MATHEMATICS RESEARCH DEVELOPMENTS

Additional books in this series can be found on Nova's website
under the Series tab.

Additional e-books in this series can be found on Nova's website
under the e-books tab.

MATHEMATICS RESEARCH DEVELOPMENTS

FOUNDATIONS OF ISO-DIFFERENTIAL CALCULUS

VOLUME 5
ISO-STOCHASTIC DIFFERENTIAL EQUATIONS

SVETLIN GEORGIEV

nova publishers
New York

Library of Congress Cataloging-in-Publication Data

ISBN: 978-1-63482-146-9

Published by Nova Science Publishers, Inc. † New York

CONTENTS

Preface

This book is intended for readers who have had a course in iso-differential calculus and theory of probability. It can be used for a senior undergraduate course.

Chapter 1 represents a short introduction to the theory of iso-probability theory. They are defined iso-probability measure, iso-probability space, random iso-variable of the first, second, third, fourth and fifth kind, iso-expected values, iso-martingales, iso-Brownian motion, iso-Wiener processes, Paley-Wiener-Zygmund integral, Itô's iso-integral, and they are deducted some of their properties.

Chapter 2 is devoted on the iso-stochastic differential equations of the first, second and third kind, and for them they are proved the general existence and uniqueness theorems. They are given some methods for solving of some classes iso-stochastic differential equations.

Chapter 3 deals with the linear iso-stochastic differential equations.

The dependence on parameters and initial data is considered in Chapter 4.

In Chapter 5 is investigated the stability of the main classes iso-stochastic differential equations.

Iso-Stratonovich iso-integral and its properties are considered in Chapter 6.

I will be very grateful to anybody who wants to inform me about errors or just misprints, or wants to express criticism or other comments, to my e-mails svetlingeorgiev1@gmail.com, sgg2000bg@yahoo.com.

Svetlin Georgiev
Paris, France
August 15, 2014

Chapter 1

A Crash Course in Basic Iso-Probability Theory

1.1 Basic Definitions

Let Ω be a nonempty set.

Definition 1.1.1. *A collection U of subsets of Ω with the properties*

(i) $\emptyset, \Omega \in U$,

(ii) *if $A_1, A_2, \ldots \in U$, then*

$$\bigcup_{k=1}^{\infty} A_k, \bigcap_{k=1}^{\infty} A_k \in U,$$

(iii) *if $A \in U$, then $A^c := \Omega \backslash A \in U$,*

will be called σ-algebra.

Below we will suppose that U is an σ-algebra of subsets of Ω. Let $\hat{T} : U \longrightarrow (0, \infty)$, $P : U \longrightarrow [0, \infty)$ be given. We note

$$\hat{P}(A) := \frac{P(A)}{\hat{T}(A)} \qquad \text{for} \qquad \forall A \in U.$$

Definition 1.1.2. *We call $\hat{P}$ an iso-probability measure provided*

(i) $\hat{P}(\emptyset) = 0$, $\hat{P}(\Omega) = \hat{I}(\Omega) := \frac{1}{\hat{T}(\Omega)}$,

(ii) *if $A_1, A_2, \ldots \in U$, then*

$$\hat{P}(\bigcup_{k=1}^{\infty} A_k) \leq \sum_{k=1}^{\infty} \hat{P}(A_k),$$

(iii) *if $A_1, A_2, \ldots \in U$ are disjoint sets, then*

$$\hat{P}(\bigcup_{k=1}^{\infty} A_k) = \sum_{k=1}^{\infty} \hat{P}(A_k).$$

Below we will suppose that $\hat{P}$ is an iso-probability measure on U.

Proposition 1.1.3. *For every $A \in U$ we have*

$$\hat{P}(A) \leq \frac{1}{\hat{T}(\Omega)}.$$

Proof. Since

$$A \bigcup A^c = U, \qquad A \bigcap A^c = \emptyset,$$

we have

$$\tfrac{1}{\hat{T}(\Omega)} = \hat{P}(\Omega) = \hat{P}(A \bigcup A^c) = \hat{P}(A) + \hat{P}(A^c),$$

whereupon, because $\hat{P}(A) \geq 0$, $\hat{P}(A^c) \geq 0$, we conclude that

$$\hat{P}(A) \leq \frac{1}{\hat{T}(\Omega)}.$$

□

Definition 1.1.4. *The triple $(\Omega, U, \hat{P})$ will be called an iso-probability space.*

Definition 1.1.5. **(i)** *A set $A \in U$ is called an event, points $\omega \in \Omega$ are sample paths.*

(ii) *$\hat{P}(A)$ is the iso-probability of the event A.*

(iii) *A property which is true except for an event of iso-probability zero is said to hold iso-almost surely(usually abbreviated "i-a.s.").*

Proposition 1.1.6. *Let A, $B \in U$, $A \subseteq B$,*

$$P(A) \leq P(B), \qquad \hat{T}(A) \geq \hat{T}(B).$$

Then

$$\hat{P}(A) \leq \hat{P}(B).$$

Proof. Since $\hat{T}(A) \geq \hat{T}(B)$, we have

$$\frac{1}{\hat{T}(A)} \leq \frac{1}{\hat{T}(B)},$$

from here and from $P(A) \leq P(B)$, we obtain

$$\hat{P}(A) = \frac{P(A)}{\hat{T}(A)} \leq \frac{P(B)}{\hat{T}(B)} = \hat{P}(B).$$

□

Corollary 1.1.7. *If P is a probability measure on U and $\hat{T}(A) \geq \hat{T}(B)$, $A \subseteq B$, A, $B \in U$, then $\hat{P}(A) \leq \hat{P}(B)$.*

Proposition 1.1.8. *Let A, $B \in U$, $A \subseteq B$, and*

$$P(A) \geq P(B), \qquad \hat{T}(A) \leq \hat{T}(B).$$

Then

$$\hat{P}(A) \geq \hat{P}(B).$$

Proof. Since $\hat{T}(A) \leq \hat{T}(B)$, then

$$\frac{1}{\hat{T}(A)} \geq \frac{1}{\hat{T}(B)}.$$

From here and from $P(A) \geq P(B)$, we get

$$\hat{P}(A) = \frac{P(A)}{\hat{T}(A)} \geq \frac{P(B)}{\hat{T}(B)} = \hat{P}(B).$$

□

Proposition 1.1.9. *Let $A \subset \Omega$. Then*

$$\hat{P}(A) + \hat{P}(A^c) = \frac{1}{\hat{T}(\Omega)}.$$

Proof. We note that

$$A \bigcup A^c = \Omega.$$

Then

$$\hat{P}(A \bigcup A^c) = \hat{P}(\Omega) = \frac{1}{\hat{T}(\Omega)}.$$

From here, since $A \bigcap A^c = \emptyset$, we get

$$\hat{P}(A \bigcup A^c) = \hat{P}(A) + \hat{P}(A^c)$$

and

$$\hat{P}(A) + \hat{P}(A^c) = \frac{1}{\hat{T}(\Omega)}.$$

□

Definition 1.1.10. *Let A, $B \subset \Omega$. If $\hat{P}(A) > 0$, then the quotient*

$$\hat{P}_A(B) = \frac{\hat{P}(A \bigcap B)}{\hat{P}(A)}$$

is defined to be conditional iso-probability of the event B under the condition A.

We can rewrite the conditional iso-probability as follows

$$\hat{P}_A(B) = \frac{\hat{P}(A\bigcap B)}{\hat{P}(A)}$$

$$= \frac{\frac{P(A\bigcap B)}{\hat{T}(A\bigcap B)}}{\frac{P(A)}{\hat{T}(A)}}$$

$$= \frac{P(A\bigcap B)\hat{T}(A)}{P(A)\hat{T}(A\bigcap B)}.$$

From the definition for conditional iso-probability, it follows

$$\hat{P}(A\bigcap B) = \hat{P}(A)\hat{P}_A(B).$$

Theorem 1.1.11. *(multiplicative theorem) For every $n \in \mathbb{N}$, $n \geq 2$, if $A_1, A_2, \ldots, A_n \in \Omega$, we have*

$$\hat{P}(A_1\bigcap A_2\bigcap \ldots \bigcap A_n) = \hat{P}(A_1)\hat{P}_{A_1}(A_2)\hat{P}_{A_1\bigcap A_2}(A_3)\bigcap \ldots \bigcap \hat{P}_{A_1\bigcap A_2\bigcap \ldots \bigcap A_{n-1}}(A_n).$$

Proof. The case $n = 2$ follows from the definition for conditional iso-probability. Now we consider the case $n = 3$. We have

$$\hat{P}(A_1\bigcap A_2\bigcap A_3) = \hat{P}(A_1\bigcap(A_2\bigcap A_3))$$

$$= \hat{P}(A_1)\hat{P}_{A_1}(A_2\bigcap A_3)$$

$$\text{now we apply the case } n = 2$$

$$= \hat{P}(A_1)\hat{P}_{A_1}(A_2)\hat{P}_{A_1\bigcap A_2}(A_3).$$

We suppose that the assertion is valid for some $n \in \mathbb{N}$, $n \geq 2$. We will prove the assertion in the case $n+1$.

Indeed,

$$\hat{P}(A_1\bigcap A_2\bigcap \ldots \bigcap A_n\bigcap A_{n+1}) = \hat{P}(A_1\bigcap(A_2\bigcap A_3\bigcap \ldots \bigcap A_{n+1}))$$

$$= \hat{P}(A_1)\hat{P}_{A_1}(A_2\bigcap A_3\bigcap \ldots \bigcap A_n\bigcap A_{n+1})$$

$$= \hat{P}(A_1)\hat{P}_{A_1}(A_2)\hat{P}_{A_1\bigcap A_2}(A_3\bigcap \ldots \bigcap A_n\bigcap A_{n+1})$$

$$\ldots$$

$$= \hat{P}(A_1)\hat{P}_{A_1}(A_2)\hat{P}_{A_1\bigcap A_2}(A_3)\ldots \hat{P}_{A_1\bigcap A_2\bigcap \ldots \bigcap A_n}(A_{n+1}).$$

□

Proposition 1.1.12. *Let $A \subset \Omega$. Then*

$$\hat{P}_A(\Omega) = 1.$$

Proof. From the definition for conditional iso-probability, we have

$$\hat{P}_A(\Omega) = \frac{\hat{P}(A \bigcap \Omega)}{\hat{P}(A)} = \frac{\hat{P}(A)}{\hat{P}(A)} = 1.$$

□

Proposition 1.1.13. *Let A, B ⊂ Ω. Then*

$$\hat{P}_A(B) \geq 0.$$

Proof. We have that

$$\hat{P}_A(B) = \frac{\hat{P}(A \bigcap B)}{\hat{P}(A)},$$

from here, since $\hat{P}(A) > 0$ and $\hat{P}(A \bigcap B) \geq 0$, we conclude that $\hat{P}_A(B) \geq 0$. □

Proposition 1.1.14. *Let A, B, C ⊂ Ω, $B \bigcap C = \emptyset$. Then*

$$\hat{P}_A(B \bigcup C) = \hat{P}_A(B) + \hat{P}_A(C).$$

Proof. We have

$$\hat{P}_A(B \bigcup C) = \frac{\hat{P}(A \bigcap (B \bigcup C))}{\hat{P}(A)}$$

$$= \frac{\hat{P}((A \bigcap B) \bigcup (A \bigcap C))}{\hat{P}(A)}$$

$$= \frac{\hat{P}(A \bigcap B) + \hat{P}(A \bigcap C)}{\hat{P}(A)}$$

$$= \hat{P}_A(B) + \hat{P}_A(C).$$

□

Theorem 1.1.15. *Let A_1, A_2, …, A_n are disjoint sets in Ω and*

$$\Omega = A_1 \bigcup A_2 \bigcup \ldots \bigcup A_n.$$

Let also, X ⊂ Ω be arbitrarily chosen. Then

$$\hat{P}(X) = \hat{P}(A_1)\hat{P}_{A_1}(X) + \cdots + \hat{P}(A_n)\hat{P}_{A_n}(X).$$

Proof. We have that

$$\Omega = A_1 \bigcup A_2 \bigcup \ldots \bigcup A_n.$$

Since

$$X = X \bigcap \Omega,$$

we get

$$X = X \bigcap \Omega = X \bigcap (A_1 \bigcup A_2 \bigcup \ldots \bigcup A_n)$$

$$= (X \bigcap A_1) \bigcup (X \bigcap A_2) \bigcup \ldots \bigcup (X \bigcap A_n).$$

Then

$$\hat{P}(X) = \hat{P}((X\bigcap A_1)\bigcup(X\bigcap A_2)\bigcup\ldots\bigcup(X\bigcap A_n))$$

$$= \hat{P}(X\bigcap A_1) + \hat{P}(X\bigcap A_2) + \cdots + \hat{P}(X\bigcap A_n)$$

$$= \hat{P}(A_1)\hat{P}_{A_1}(X) + \hat{P}(A_2)\hat{P}_{A_2}(X) + \cdots + \hat{P}(A_n)\hat{P}_{A_n}(X).$$

□

Proposition 1.1.16. *Let A, $B \subset \Omega$, $\hat{P}(A) > 0$, $\hat{P}(B) > 0$. Then*

$$\hat{P}_B(A) = \frac{\hat{P}(A)\hat{P}_A(B)}{\hat{P}(B)}.$$

Proof. We have

$$\hat{P}_A(B) = \frac{\hat{P}(A\bigcap B)}{\hat{P}(A)},$$

whereupon

$$\hat{P}(A\bigcap B) = \hat{P}_A(B)\hat{P}(A).$$

From here,

$$\hat{P}_B(A) = \frac{\hat{P}(A\bigcap B)}{\hat{P}(B)} = \frac{\hat{P}(A)\hat{P}_A(B)}{\hat{P}(B)}.$$

□

Definition 1.1.17. *The events A_1, A_2, …, A_m will be called mutually independent if*

$$\hat{P}(A_1\bigcap A_2\bigcap\ldots\bigcap A_m) = \hat{T}(\Omega)\hat{P}(A_1)\hat{P}(A_2)\ldots\hat{P}(A_m).$$

Proposition 1.1.18. *Let A_1, $A_2 \subset \Omega$ be mutually independent. Then*

$$\hat{P}(A_1\bigcap A_2^c) = \hat{T}(\Omega)\hat{P}(A_1)\hat{P}(A_2^c).$$

Proof. Firstly, we will note

$$\hat{P}(A_1) = \hat{P}(A_1\bigcap\Omega)$$

$$= \hat{P}(A_1\bigcap(A_2\bigcup A_2^c))$$

$$= \hat{P}((A_1\bigcap A_2)\bigcup(A_1\bigcap A_2^c))$$

$$= \hat{P}(A_1\bigcap A_2) + \hat{P}(A_1\bigcap A_2^c),$$

i.e.,

$$\hat{P}(A_1\bigcap A_2^c) = \hat{P}(A_1) - \hat{P}(A_1\bigcap A_2),$$

whereupon

$$\hat{P}(A_1\bigcap A_2^c) = \hat{P}(A_1) - \hat{T}(\Omega)\hat{P}(A_1)\hat{P}(A_2)$$

$$= \hat{T}(\Omega)\hat{P}(A_1)\left(\frac{1}{\hat{T}(\Omega)} - \hat{P}(A_2)\right)$$

$$= \hat{T}(\Omega)\hat{P}(A_1)\hat{P}(\Omega\backslash A_2)$$

$$= \hat{T}(\Omega)\hat{P}(A_1)\hat{P}(A_2^c).$$

□

As in above, one can prove the following assertion.

Proposition 1.1.19. *Let A_1, $A_2 \subset \Omega$ be mutually independent. Then*

$$\hat{P}(A_1^c\bigcap A_2) = \hat{T}(\Omega)\hat{P}(A_1^c)\hat{P}(A_2).$$

Proposition 1.1.20. *Let A_1, $A_2 \subset \Omega$ be mutually independent. Then*

$$\hat{P}(A_1^c\bigcap A_2^c) = \hat{T}(\Omega)\hat{P}(A_1^c)\hat{P}(A_2^c).$$

Proof. We have

$$\hat{P}(A_1^c) = \hat{P}(A_1^c\bigcap\Omega)$$

$$= \hat{P}(A_1^c\bigcap(A_2^c\bigcup A_2))$$

$$= \hat{P}((A_1^c\bigcap A_2^c)\bigcup(A_1^c\bigcap A_2))$$

$$= \hat{P}(A_1^c\bigcap A_2^c) + \hat{P}(A_1^c\bigcap A_2)$$

$$= \hat{P}(A_1^c\bigcap A_2) + \hat{T}(\Omega)\hat{P}(A_1^c)\hat{P}(A_2),$$

whereupon

$$\hat{T}(\Omega)\hat{P}(A_1^c)\left(\frac{1}{\hat{T}(\Omega)} - \hat{P}(A_2)\right) = \hat{P}(A_1^c\bigcap A_2^c),$$

or

$$\hat{P}(A_1^c\bigcap A_2^c) = \hat{T}(\Omega)\hat{P}(A_1^c)\hat{P}(A_2^c).$$

□

Definition 1.1.21. *Let A_1, A_2, ..., A_n, ... be events. Then the event*

$$\bigcap_{n=1}^{\infty}\bigcup_{m=n}^{\infty} A_m = \left\{\omega\in\Omega : \omega \text{ belongs to infinitely many of the } A_n\right\}$$

*is called A_n infinitely often, abbreviated A_n*i.o., *i.e.,*

$$A_n\text{i.o.} = \bigcap_{n=1}^{\infty}\bigcup_{m=n}^{\infty} A_m.$$

Lemma 1.1.22. *(iso-Borel-Cantelli lemma) Let A_1, A_2, ..., A_n, ..., be events such that $\hat{T}(A_n) \geq Q$ for every $n \in \mathbb{N}$, where Q is a positive constant. Let also, $P(A) \leq P(B)$ and $\hat{T}(A) \geq \hat{T}(B)$ for every A, $B \in U$, $A \subset B$. If $\sum_{n=1}^{\infty} P(A_n) < \infty$, then*

$$\hat{P}(A_n \text{i.o.}) = 0.$$

Proof. By definition we have

$$A_n \text{i.o.} = \bigcap_{n=1}^{\infty} \bigcup_{m=n}^{\infty} A_m.$$

Therefore

$$A_n \text{i.o.} \subset \bigcup_{m=n}^{\infty} A_m \qquad \text{for} \qquad \forall n \in \mathbb{N}.$$

Hence,

$$P(A_n \text{i.o.}) \leq P\left(\cup_{m=n}^{\infty} A_m\right),$$

$$\hat{T}(A_n \text{i.o.}) \geq \hat{T}\left(\cup_{m=n}^{\infty} A_m\right),$$

$$\frac{1}{\hat{T}(A_n \text{i.o.})} \leq \frac{1}{\hat{T}(\cup_{m=n}^{\infty} A_m)},$$

and

$$\begin{aligned}
\hat{P}(A_n \text{i.o.}) &= \frac{P(A_n \text{i.o.})}{\hat{T}(A_n \text{i.o.})} \\
&\leq \frac{P(\cup_{m=n}^{\infty} A_m)}{\hat{T}(\cup_{m=n}^{\infty} A_m)} \\
&= \hat{P}\left(\cup_{m=n}^{\infty} A_m\right) \\
&\leq \sum_{m=n}^{\infty} \hat{P}(A_m) \\
&= \sum_{m=n}^{\infty} \frac{P(A_m)}{\hat{T}(A_m)} \\
&\leq \frac{1}{Q} \sum_{m=n}^{\infty} P(A_m) \\
&\longrightarrow_{n \longrightarrow \infty},
\end{aligned}$$

because $\sum_{n=1}^{\infty} P(A_n) < \infty$. □

Definition 1.1.23. *A sequence $\{\hat{X}_k\}_{k=1}^{\infty}$ of random iso-variables converges in iso-probability to an iso-random variable $\hat{X}$, provided*

$$\lim_{k \longrightarrow \infty} \hat{P}(|\hat{X}_k - \hat{X}|) > \varepsilon$$

for every $\varepsilon > 0$.

Theorem 1.1.24. *Let $Q_1 \geq \hat{T}(A) \geq Q$ for every $A \in U$ and for some constants $Q, Q_1 > 0$. Let also, $P(A) \leq P(B)$ and $\hat{T}(A) \geq \hat{T}(B)$ for every A, $B \in U$, $A \subset B$. If $\hat{X}_k \longrightarrow \hat{X}$ in iso-probability, then there exists a subsequence $\{\hat{X}_{k_j}\}_{j=1}^{\infty} \subset \{\hat{X}_k\}_{k=1}^{\infty}$ such that*

$$\hat{X}_{k_j}(\omega) \longrightarrow \hat{X}(\omega)$$

for almost every ω.

Proof. For each positive integer j we choose k_j so that

$$\hat{P}(|\hat{X}_{k_j} - \hat{X}| > \frac{1}{j}) \leq \frac{1}{j^2}$$

and also

$$k_{j-1} < k_j < \cdots, \qquad k_j \longrightarrow \infty.$$

Hence,

$$\frac{1}{Q_1} P(|\hat{X}_{k_j} - \hat{X}| > \frac{1}{j}) \leq \frac{P(|\hat{X}_{k_j} - \hat{X}| > \frac{1}{j})}{\hat{T}(|\hat{X}_{k_j} - \hat{X}| > \frac{1}{j})} \leq \frac{1}{j^2}$$

or

$$P(|\hat{X}_{k_j} - \hat{X}| > \frac{1}{j}) \leq \frac{Q_1}{j^2}.$$

Let

$$A_j := \{|\hat{X}_{k_j} - \hat{X}| > \frac{1}{j}\}.$$

Then

$$\sum_{j=1}^{\infty} P(A_j) \leq Q_1 \sum_{j-1}^{\infty} \frac{1}{j^2} < \infty.$$

From here and from the iso-Borel-Cantelli lemma it follows that

$$\hat{P}(A_n \text{i.o.}) = 0,$$

i.e.,

$$\hat{X}_{k_j}(\omega) \longrightarrow_{j \longrightarrow \infty} \hat{X}(\omega)$$

for almost every ω. □

1.2 Random Iso-variables

Let $\mathcal{B}$ denotes the collection of Borel sets of $\mathbb{R}^n$, i.e., $\mathcal{B}$ is the smallest σ-algebra of subsets of $\mathbb{R}^n$ containing all open sets.

Definition 1.2.1. *For $\omega \in \Omega$ we define the operators*

$$\hat{\imath}(\omega) := \frac{\omega}{\hat{T}(\omega)}, \qquad \hat{m}(\omega) := \omega \hat{T}(\omega).$$

Definition 1.2.2. *Let $X : \Omega \longrightarrow \mathbb{R}^n$ be given.*

1. *Then the map* $\hat{X}^{\wedge\wedge} : \Omega \longrightarrow \mathbb{R}^n$ *defined by*

$$\hat{X}^{\wedge\wedge}(\omega) := \frac{X(\omega)}{\hat{T}(\omega)}, \qquad \omega \in \Omega,$$

will be called an n-dimensional random iso-variable of the first kind if for each $B \in \mathcal{B}$, *we have*

$$\hat{X}^{\wedge\wedge -1}(B) \in U.$$

We equivalently say that $\hat{X}^{\wedge\wedge}$ *is* U*-iso-measurable of the first kind.*

2. *Let also,* $\hat{m} : \Omega \longrightarrow \Omega$. *Then the map* $\hat{X}^{\wedge} : \Omega \longrightarrow \mathbb{R}^n$ *defined by*

$$\hat{X}^{\wedge}(\omega) := \frac{X(\omega\hat{T}(\omega))}{\hat{T}(\omega)}, \qquad \omega \in \Omega,$$

will be called an random iso-variable of the second kind if for each $B \in \mathcal{B}$

$$\hat{X}^{\wedge -1}(B) \in U.$$

We equivalently say $\hat{X}^{\wedge}$ *is* U*-iso-measurable of the first kind.*

3. *Let also,* $\hat{I} : \Omega \longrightarrow \Omega$. *Then the map* $\hat{\hat{X}} : \Omega \longrightarrow \mathbb{R}^n$ *defined by*

$$\hat{\hat{X}}(\omega) := \frac{X\left(\frac{\omega}{\hat{T}(\omega)}\right)}{\hat{T}(\omega)}, \qquad \omega \in \Omega,$$

will be called an random iso-variable of the third kind if for each $B \in \mathcal{B}$ *we have*

$$\hat{\hat{X}}^{-1}(B) \in U.$$

We equivalently say $\hat{\hat{X}}$ *is* U*-iso-measurable of the third kind.*

4. *Let also,* $\hat{m} : \Omega \longrightarrow \Omega$. *Then the map* $X^{\wedge} : \Omega \longrightarrow \mathbb{R}^n$ *defined by*

$$X^{\wedge}(\omega) := X(\omega\hat{T}(\omega)), \qquad \omega \in \Omega,$$

will be called an iso-variable of the fourth kind if for each $B \in \mathcal{B}$ *we have*

$$X^{\wedge -1}(B) \in U.$$

We equivalently say $X^{\wedge}$ *is* U*-iso-measurable of the fourth kind.*

5. *Let also,* $\hat{I} : \Omega \longrightarrow \Omega$. *Then the map* $X^{\vee} : \Omega \longrightarrow \mathbb{R}^n$ *defined by*

$$X^{\vee}(\omega) := X\left(\frac{\omega}{\hat{T}(\omega)}\right), \qquad \omega \in \Omega,$$

will be called an iso-variable of the fifth kind if for each $B \in \mathcal{B}$ *we have*

$$X^{\vee -1}(B) \in U.$$

We equivalently say $X^{\vee}$ *is* U*-iso-measurable of the fifth kind.*

Below with $\hat{X}$ we will denote an random iso-variable of the first, second, third, fourth or fifth kind.

Remark 1.2.3. *We will note that there is a possibility X to be a random variable and there is a possibility X to be not a random variable, and at the same time $\hat{X}$ to be a random iso-variable.*

We denote $\hat{P}(\hat{X}^{-1}(B))$ as $\hat{P}(\hat{X} \in B)$, the iso-probability that $\hat{X}$ is in B. Also, we usually write $\hat{X}$and not $\hat{X}(\omega)$.

Example 1.2.4. *Let $A \in U$,*

$$\chi_A(\omega) := \begin{cases} 1 & \text{if} \quad \omega \in A \\ 0 & \text{if} \quad \omega \notin A, \end{cases} \qquad \hat{T}_A(\omega) = 1 \quad \text{for} \quad \text{every} \quad \omega \in \Omega.$$

Then $\hat{\chi}_A$ is a random iso-variable.

Example 1.2.5. *(generalization of the previous example) If $A_1,\ A_2,\ \ldots,\ A_m \in U,\ \Omega = \bigcup_{i=1}^m A_i$, and $a_1,\ a_2,\ \ldots,\ a_m$ are real numbers, $\hat{T}(\omega) = 1$ for every $\omega \in \Omega$. Then*

$$\hat{X} = \sum_{i=1}^m \hat{a}_i \hat{\times} \hat{\chi}_{A_i}, \qquad \hat{X} = \sum_{i=1}^m \hat{a}_i \hat{\chi}_{A_i}, \qquad \hat{X} = \sum_{i=1}^m a_i \hat{\times} \hat{\chi}_{A_i}, \qquad \hat{X} = \sum_{i=1}^m a_i \hat{\chi}_{A_i}$$

are random iso-variables.

Definition 1.2.6. *If a random iso-variable $\hat{Y}$ is an iso-function of $\hat{X}$, that is, if*

$$\hat{Y} = \tilde{\Phi}(\hat{X}),$$

where $\tilde{\Phi}$ is an iso-function of first, second, third, fourth or fifth kind, then the random iso-variable $\hat{Y}$ will be called $U(\hat{X})$-iso-measurable.

Definition 1.2.7. *A collection $\{\hat{X}_t : t \geq 0\}$ of random iso-variables is called an iso-stochastic process.*

For each point $\omega \in \Omega$, the mapping $t \longrightarrow \hat{X}_t$ is the corresponding iso-sample path.

1.3 Iso-Expected Value. Iso-Variance

Let $\hat{X} : \Omega \longrightarrow \mathbb{R}$ be a real random iso-variable. Let also, $\hat{T} : U \longrightarrow (0,\infty)$ and $\hat{T} : \Omega \longrightarrow (0,\infty)$ is differentiable.

Definition 1.3.1. *We define iso-integral as follows*

$$\hat{\int}_\Omega \hat{X} \hat{\times} \hat{d}\hat{P}.$$

For the iso-integral we have the following representation

$$\hat{\int}_{\Omega}\hat{X}\hat{\times}\hat{d}\hat{P} = \int_{\Omega}\hat{T}^{-1}(\omega)\hat{X}(\omega)\hat{T}(\omega)\hat{T}(\omega)d\hat{P}$$

$$= \int_{\Omega}\hat{X}(\omega)\hat{T}(\omega)d\frac{P(\omega)}{\hat{T}(\omega)}$$

$$= \int_{\Omega}\hat{X}(\omega)dP(\omega) - \int_{\Omega}\hat{X}(\omega)\frac{\hat{T}'(\omega)}{\hat{T}(\omega)}P(\omega)d\omega.$$

1. If $\hat{X} = \hat{X}^{\wedge\wedge}$, then

$$\hat{\int}_{\Omega}\hat{X}\hat{\times}\hat{d}\hat{P} = \int_{\Omega}\frac{X(\omega)}{\hat{T}(\omega)}dP(\omega) - \int_{\Omega}\frac{X(\omega)}{\hat{T}(\omega)}\frac{\hat{T}'(\omega)}{\hat{T}(\omega)}P(\omega)d\omega$$

$$= \int_{\Omega}\frac{X(\omega)}{\hat{T}(\omega)}dP(\omega) - \int_{\Omega}X(\omega)\frac{\hat{T}'(\omega)}{\hat{T}^2(\omega)}P(\omega)d\omega.$$

2. If $\hat{X} = \hat{X}^{\wedge}$, then

$$\hat{\int}_{\Omega}\hat{X}\hat{\times}\hat{d}\hat{P} = \int_{\Omega}\frac{X(\omega\hat{T}(\omega))}{\hat{T}(\omega)}dP(\omega) - \int_{\Omega}\frac{X(\omega\hat{T}(\omega))}{\hat{T}(\omega)}\frac{\hat{T}'(\omega)}{\hat{T}(\omega)}P(\omega)d\omega$$

$$= \int_{\Omega}\frac{X(\omega\hat{T}(\omega))}{\hat{T}(\omega)}dP(\omega) - \int_{\Omega}X(\omega\hat{T}(\omega))\frac{\hat{T}'(\omega)}{\hat{T}^2(\omega)}P(\omega)d\omega.$$

3. If $\hat{X} = \hat{\hat{X}}$, then

$$\hat{\int}_{\Omega}\hat{X}\hat{\times}\hat{d}\hat{P} = \int_{\Omega}\frac{X\left(\frac{\omega}{\hat{T}(\omega)}\right)}{\hat{T}(\omega)}dP(\omega) - \int_{\Omega}\frac{X\left(\frac{\omega}{\hat{T}(\omega)}\right)}{\hat{T}(\omega)}\frac{\hat{T}'(\omega)}{\hat{T}(\omega)}P(\omega)d\omega$$

$$= \int_{\Omega}\frac{X\left(\frac{\omega}{\hat{T}(\omega)}\right)}{\hat{T}(\omega)}dP(\omega) - \int_{\Omega}X\left(\frac{\omega}{\hat{T}(\omega)}\right)\frac{\hat{T}'(\omega)}{\hat{T}^2(\omega)}P(\omega)d\omega.$$

4. If $\hat{X} = X^{\wedge}$, then

$$\hat{\int}_{\Omega}\hat{X}\hat{\times}\hat{d}\hat{P} = \int_{\Omega}X(\omega\hat{T}(\omega))dP(\omega) - \int_{\Omega}X(\omega\hat{T}(\omega))\frac{\hat{T}'(\omega)}{\hat{T}(\omega)}P(\omega)d\omega.$$

5. If $\hat{X} = X^{\vee}$, then

$$\hat{\int}_{\Omega}\hat{X}\hat{\times}\hat{d}\hat{P} = \int_{\Omega}X\left(\frac{\omega}{\hat{T}(\omega)}\right)dP(\omega) - \int_{\Omega}X\left(\frac{\omega}{\hat{T}(\omega)}\right)\frac{\hat{T}'(\omega)}{\hat{T}(\omega)}P(\omega)d\omega.$$

Now we suppose that $\hat{X} : \Omega \longrightarrow \mathbb{R}^n$ be a vector-valued random iso-variable, $\hat{X} = (\hat{X}^1, \hat{X}^2, \dots, \hat{X}^n)$. Then

$$\hat{\int}_{\Omega}\hat{X}\hat{\times}\hat{d}\hat{P} = \left(\hat{\int}_{\Omega}\hat{X}^1\hat{\times}\hat{d}\hat{P}, \hat{\int}_{\Omega}\hat{X}^2\hat{\times}\hat{d}\hat{P}, \dots, \hat{\int}_{\Omega}\hat{X}^n\hat{\times}\hat{d}\hat{P}\right).$$

Definition 1.3.2. *We call*

$$\hat{E}(\hat{X}) := \hat{\int}_{\Omega} \hat{X} \hat{\times} \hat{d}\hat{P}$$

the iso-expected value of $\hat{X}$.

Definition 1.3.3. *We call*

$$\hat{V}(\hat{X}) := \hat{\int}_{\Omega} \left| \hat{X} - \frac{\hat{E}(\hat{X})}{\hat{T}} \right|^2 \hat{\times} \hat{d}\hat{P}$$

the iso-variance of $\hat{X}$. *Here* $|\cdot|$ *denotes the Euclidean norm.*

We observe

$$\hat{X} - \hat{E}(\hat{X}) = (\hat{X}^1 - \hat{E}^1(\hat{X}), \hat{X}^2 - \hat{E}^2(\hat{X}), \ldots, \hat{X}^n - \hat{E}^n(\hat{X})),$$

$$|\hat{X} - \hat{E}(\hat{X})|^2 = \Sigma_{i=1}^n \left(\hat{X}^i - \hat{E}^i(\hat{X})\right)^2$$

$$= \Sigma_{i=1}^n \hat{X}^{i2} - 2\Sigma_{i=1}^n \hat{X}^i \hat{E}^i(\hat{X}) + \Sigma_{i=1}^n \hat{E}^{i2}(\hat{X}),$$

$$\hat{V}(\hat{X}) = \hat{\int}_{\Omega} |\hat{X} - \hat{E}(\hat{X})|^2 \hat{\times} \hat{d}\hat{P}$$

$$= \Sigma_{i=1}^n \hat{\int}_{\Omega} \hat{X}^{i2} \hat{\times} \hat{d}\hat{P} - 2\Sigma_{i=1}^n \hat{E}^i(\hat{X}) \hat{\int}_{\omega} \hat{X}^i \hat{\times} \hat{d}\hat{P} + \Sigma_{i=1}^n \hat{E}^{i2}(\hat{X}) \hat{\int}_{\Omega} 1 \hat{\times} \hat{d}\hat{P}$$

$$= \hat{\int}_{\Omega} |\hat{X}|^2 \hat{\times} \hat{d}\hat{P} - 2\Sigma_{i=1}^n \hat{E}^{i2}(\hat{X}) + \Sigma_{i=1}^n \hat{E}^{i2}(\hat{X}) \hat{\int}_{\Omega} 1 \hat{\times} \hat{d}\hat{P}$$

$$= \hat{\int}_{\Omega} |\hat{X}|^2 \hat{\times} \hat{d}\hat{P} - 2|\hat{E}(\hat{X})|^2 + |\hat{E}(\hat{X})|^2 \hat{\int}_{\Omega} 1 \hat{\times} \hat{d}\hat{P}.$$

Lemma 1.3.4. *(iso-Chebychev's inequality) If* $\dot{X}$ *be a random iso-variable and* $1 \le p < \infty$, *then*

$$P(|\hat{X}|^p \ge \lambda) \le \frac{1}{\lambda^p} \hat{E}(|\hat{X}|^p) + \frac{1}{\lambda^p} \int_{\Omega} |\hat{X}(\omega)|^p \frac{\hat{T}'(\omega)}{\hat{T}(\omega)} P(\omega) d\omega$$

for all $\lambda > 0$.

In addition, if $\hat{T}(|\hat{X}| \ge \lambda)$ *for some* $\lambda > 0$, *we have*

$$\hat{P}(|\hat{X}|^p \ge \lambda) \le \frac{1}{\lambda^p} \hat{E}(|\hat{X}|^p) + \frac{1}{\lambda^p} \int_{\Omega} |\hat{X}(\omega)|^p \frac{\hat{T}'(\omega)}{\hat{T}(\omega)} P(\omega) d\omega.$$

Proof. Let $\lambda > 0$ be arbitrarily chosen. Then we have

$$\hat{E}(|\hat{X}|^p) = \int_{\Omega} |\hat{X}(\omega)|^p dP(\omega) - \int_{\Omega} |\hat{X}(\omega)|^p \tfrac{\hat{T}'(\omega)}{\hat{T}(\omega)} P(\omega) d\omega$$

$$\ge \int_{|\hat{X}| \ge \lambda} |\hat{X}(\omega)|^p dP(\omega) - \int_{\Omega} |\hat{X}(\omega)|^p \tfrac{\hat{T}'(\omega)}{\hat{T}(\omega)} P(\omega) d\omega$$

$$\ge \lambda^p \int_{|\hat{X}| \ge \lambda} dP(\omega) - \int_{\Omega} |\hat{X}(\omega)|^p \tfrac{\hat{T}'(\omega)}{\hat{T}(\omega)} P(\omega) d\omega$$

$$= \lambda^p P(|\hat{X}| \ge \lambda) - \int_{\Omega} |\hat{X}(\omega)|^p \tfrac{\hat{T}'(\omega)}{\hat{T}(\omega)} P(\omega) d\omega,$$

whereupon

$$\lambda^p P(|\hat{X}| \geq \lambda) \leq \hat{E}(|\hat{X}|^p) - \int_\Omega |\hat{X}(\omega)|^p \frac{\hat{T}'(\omega)}{\hat{T}(\omega)} P(\omega) d\omega$$

or

$$P(|\hat{X}| \geq \lambda) \leq \frac{1}{\lambda^p}\hat{E}(|\hat{X}|^p) - \frac{1}{\lambda^p}\int_\Omega |\hat{X}(\omega)|^p \frac{\hat{T}'(\omega)}{\hat{T}(\omega)} P(\omega) d\omega.$$

Let now $\hat{T}(|\hat{X}| \geq \lambda) \geq 1$ for some $\lambda > 0$. Then

$$\begin{aligned} \hat{P}(|\hat{X}| \geq \lambda) &= \tfrac{P(|\hat{X}|\geq\lambda)}{\hat{T}(|\hat{X}|\geq\lambda)} \\ &\leq P(|\hat{X}| \geq \lambda) \\ &\leq \tfrac{1}{\lambda^p}\hat{E}(|\hat{X}|^p) - \tfrac{1}{\lambda^p}\int_\Omega |\hat{X}(\omega)|^p \tfrac{\hat{T}'(\omega)}{\hat{T}(\omega)} P(\omega) d\omega. \end{aligned}$$

□

1.4 Distribution Iso-functions

Suppose $\hat{X} : \Omega \longrightarrow \mathbb{R}^n$ be a random iso-variable.

For

$$x = (x_1, x_2, \dots, x_n) \in \mathbb{R}^n, \qquad y = (y_1, y_2, \dots, y_n) \in \mathbb{R}^n,$$

we will write

$$x \leq y$$

if

$$x_i \leq y_i \qquad \text{for} \qquad \forall i = 1, 2, \dots, m.$$

Definition 1.4.1. **1.** *The distribution iso-function of $\hat{X}$ is the function $F_{\hat{X}} : \mathbb{R}^n \longrightarrow \left[0, \frac{1}{\hat{T}(\Omega)}\right]$ defined by*

$$F_{\hat{X}}(x) := \hat{P}(\hat{X} \leq x) \qquad \text{for} \qquad \text{all} \qquad x \in \mathbb{R}^n.$$

2. *If $\hat{X}_1$, $\hat{X}_2$, ..., $\hat{X}_m : \Omega \longrightarrow \mathbb{R}^n$ are random iso-variables. Their joint distribution iso-function is*

$$F_{\hat{X}_1\hat{X}_2\dots\hat{X}_m} : (\mathbb{R}^n)^m \longrightarrow \left[0, \frac{1}{\hat{T}(\Omega)}\right],$$

$$F_{\hat{X}_1\hat{X}_2\dots\hat{X}_m}(x_1, x_2, \dots, x_m) := \hat{P}(\hat{X}_1 \leq x_1, \hat{X}_2 \leq x_2, \dots, \hat{X}_m \leq x_m)$$

for all $x_i \in \mathbb{R}^n$, $i = 1, 2, \dots, m$.

Now we suppose that $\tilde{T}$, $\hat{T}_i : \mathbb{R} \longrightarrow (0, \infty)$ are differentiable functions, $i = 1, 2, \dots, n$, also, $\overline{T} : \mathbb{R}^n \longrightarrow (0, \infty)$.

Definition 1.4.2. *Suppose that $\hat{X} : \Omega \longrightarrow \mathbb{R}^n$ be a random iso-variable and $F = F_{\hat{X}}$ is its distribution iso-function. If there exists a nonnegative, integrable iso-function of the first, second, third, fourth and fifth kind $\hat{f} : \mathbb{R}^n \longrightarrow \mathbb{R}$ such that*

$$F(x) = F(x_1, x_2, \ldots, x_n) = \hat{\int}_{-\infty}^{x_1} \hat{\int}_{-\infty}^{x_2} \cdots \hat{\int}_{-\infty}^{x_n} \hat{f}(y_1, y_2, \ldots, y_n) \hat{\times} \hat{d}\hat{y}_1 \hat{\times} \ldots \hat{\times} \hat{d}\hat{y}_2 \hat{\times} \hat{d}\hat{y}_1,$$

then $\hat{f}$ is called the density iso-function for $\hat{X}$.

We can rewrite the density iso-function as follows.

$$F(x) = \int_{-\infty}^{x_1} \int_{-\infty}^{x_2} \cdots \int_{-\infty}^{x_n} \hat{T}_1^{-1}(y_1)\hat{T}_2^{-1}(y_2) \ldots \hat{T}_n^{-1}(y_n)\hat{f}(y_1, y_2, \ldots, y_n)$$

$$\times \hat{T}_1(y_1)\hat{T}_2(y_2) \ldots \hat{T}_n(y_n) \hat{d}\hat{y}_n \ldots \hat{d}\hat{y}_2 \hat{d}\hat{y}_1$$

$$= \int_{-\infty}^{x_1} \int_{-\infty}^{x_2} \cdots \int_{-\infty}^{x_n} \hat{f}(y_1, y_2, \ldots, y_n) \hat{d}\hat{y}_n \ldots \hat{d}\hat{y}_2 \hat{d}\hat{y}_1$$

$$= \int_{-\infty}^{x_1} \int_{-\infty}^{x_2} \cdots \int_{-\infty}^{x_n} \hat{f}(y_1, y_2, \ldots, y_n)\hat{T}(y_1)\hat{T}(y_2) \ldots \hat{T}(y_n) d\hat{y}_n \ldots d\hat{y}_2 d\hat{y}_1$$

$$= \int_{-\infty}^{x_1} \int_{-\infty}^{x_2} \cdots \int_{-\infty}^{x_n} \hat{f}(y_1, y_2, \ldots, y_n) \prod_{i=1}^{n} \left(1 - \frac{\hat{T}_i'(y_i)}{\hat{T}_i(y_i)}\right) dy_n \ldots dy_2 dy_1.$$

1. If $\hat{f} = \hat{f}^{\wedge\wedge}$, then

$$F(x) = \int_{-\infty}^{x_1} \int_{-\infty}^{x_2} \cdots \int_{-\infty}^{x_n} \frac{f(y_1, y_2, \ldots, y_n)}{\overline{T}(y_1, y_2, \ldots, y_n)} \prod_{i=1}^{n} \left(1 - \frac{\hat{T}_i'(y_i)}{\hat{T}_i(y_i)}\right) dy_n \ldots dy_2 dy_1.$$

2. If $\hat{f} = \hat{f}^{\wedge}$, then

$$F(x) = \int_{-\infty}^{x_1} \int_{-\infty}^{x_2} \cdots \int_{-\infty}^{x_n} \frac{f(y_1\hat{T}_1(y_1), y_2\hat{T}_2(y_2), \ldots, y_n\hat{T}_n(y_n))}{\overline{T}(y_1, y_2, \ldots, y_n)}$$

$$\times \prod_{i=1}^{n} \left(1 - \frac{\hat{T}_i'(y_i)}{\hat{T}_i(y_i)}\right) dy_n \ldots dy_2 dy_1.$$

3. If $\hat{f} = \hat{\hat{f}}$, then

$$F(x) = \int_{-\infty}^{x_1} \int_{-\infty}^{x_2} \cdots \int_{-\infty}^{x_n} \frac{f\left(\frac{y_1}{\hat{T}_1(y_1)}, \frac{y_2}{\hat{T}_2(y_2)}, \ldots, \frac{y_n}{\hat{T}_n(y_n)}\right)}{\overline{T}(y_1, y_2, \ldots, y_n)}$$

$$\times \prod_{i=1}^{n} \left(1 - \frac{\hat{T}_i'(y_i)}{\hat{T}_i(y_i)}\right) dy_n \ldots dy_2 dy_1.$$

4. If $\hat{f} = f^{\wedge}$, then

$$F(x) = \int_{-\infty}^{x_1} \int_{-\infty}^{x_2} \cdots \int_{-\infty}^{x_n} f(y_1\hat{T}_1(y_1), y_2\hat{T}_2(y_2), \ldots, y_n\hat{T}_n(y_n))$$

$$\times \prod_{i=1}^{n} \left(1 - \frac{\hat{T}_i'(y_i)}{\hat{T}_i(y_i)}\right) dy_n \ldots dy_2 dy_1.$$

5. If $\hat{f} = f^{\vee}$, then

$$F(x) = \int_{-\infty}^{x_1}\int_{-\infty}^{x_2}\cdots\int_{-\infty}^{x_n} f\left(\frac{y_1}{\hat{T}_1(y_1)}, \frac{y_2}{\hat{T}_2(y_2)}, \ldots, \frac{y_n}{\hat{T}_n(y_n)}\right)$$

$$\times\prod_{i=1}^{n}\left(1 - \frac{\hat{T}_i'(y_i)}{\hat{T}_i(y_i)}\right) dy_n \ldots dy_2 dy_1.$$

Remark 1.4.3. *It follows that*

$$\hat{P}(\hat{X} \in B) = \hat{\int_B} \hat{f}(x)\hat{\times}\hat{d}\hat{x}$$

for every $B \in \mathcal{B}$.

Example 1.4.4. *If* $\hat{X} : \Omega \longrightarrow \mathbb{R}$ *has density*

1.

$$\hat{f}(x) = \frac{1}{\sqrt{2\pi\sigma^2}} e^{-\frac{|x-m|^2}{2\sigma^2}} \frac{1}{\tilde{T}(x)},$$

we say $\hat{X}$ *has a Gaussian or normal iso-distribution of the first kind. In this case we will write*

$$\hat{X} \quad \text{is} \quad \text{an} \quad N^1(m, \sigma^2) \quad \text{random} \quad \text{iso} - \text{variable.}$$

2.

$$\hat{f}(x) = \frac{1}{\sqrt{2\pi\sigma^2}} e^{-\frac{|x\tilde{T}(x)-m|^2}{2\sigma^2}} \frac{1}{\tilde{T}(x)},$$

we say $\hat{X}$ *has a Gaussian or normal iso-distribution of the second kind. In this case we will write*

$$\hat{X} \quad \text{is} \quad \text{an} \quad N^2(m, \sigma^2) \quad \text{random} \quad \text{iso} - \text{variable.}$$

3.

$$\hat{f}(x) = \frac{1}{\sqrt{2\pi\sigma^2}} e^{-\frac{\left|\frac{x}{\tilde{T}(x)}-m\right|^2}{2\sigma^2}} \frac{1}{\tilde{T}(x)},$$

we say $\hat{X}$ *has a Gaussian or normal iso-distribution of the third kind. In this case we will write*

$$\hat{X} \quad \text{is} \quad \text{an} \quad N^3(m, \sigma^2) \quad \text{random} \quad \text{iso} - \text{variable.}$$

4.

$$\hat{f}(x) = \frac{1}{\sqrt{2\pi\sigma^2}} e^{-\frac{|x\tilde{T}(x)-m|^2}{2\sigma^2}},$$

we say $\hat{X}$ *has a Gaussian or normal iso-distribution of the fifth kind. In this case we will write*

$$\hat{X} \quad \text{is} \quad \text{an} \quad N^5(m, \sigma^2) \quad \text{random} \quad \text{iso} - \text{variable.}$$

5.

$$\hat{f}(x) = \frac{1}{\sqrt{2\pi\sigma^2}} e^{-\frac{\left|\frac{x}{\hat{T}(x)}-m\right|^2}{2\sigma^2}},$$

we say $\hat{X}$ has a Gaussian or normal iso-distribution of the fifth kind. In this case we will write

$$\hat{X} \quad \text{is} \quad \text{an} \quad N^5(m,\sigma^2) \quad \text{random} \quad \text{iso}-\text{variable}.$$

If $x \in \mathbb{R}$, then we will write

$$\hat{x} = \frac{x}{\tilde{T}(x)}, \qquad x^{\vee} = x\tilde{T}(x).$$

If $x \in \mathbb{R}^n$, $n > 1$, then we will write

$$\hat{x} = \left(\frac{x_1}{\hat{T}_1(x_1)}, \frac{x_2}{\hat{T}_2(x_2)}, \ldots, \frac{x_n}{\hat{T}_n(x_n)}\right), \qquad x^{\vee} = \left(x_1\hat{T}_1(x_1), x_2\hat{T}_2(x_2), \ldots, x_n\hat{T}_n(x_n)\right).$$

Definition 1.4.5. *Let $A \in U$. Then the indicator function of is defined as follows*

$$\chi_A(\omega) := \begin{cases} 1 & \text{if} \quad \omega \in A \\ 0 & \text{if} \quad \omega \notin A. \end{cases}$$

Definition 1.4.6. *If A_1, A_2, ..., $A_n \in U$, with $\Omega = \bigcup_{i=1}^{m} A_i$, and a_1, a_2, ..., a_m are real numbers, then*

$$X = \sum_{i=1}^{m} a_i \chi_{A_i}$$

is called a simple function.

Lemma 1.4.7. *Let $\hat{X} : \Omega \longrightarrow \mathbb{R}^n$ be a random iso-variable, and assume that its distribution iso-function F has the density iso-function $\hat{f}$. Suppose that $g : \mathbb{R}^n \longrightarrow \mathbb{R}$ and $g(\hat{X})$ is integrable. Then*

$$\hat{E}\left(\frac{g(\hat{X})}{\hat{T}}\right) = \int_{\mathbb{R}^n} g(y)\hat{f}(y)\prod_{i=1}^{n}\left(1 - \frac{\hat{T}_i'(y_i)}{\hat{T}_i(y_i)}\right)dy.$$

In particular,

$$\hat{E}\left(\frac{\hat{X}}{\hat{T}}\right) = \int_{\mathbb{R}^n} y\hat{f}(y)\prod_{i=1}^{n}\left(1 - \frac{\hat{T}_i'(y_i)}{\hat{T}_i(y_i)}\right)dy.$$

Proof. Suppose first g is a simple function on $\mathbb{R}^n$,

$$g = \sum_{i=1}^{m} b_i \chi_{B_i}, \qquad B_i \in \mathcal{B}.$$

Then

$$\int_{\mathbb{R}^n} g(y)\hat{f}(y)\prod_{i=1}^n \left(1-\frac{\hat{T}_i'(y_i)}{\hat{T}_i(y_i)}\right)dy = \int_{\mathbb{R}^n} \sum_{j=1}^m b_j\chi_{B_j}\hat{f}(y)\prod_{i=1}^n \left(1-\frac{\hat{T}_i'(y_i)}{\hat{T}_i(y_i)}\right)dy$$

$$= \sum_{j=1}^m b_j \int_{B_j} \hat{f}(y)\prod_{i=1}^n \left(1-\frac{\hat{T}_i'(y_i)}{\hat{T}_i(y_i)}\right)dy$$

$$= \sum_{j=1}^m b_j\hat{P}(\hat{X}\in B_j).$$

On the other hand,

$$\hat{E}\left(\frac{g(\hat{X})}{\hat{T}}\right) = \hat{\int}_\Omega \frac{g(\hat{X})}{\hat{T}}\hat{\times}d\hat{P}$$

$$= \int_\Omega \sum_{j=1}^m b_j\chi_{B_j}(\hat{X})d\hat{P}$$

$$= \sum_{j=1}^m b_j \int_\Omega \chi_{B_j}(\hat{X})d\hat{P}$$

$$= \sum_{j=1}^m b_j\hat{P}(\hat{X}\in B_j).$$

Therefore the formula holds for all simple functions g. Hence, by approximation, it holds for general functions g. □

Definition 1.4.8. *Let $\hat{X}_i : \Omega \longrightarrow \mathbb{R}^n$ be random iso-variables, $i = 1,2,\ldots$. We say that the random iso-variables $\hat{X}_1$, $\hat{X}_2$, ..., are independent if for all integer $k \geq 2$ and all choices of Borel sets B_1, B_2, ..., $B_k \subset \mathbb{R}^n$ we have*

$$\hat{P}(\hat{X}_1\in B_1,\hat{X}_2\in B_2,\ldots,\hat{X}_k\in B_k) = \hat{P}(\hat{X}_1\in B_1)\hat{P}(\hat{X}_2\in B_2)\ldots\hat{P}(\hat{X}_k\in B_k).$$

This is equivalent to say that σ-algebras $\{U(\hat{X}_i)\}_{i=1}^\infty$ are independent.

Theorem 1.4.9. *Let $\hat{X}_1$, $\hat{X}_2$, ... be independent random iso-variables, $(i_1,i_2,\ldots)$, $(j_1,j_2,\ldots)$ disjoint sets of integers. Then the fields*

$$f = f(\hat{X}_{i_1},\hat{X}_{i_2},\ldots), \qquad g = g(\hat{X}_{j_1},\hat{X}_{j_2},\ldots)$$

are independent.

Proof. Let $D \in g$ be arbitrarily chosen,

$$D = \{\hat{X}_{j_1}\in B_1,\hat{X}_{j_2}\in B_2,\ldots,\hat{X}_{j_m}\in B_m\},$$

$B_k \in \mathcal{B}$, $k = 1,2,\ldots,m$.

For arbitrarily chosen $A \in f$ we define

$$\hat{P}_1(A) = \hat{P}(A\bigcap D), \qquad \hat{P}_2(A) = \hat{P}(A)\hat{P}(D),$$

$$\hat{P}_2(D) = \hat{P}(A\bigcap D), \qquad \hat{P}_3(D) = \hat{P}(A)\hat{P}(D).$$

Let now

$$C = \{\hat{X}_{j_1}\in E_1,\hat{X}_{j_2}\in E_2,\ldots,\hat{X}_{j_n}\in E_n\},$$

$E_i \in \mathcal{B}$, $i = 1, 2, \ldots, n$. We note that

$$\hat{P}_1(C) = \hat{P}(C \bigcap D)$$

$$= \hat{P}(\bigcap_{l=1}^{n} \{\hat{X}_{i_l} \in E_l\} \bigcap_{k=1}^{m} \{\hat{X}_{j_k} \in B_k\})$$

$$(1) \qquad = \prod_{l=1}^{n} \hat{P}(\hat{X}_{i_l} \in E_l) \prod_{k=1}^{m} \hat{P}(\hat{X}_{j_k} \in B_k)$$

$$= \hat{P}(A)\hat{P}(D)$$

$$= \hat{P}_2(C).$$

Thus

$$\hat{P}_1 = \hat{P}_2 \qquad \text{on} \qquad C.$$

Because C is closed under $\bigcap$, we have that $U(C) = f$, i.e.,

$$\hat{P}_1 = \hat{P}_2 \qquad \text{on} \qquad f.$$

Also, form (1), we have

$$\hat{P}_3(D) = \hat{P}_4(D),$$

which implies, because D is closed under $\bigcap$,

$$\hat{P}_3 = \hat{P}_4 \qquad \text{on} \qquad g.$$

Consequently, for any $A_1 \in f$ and $A_2 \in g$ we have

$$\hat{P}(A_1 \bigcap A_2) = \hat{P}(A_1)\hat{P}(A_2).$$

□

Theorem 1.4.10. *The random iso-variable* $\hat{X}_1$, $\hat{X}_2$, …, $\hat{X}_m : \Omega \longrightarrow \mathbb{R}^n$ *are independent if and only if*

$$(2) \qquad F_{\hat{X}_1, \hat{X}_2, \ldots, \hat{X}_m} = F_{\hat{X}_1}(x_1) F_{\hat{X}_2}(x_2) \ldots F_{\hat{X}_m}(x_m)$$

for all $x_i \in \mathbb{R}^n$, $i = 1, 2, \ldots, m$.

If the random iso-variable have iso-densities, then (2) *is equivalent to*

$$\hat{f}_{\hat{X}_1, \hat{X}_2, \ldots, \hat{X}_m} = \hat{f}_{\hat{X}_1}(x_1) \hat{f}_{\hat{X}_2}(x_2) \ldots \hat{f}_{\hat{X}_m}(x_m)$$

for all $x_i \in \mathbb{R}^n$, $i = 1, 2, \ldots, m$, *where the functions* $\hat{f}$ *are the appropriate iso-densities.*

Proof. 1. Let $\hat{X}_1$, $\hat{X}_2$, …, $\hat{X}_m$ are independent. Then

$$F_{\hat{X}_1, \hat{X}_2, \ldots, \hat{X}_m}(x_1, x_2, \ldots, x_m) = \hat{P}(\hat{X}_1 \le x_1, \hat{X}_2 \le x_2, \ldots, \hat{X}_m \le x_m)$$

$$= \hat{P}(\hat{X}_1 \le x_1) \hat{P}(\hat{X}_2 \le x_2) \ldots \hat{P}(\hat{X}_m \le x_m)$$

$$= F_{\hat{X}_1}(x_1) F_{\hat{X}_2}(x_2) \ldots F_{\hat{X}_m}(x_m).$$

2. Let B_1, B_2, ..., $B_m \subset \mathbb{R}^n$ are arbitrarily chosen Borel sets. Let also, $A_i = \hat{X}_i(B_i^{-1})$ and $2 \le l \le m$. Then we have

$$\hat{P}(A_1 \bigcap A_2 \bigcap \ldots \bigcap A_l) = \hat{P}(\hat{X}_1 \in B_1, \hat{X}_2 \in B_2, \ldots, \hat{X}_l \in B_l)$$

$$= F_{\hat{X}_1,\hat{X}_2,\ldots,\hat{X}_l}(x_1,x_2,\ldots,x_l)$$

$$= F_{\hat{X}_1}(x_1)F_{\hat{X}_2}(x_2)\ldots F_{\hat{X}_l}(x_l)$$

$$= \hat{P}(\hat{X}_1 \in B_1)\hat{P}(\hat{X}_2 \in B_2)\ldots\hat{P}(\hat{X}_k \in B_k)$$

$$= \hat{P}(A_1)\hat{P}(A_2)\ldots\hat{P}(A_l).$$

Therefore $\hat{X}_1$, $\hat{X}_2$, ..., $\hat{X}_m$ are independent iso-variables.

□

Exercise 1.4.11. *Let $\hat{X}_1$, $\hat{X}_2$, ..., $\hat{X}_m : \Omega \longrightarrow \mathbb{R}^n$ are independent iso-variables. Let also, a_1, a_2, ..., a_m are arbitrarily chosen real constants. Prove that*

$$\hat{X}_1 + a_1, \quad \hat{X}_2 + a_2, \quad \ldots, \quad \hat{X}_m + a_m$$

are independent iso-variables.

Exercise 1.4.12. *Let $\hat{X}_1$, $\hat{X}_2$, ..., $\hat{X}_m : \Omega \longrightarrow \mathbb{R}^n$ are independent iso-variables. Let also, a_1, a_2, ..., a_m are arbitrarily chosen positive real constants. Prove that*

$$a_1\hat{X}_1, \quad a_2\hat{X}_2, \quad \ldots, \quad a_m\hat{X}_m$$

are independent iso-variables.

Theorem 1.4.13. *Let $\hat{X}_1$, $\hat{X}_2$, ..., $\hat{X}_m$ are independent, real-valued iso-variables, with*

$$\hat{E}\left(\frac{|\hat{X}_i|}{\hat{T}}\right) < \infty, \qquad i = 1,2,\ldots,m.$$

Then

$$\hat{E}\left(\frac{|\hat{X}_1\hat{X}_2\ldots\hat{X}_m|}{\hat{T}}\right) < \infty$$

and

$$\hat{E}\left(\frac{|\hat{X}_1\hat{X}_2\ldots\hat{X}_m|}{\hat{T}}\right) = \hat{E}\left(\frac{|\hat{X}_1|}{\hat{T}}\right)\hat{E}\left(\frac{|\hat{X}_2|}{\hat{T}}\right)\ldots\hat{E}\left(\frac{|\hat{X}_m|}{\hat{T}}\right).$$

Proof. Suppose that $\hat{X}_i$ is bounded and has an iso-density. Then

$$\hat{E}\left(\frac{\hat{X}_1\hat{X}_2\ldots\hat{X}_m}{\hat{T}}\right) = \int_{\mathbb{R}^m} x_1x_2\ldots x_m \hat{f}_{\hat{X}_1,\hat{X}_2,\ldots,\hat{X}_m}(x_1,x_2,\ldots,x_m)\prod_{l=1}^m\left(1 - \frac{\hat{T}_l'(x_l)}{\hat{T}_l(x_l)}\right)dx$$

$$= \int_{\mathbb{R}^m} x_1x_2\ldots x_m \hat{f}_{\hat{X}_1}(x_1)\hat{f}_{\hat{X}_2}(x_2)\ldots\hat{f}_{\hat{X}_m}(x_m)\prod_{l=1}^m\left(1 - \frac{\hat{T}_l'(x_l)}{\hat{T}_l(x_l)}\right)dx$$

$$= \left(\int_{\mathbb{R}} x_1\hat{f}_{X_1}(x_1)\left(1 - \frac{\hat{T}_1'(x_1)}{\hat{T}_1(x_1)}\right)dx_1\right)\ldots\left(\int_{\mathbb{R}} x_m\hat{f}_{X_m}(x_m)\left(1 - \frac{\hat{T}_m'(x_m)}{\hat{T}_m(x_m)}\right)dx_m\right)$$

$$= \hat{E}\left(\frac{\hat{X}_1}{\hat{T}}\right)\ldots\hat{E}\left(\frac{\hat{X}_m}{\hat{T}}\right).$$

□

Lemma 1.4.14. *Let $\hat{X}:\Omega \longrightarrow \mathbb{R}^n$ be random iso-variable and $m=\hat{E}(\hat{X})$. Then*

$$\hat{E}\left(\hat{X}-\frac{m}{\hat{T}}\right)=0.$$

Proof. We have

$$\hat{E}\left(\hat{X}-\frac{m}{\hat{T}}\right)=\hat{\int}_{\Omega}\left(\hat{X}-\frac{m}{\hat{T}}\right)\hat{\times}\hat{d}\hat{P}$$

$$=\hat{\int}_{\Omega}\hat{X}\hat{\times}\hat{d}\hat{P}-\hat{\int}_{\Omega}\frac{m}{\hat{T}}\hat{\times}\hat{d}\hat{P}$$

$$=\hat{\int}_{\Omega}\hat{X}\hat{\times}\hat{d}\hat{P}-m\int_{\Omega}\hat{T}^{-1}\frac{1}{\hat{T}}\hat{T}\hat{T}d\hat{P}$$

$$=\hat{\int}_{\Omega}\hat{X}\hat{\times}\hat{d}\hat{P}-m\int_{\Omega}d\hat{P}$$

$$=m-m\hat{P}(\Omega)$$

$$=m-m$$

$$=0.$$

□

Theorem 1.4.15. *If*

$$\hat{X}_1+\frac{a_1}{\hat{T}},\quad \hat{X}_+\frac{a_2}{\hat{T}},\quad \ldots,\quad \hat{X}_m+\frac{a_m}{\hat{T}}$$

are independent, real-valued iso-variables for all $(a_1,a_2,\ldots,a_m)\in\mathbb{R}^m$, with $\hat{V}(\hat{X}_i)<\infty$, $i=1,2,\ldots,m$, then

$$\hat{V}(\hat{X}_1+\hat{X}_2+\cdots+\hat{X}_m)=\hat{V}(\hat{X}_1)+\hat{V}(\hat{X}_2)+\cdots+\hat{V}(\hat{V}_m).$$

Proof. Firstly, we observe that, if $\hat{E}(\hat{X}_i)=m_i$, $i=1,2,\ldots,m$, $2\le l\le m$,

$$\hat{E}(\hat{X}_1+\hat{X}_2+\cdots+\hat{X}_m)=\hat{\int}_{\Omega}(\hat{X}_1+\hat{X}_2+\cdots+\hat{X}_l)\hat{\times}\hat{d}\hat{P}$$

$$=\hat{\int}_{\Omega}\hat{X}_1\hat{\times}\hat{d}\hat{P}+\hat{\int}_{\Omega}\hat{X}_2\hat{\times}\hat{d}\hat{P}+\cdots+\hat{\int}_{\Omega}\hat{X}_l\hat{\times}\hat{d}\hat{P}$$

$$=\hat{E}(\hat{X}_1)+\hat{E}(\hat{X}_2)+\cdots+\hat{E}(\hat{X}_l)$$

$$=m_1+m_2+\cdots+m_l.$$

We will prove the assertion in the case $m=2$.

We have

$$\hat{V}(\hat{X}_1+\hat{X}_2)=\hat{\int}_{\Omega}\left(\hat{X}_1+\hat{X}_2-\frac{\hat{E}(\hat{X}_1+\hat{X}_2)}{\hat{T}}\right)^2\hat{\times}\hat{d}\hat{P}$$

$$=\hat{\int}_{\Omega}\left(\hat{X}_1+\hat{X}_2-\frac{m_1+m_2}{\hat{T}}\right)^2\hat{\times}\hat{d}\hat{P}$$

$$=\hat{\int}_{\Omega}\left(\hat{X}_1-\frac{m_1}{\hat{T}}+\hat{X}_2-\frac{m_2}{\hat{T}}\right)^2\hat{\times}\hat{d}\hat{P}$$

$$=\hat{\int}_{\Omega}\left(\hat{X}_1-\frac{m_1}{\hat{T}}\right)^2\hat{\times}\hat{d}\hat{P}+\hat{\int}_{\Omega}\left(\hat{X}_2-\frac{m_2}{\hat{T}}\right)^2\hat{\times}\hat{d}\hat{P}+2\hat{\int}_{\Omega}\left(\hat{X}_1-\frac{m_1}{\hat{T}}\right)\left(\hat{X}_2-\frac{m_2}{\hat{T}}\right)\hat{\times}\hat{d}\hat{P}$$

$$=\hat{V}(\hat{X}_1)+\hat{V}(\hat{X}_2)-2\hat{\int}_{\Omega}\left(\hat{X}_1-\frac{m_1}{\hat{T}}\right)\hat{\times}\hat{d}\hat{P}\hat{\int}_{\Omega}\left(\hat{X}_2-\frac{m_2}{\hat{T}}\right)\hat{\times}\hat{d}\hat{P}$$

$$=\hat{V}(\hat{X}_1)+\hat{V}(\hat{X}_2)-2\hat{E}\left(\hat{X}_1-\frac{m_1}{\hat{T}}\right)\hat{E}\left(\hat{X}_2-\frac{m_2}{\hat{T}}\right)$$

$$=\hat{V}(\hat{X}_1)+\hat{V}(\hat{X}_2).$$

□

Definition 1.4.16. *Let $\hat{X}$ be an $\mathbb{R}^n$-valued random iso-variable. Then*

$$\hat{\phi}_{\hat{X}}(\lambda):=\hat{E}\left(\frac{e^{i\lambda\cdot\hat{X}}}{\hat{T}}\right)$$

is the characteristic iso-function of $\hat{X}$.

Theorem 1.4.17. *Let $\hat{X}_1$, $\hat{X}_2$, ..., $\hat{X}_m$ be independent random iso-variables. Then for each $\lambda\in\mathbb{R}^n$*

$$\hat{\phi}_{\hat{X}_1+\hat{X}_2+\cdots+\hat{X}_m}(\lambda)=\hat{\phi}_{\hat{X}_1}(\lambda)\hat{\phi}_{\hat{X}_2}(\lambda)\ldots\hat{\phi}_{\hat{X}_m}(\lambda).$$

Proof. We have

$$\hat{\phi}_{\hat{X}_1+\hat{X}_2+\cdots+\hat{X}_m}(\lambda)=\hat{E}\left(\frac{e^{i\lambda\cdot(\hat{X}_1+\hat{X}_2+\cdots+\hat{X}_m)}}{\hat{T}}\right)$$

$$=\hat{E}\left(\frac{e^{i\lambda\cdot\hat{X}_1+i\lambda\cdot\hat{X}_2+\cdots+i\lambda\cdot\hat{X}_m}}{\hat{T}}\right)$$

$$=\hat{E}\left(\frac{e^{i\lambda\cdot\hat{X}_1}e^{i\lambda\cdot\hat{X}_2}\ldots e^{i\lambda\cdot\hat{X}_m}}{\hat{T}}\right)$$

$$=\hat{E}\left(\frac{e^{i\lambda\cdot\hat{X}_1}}{\hat{T}}\right)\hat{E}\left(\frac{e^{i\lambda\cdot\hat{X}_2}}{\hat{T}}\right)\ldots\hat{E}\left(\frac{e^{i\lambda\cdot\hat{X}_m}}{\hat{T}}\right)$$

$$=\hat{\phi}_{\hat{X}_1}(\lambda)\hat{\phi}_{\hat{X}_2}(\lambda)\ldots\hat{\phi}_{\hat{X}_m}(\lambda).$$

□

Exercise 1.4.18. *Let $\hat{X}$ be a real-valued random iso-variable. Prove that*

$$\hat{\phi}^{(k)}_{\hat{X}}(0) = i^k \hat{E}\left(\frac{\hat{X}^k}{\hat{T}}\right), \qquad k = 0, 1, \ldots.$$

Exercise 1.4.19. *Let $\hat{X}$ and $\hat{Y}$ be real-valued random iso-variables and*

$$\hat{\phi}_{\hat{X}}(\lambda) = \hat{\phi}_{\hat{Y}}(\lambda) \qquad \text{for} \qquad \text{all} \qquad \lambda \in \mathbb{R}.$$

Prove that

$$F_{\hat{X}}(x) = F_{\hat{Y}}(x) \qquad \text{for} \qquad \text{all} \qquad x \in \mathbb{R}.$$

Definition 1.4.20. *A sequence $\hat{X}_1$, $\hat{X}_2$, ..., $\hat{X}_n$, ... of real-valued random iso-variables is called independent iso-distributed if*

$$F_{\hat{X}_1}(x) = F_{\hat{X}_2}(x) = \ldots = F_{\hat{X}_n}(x) = \ldots \qquad \text{for} \qquad \text{all} \qquad x.$$

Theorem 1.4.21. *(iso-strong law of large numbers) Let $P(A) \leq P(B)$, $\hat{T}(A) \geq \hat{T}(B)$, $\hat{T}(A) \geq Q$ for all A, $B \in U$, $A \subset B$ and for some positive constant Q. Let also, $\hat{X}_1$, $\hat{X}_2$, ..., $\hat{X}_n$, ... be a sequence of independent, identically iso-distributed, integrable iso-variables defined on the same iso-probability space,*

$$\left| \int_\Omega \left(\sum_{i=1}^n \hat{X}_i(\omega)\right)^4 \frac{\hat{T}'(\omega)}{\hat{T}(\omega)} P(\omega) d\omega \right| \leq M$$

for all $n \subset \mathbb{N}$, $\hat{E}(\hat{X}_i^2) < \infty$, $\hat{E}(\hat{X}_i^4) < \infty$, $i = 1, 2, \ldots$. Write $m := \hat{E}(\hat{X}_i)$, $i = 1, 2, \ldots$. Then

$$\hat{P}\left(\lim_{n \longrightarrow \infty} \frac{\hat{X}_1 + \hat{X}_2 + \cdots + \hat{X}_n}{n} = \frac{m}{\hat{T}}\right) = \frac{1}{\hat{T}(\Omega)}.$$

Proof. Since $\hat{X}_1$, $\hat{X}_2$, ..., $\hat{X}_n$, ... are independent, identically iso-distributed random iso-variables, we have that

$$\hat{E}(\hat{X}_i^l) = \hat{E}(\hat{X}_m^l)$$

for all i, l, $m \in \mathbb{N}$.

Without loss of generality we assume that $m = 0$. Otherwise, we will consider the random iso-variable $\hat{X}_i - \frac{m}{\hat{T}}$, $i = 1, 2, \ldots$. Therefore we have to prove that

$$\hat{P}\left(\lim_{n \longrightarrow \infty} \frac{\hat{X}_1 + \hat{X}_2 + \cdots + \hat{X}_n}{n} = 0\right) = \frac{1}{\hat{T}(\Omega)}.$$

We have

$$\hat{E}\left(\left(\sum_{i=1}^n \hat{X}_i\right)^4\right) = \sum_{i,j,k,l=1}^n \hat{E}(\hat{X}_i \hat{X}_j \hat{X}_k \hat{X}_l).$$

Also, if $i \neq j, k$ and l, we have

$$\hat{E}(\hat{X}_i \hat{X}_j \hat{X}_k \hat{X}_l) = \hat{E}(\hat{X}_i) \hat{E}(\hat{X}_j \hat{X}_k \hat{X}_l)$$

$$= 0 \cdot \hat{E}(\hat{X}_j \hat{X}_k \hat{X}_l)$$

$$= 0.$$

Hence,

$$\hat{E}\left(\left(\sum_{i=1}^n \hat{X}_i\right)^4\right) = \sum_{i=1}^n \hat{E}(\hat{X}_i^4) + 4\sum_{i,j=1,i<j}^n \hat{E}(\hat{X}_i^2\hat{X}_j^2)$$

$$= \sum_{i=1}^n \hat{E}(\hat{X}_i^4) + 4\sum_{i,j=1,i<j}^n \hat{E}(\hat{X}_i^2)\hat{E}(\hat{X}_j^2)$$

$$= \sum_{i=1}^n \hat{E}(\hat{X}_i^4) + 4\sum_{i,j=1,i<j}^n \hat{E}(\hat{X}_1^2)^2$$

$$= n\hat{E}(\hat{X}_1^4) + 4\tfrac{n(n-1)}{2}\hat{E}(\hat{X}_1^2)$$

$$\leq Cn^2,$$

i.e.,

$$\hat{E}\left(\left(\sum_{i=1}^n \hat{X}_i\right)^4\right) \leq Cn^2,$$

for some positive constant C.

On the other hand, using iso-Chebychev inequality, we have

$$\hat{P}\left(\left|\tfrac{1}{n}\sum_{i=1}^n \hat{X}_i\right| \geq \varepsilon\right) = \hat{P}\left(\left|\sum_{i=1}^n \hat{X}_i\right| \geq \varepsilon n\right)$$

$$\leq \tfrac{1}{\varepsilon^4 n^4}\hat{E}\left(\left(\sum_{i=1}^n \hat{X}_i\right)^4\right) + \tfrac{1}{\varepsilon^4 n^4}\int_\Omega \left|\sum_{i=1}^n \hat{X}_i\right|^4 \tfrac{\hat{T}'(\omega)}{\hat{T}(\omega)}P(\omega)d\omega$$

$$\leq \tfrac{Cn^2}{\varepsilon^4 n^4} + \tfrac{M}{\varepsilon^4 n^4}$$

$$= \tfrac{C}{\varepsilon^4 n^2} + \tfrac{M}{\varepsilon^4 n^4},$$

whereupon

$$\sum_{n=1}^{\infty} \hat{P}\left(\left|\frac{1}{n}\sum_{i=1}^n \hat{X}_i\right| \geq \varepsilon\right) \leq \frac{C}{\varepsilon^4}\sum_{n=1}^{\infty}\frac{1}{n^4} + M\sum_{n=1}^{\infty}\frac{1}{n^4} < \infty.$$

From here and from the iso-Borel-Cantelli lemma it follows that

(3) $\quad \hat{P}\left(\left|\tfrac{1}{n}\sum_{i=1}^n \hat{X}_i\right| \geq \varepsilon \text{i.o.}\right) = 0.$

Let now $\varepsilon = \frac{1}{k}$. Then, using (3), we get

$$\limsup_{k\longrightarrow\infty}\left|\frac{1}{n}\sum_{i=1}^n \hat{X}_i(\omega)\right| \leq \frac{1}{k}$$

except possible for $\omega \in B_k$ such that $\hat{P}(B_k) = 0$.

We put

$$B := \bigcup_{k=1}^{\infty} B_k.$$

Then

$$\hat{P}(B) = \hat{P}(\bigcup_{k=1}^{\infty})$$

$$\leq \sum_{k=1}^{\infty} \hat{P}(B_k)$$

$$= 0.$$

Therefore

$$\lim_{n\longrightarrow\infty} \frac{1}{n}\sum_{i=1}^{n} \hat{X}_i(\omega) = 0$$

for $\omega \notin B$.

Consequently,

$$\hat{P}\left(\lim_{n\longrightarrow\infty} \frac{1}{n}\sum_{i=1}^{n} \hat{X}_i(\omega)\right) = \frac{1}{\hat{T}(\Omega)}.$$

□

Theorem 1.4.22. *Suppose that the real-valued random iso-variables $\hat{X}_1$, $\hat{X}_2$, ..., $\hat{X}_n$, ... are independent and identically iso-distributed, with*

$$\begin{cases} \hat{P}\left(\hat{X}_i = \frac{1}{\hat{T}}\right) = p \\ \hat{P}\left(\hat{X}_i = 0\right) = q, \end{cases}$$

for p, $q \geq 0$, $p+q = \frac{1}{\hat{T}(\Omega)}$,

$$m_i = \int_{\hat{X}_i=\frac{1}{\hat{T}}} \frac{1}{\hat{T}} d\hat{P}, \qquad k_i = \int_{\hat{X}_i=0} \frac{1}{\hat{T}} d\hat{P}.$$

Then

$$\hat{E}(\hat{X}_1 + \hat{X}_2 + \cdots + \hat{X}_n) = np,$$

$$\hat{V}(\hat{X}_1 + \hat{X}_2 + \cdots + \hat{X}_n) = (1-p)^2 \sum_{i=1}^{n} m_i + p^2 \sum_{i=1}^{n} k_i.$$

Proof. We have

$$\hat{E}(\hat{X}_i) = \hat{\int}_{\Omega} \hat{X}_i \hat{\times} \hat{d}\hat{P}$$

$$= \hat{\int}_{\hat{X}_i=\frac{1}{\hat{T}}} \hat{X}_i \hat{\times} \hat{d}\hat{P} + \hat{\int}_{\hat{X}_i=0} \hat{X}_i \hat{\times} \hat{d}\hat{P}$$

$$= \int_{\hat{X}_i=\frac{1}{\hat{T}}} \frac{1}{\hat{T}} \frac{1}{\hat{T}} \hat{T}\hat{T} d\hat{P}$$

$$= \int_{\hat{X}_i=\frac{1}{\hat{T}}} d\hat{P}$$

$$= \hat{P}\left(\hat{X}_i = \frac{1}{\hat{T}}\right)$$

$$= p.$$

Therefore

$$\hat{E}(\hat{X}_1+\hat{X}_2+\cdots+\hat{X}_n)=\hat{E}(\hat{X}_1)+\hat{E}(\hat{X}_2)+\cdots+\hat{E}(\hat{X}_n)$$

$$=p+p+\cdots+p$$

$$=np.$$

For the iso-variance we have

$$\hat{V}(\hat{X}_i)=\hat{\int}_{\Omega}\left(\hat{X}_i-\frac{\hat{E}(\hat{X}_i)}{\hat{T}}\right)^2\hat{\times}\hat{d}\hat{P}$$

$$=\hat{\int}_{\Omega}\left(\hat{X}_i-\frac{p}{\hat{T}}\right)^2\hat{\times}\hat{d}\hat{P}$$

$$=\hat{\int}_{\Omega}\left(\hat{X}_i^2-2\frac{p}{\hat{T}}\hat{X}_i+\frac{p^2}{\hat{T}^2}\right)\hat{\times}\hat{d}\hat{P}$$

$$=\hat{\int}_{\hat{X}_i=\frac{1}{\hat{T}}}\left(\hat{X}_i^2-2\frac{p}{\hat{T}}\hat{X}_i+\frac{p^2}{\hat{T}^2}\right)\hat{\times}\hat{d}\hat{P}+\hat{\int}_{\hat{X}_i=0}\left(\hat{X}_i^2-2\frac{p}{\hat{T}}\hat{X}_i+\frac{p^2}{\hat{T}^2}\right)\hat{\times}\hat{d}\hat{P}$$

$$=\hat{\int}_{\hat{X}_i=\frac{1}{\hat{T}}}\left(\frac{1}{\hat{T}^2}-2\frac{p}{\hat{T}^2}+\frac{p^2}{\hat{T}^2}\right)\hat{\times}\hat{d}\hat{P}+p^2\hat{\int}_{\hat{X}_i=0}\frac{1}{\hat{T}^2}\hat{\times}\hat{d}\hat{P}$$

$$=(1-p)^2\int_{\hat{X}_i=\frac{1}{\hat{T}}}\frac{1}{\hat{T}}\frac{1}{\hat{T}^2}\hat{T}\hat{T}d\hat{P}+p^2\int_{\hat{X}_i=0}\frac{1}{\hat{T}}\frac{1}{\hat{T}^2}\hat{T}\hat{T}d\hat{P}$$

$$=(1-p)^2\int_{\hat{X}_i=\frac{1}{\hat{T}}}\frac{1}{\hat{T}}d\hat{P}+p^2\int_{\hat{X}_i=0}\frac{1}{\hat{T}}d\hat{P}$$

$$=(1-p)^2m_i+p^2k_i.$$

Therefore

$$\hat{V}(\hat{X}_1+\hat{X}_2+\cdots+\hat{X}_n)=\hat{V}(\hat{X}_1)+\hat{V}(\hat{X}_2)+\cdots+\hat{V}(\hat{X}_n)$$

$$=(1-p)^2m_1+p^2k_1+(1-p)^2m_2+p^2k_2+\cdots+(1-p)^2m_n+p^2k_n$$

$$=(1-p)^2\sum_{i=1}^n m_i+p^2\sum_{i=1}^n k_i.$$

□

Corollary 1.4.23. *Suppose that the real-valued random iso-variables $\hat{X}_1$, $\hat{X}_2$, …, $\hat{X}_n$, … are independent and identically iso-distributed, with*

$$\begin{cases}\hat{P}\left(\hat{X}_i=\frac{1}{\hat{T}}\right)=p\\ \hat{P}\left(\hat{X}_i=0\right)=q,\end{cases}$$

for $p,\ q\geq 0$, $p+q=\frac{1}{\hat{T}(\Omega)}$,

$$m=\int_{\hat{X}_i=\frac{1}{\hat{T}}}\frac{1}{\hat{T}}d\hat{P},\qquad r=\int_{\hat{X}_i=0}\frac{1}{\hat{T}}d\hat{P},\qquad i=1,2,\ldots,$$

$$l=m(1-p)^2+rp^2.$$

Then

$$\hat{E}(\hat{X}_1+\hat{X}_2+\cdots+\hat{X}_n)=np,$$

$$\hat{V}(\hat{X}_1+\hat{X}_2+\cdots+\hat{X}_n)=nl.$$

Proof. Here we have that $m_i=m$, $k_i=r$, $i=1,2,\ldots$. Then from the previous theorem it follows that

$$\hat{V}(\hat{X}_1+\hat{X}_2+\cdots+\hat{X}_n)=(1-p)^2\sum_{i=1}^n m_i+p^2\sum_{i=1}^n k_i$$

$$=(1-p)^2\sum_{i=1}^n m+p^2\sum_{i=1}^n r$$

$$=nm(1-p)^2+nrp^2$$

$$=n\left(m(1-p)^2+rp^2\right)$$

$$=nl.$$

□

Theorem 1.4.24. *Suppose that the real-valued random iso-variables $\hat{X}_1$, $\hat{X}_2$, …, $\hat{X}_n$, … are independent and identically iso-distributed, with*

$$\begin{cases}\hat{P}\left(\hat{X}_i=\frac{1}{\hat{T}}\right)=p\\ \hat{P}\left(\hat{X}_i=0\right)=q,\end{cases}$$

for $p,\ q\geq 0$, $p+q=\frac{1}{\hat{T}(\Omega)}$,

$$m=\int_{\hat{X}_i=\frac{1}{\hat{T}}}\frac{1}{\hat{T}}d\hat{P},\qquad r=\int_{\hat{X}_i=0}\frac{1}{\hat{T}}d\hat{P},\qquad i=1,2,\ldots,$$

$$l=m(1-p)^2+rp^2.$$

Define the sums

$$S_n:=\hat{X}_1+\hat{X}_2+\cdots+\hat{X}_n.$$

Then for all $-\infty<a<b<\infty$ we have

$$\lim_{n\longrightarrow\infty}\hat{P}\left(a\leq\frac{S_n-np}{\sqrt{nr}}\leq b\right)=\sqrt{\frac{r}{pq}}\frac{1}{\hat{T}^2(\Omega)\sqrt{2\pi}}\int_a^b e^{-\frac{r}{2pq\hat{T}^2(\Omega)}x^2}dx.$$

Proof. Let

$$S_n^*=\frac{S_n-np}{\sqrt{nr}}$$

and we consider S_n^* as a random iso-variable taking on the values

$$x_k=\frac{k-n\hat{T}(\Omega)p}{\sqrt{nr}},\qquad k=0,1,2,\ldots,n,$$

with iso-probability

$$p_n(k) = \begin{pmatrix} n \\ k \end{pmatrix} p^k q^{n-k} \hat{T}^{n-1}(\Omega),$$

i.e.,

$$\hat{P}(S_n^*(x_k)) = \begin{pmatrix} n \\ k \end{pmatrix} p^k q^{n-k} \hat{T}^{n-1}(\Omega).$$

We have

$$\hat{P}(\Omega) = \Sigma_{k=0}^{n} \hat{P}(S_n^*(x_k))$$

$$= \hat{T}^{n-1}(\Omega) \Sigma_{k=0}^{n} \begin{pmatrix} n \\ k \end{pmatrix} p^k q^{n-k}$$

$$= \hat{T}^{n-1}(\Omega)(p+q)^n$$

$$= \hat{T}^{n-1}(\Omega) \frac{1}{\hat{T}^n(\Omega)}$$

$$= \frac{1}{\hat{T}(\Omega)}.$$

We observe that

$$x_{k+1} - x_k = \frac{k+1-n\hat{T}(\Omega)p}{\sqrt{nr}} - \frac{k-n\hat{T}(\Omega)p}{\sqrt{nr}}$$

$$= \frac{k+1-n\hat{T}(\Omega)p-k+n\hat{T}(\Omega)p}{\sqrt{nr}}$$

$$= \frac{1}{\sqrt{nr}}.$$

Thus, the points x_k divide the interval

$$\left[-\frac{n\hat{T}(\Omega)p}{\sqrt{nr}}, \frac{n\hat{T}(\Omega)q}{\sqrt{nr}} \right]$$

into n subintervals of length

$$h := \frac{1}{\sqrt{nr}}.$$

We note that

$$k - n\hat{T}(\Omega)p = x_k \sqrt{nr} \qquad \Longrightarrow$$

$$k = n\hat{T}(\Omega)p + x + k\sqrt{nr}.$$

Hence, if n goes infinity and at the same time k changes so that $|x_k|$ is bounded, then

$$k \longrightarrow_{n \longrightarrow \infty} \infty$$

and

$$\begin{aligned}
n-k &= n-\left(n\hat{T}(\Omega)p+x+k\sqrt{nr}\right)\\
&= n-n\hat{T}(\Omega)p-x_k\sqrt{nr}\\
&= n\hat{T}(\Omega)\left(\frac{1}{\hat{T}(\Omega)}-p\right)-x_k\sqrt{nr}\\
&= n\hat{T}(\Omega)q-x_k\sqrt{nr}\\
&\longrightarrow_{n\longrightarrow\infty}\infty.
\end{aligned}$$

From the Stirling's formula we have that

$$n! = e^{-n}n^n\sqrt{2\pi n}(1+o(1)), \qquad \text{as} \qquad n\longrightarrow\infty.$$

Then

$$\begin{aligned}
p_n(k) &= \hat{T}^{n-1}(\Omega)\left(\begin{array}{c} n\\ k\end{array}\right)p^kq^{n-k}\\
&= \hat{T}^{n-1}(\Omega)\frac{n!}{k!(n-k)!}p^kq^{n-k}\\
&= \hat{T}^{n-1}(\Omega)\frac{e^{-n}n^n\sqrt{2\pi n}(1+o(1))p^kq^{n-k}}{e^{-k}k^k\sqrt{2\pi k}(1+o(1))e^{-(n-k)}(n-k)^{n-k}\sqrt{2\pi(n-k)}(1+o(1))}\\
&= \hat{T}^{n-1}(\Omega)\frac{e^{-n}n^n\sqrt{2\pi n}p^kq^{n-k}}{e^{-k}k^k\sqrt{2\pi k}e^{-n}e^k(n-k)^{n-k}\sqrt{2\pi(n-k)}}(1+o(1))\\
&= \hat{T}^{n-1}(\Omega)\frac{1}{\sqrt{2\pi}}\sqrt{\frac{n}{k(n-k)}}\left(\frac{np}{k}\right)^k\left(\frac{nq}{n-k}\right)^{n-k}(1+o(1)),
\end{aligned}$$

i.e.,

$$p_n(k) = \hat{T}^{n-1}(\Omega)\frac{1}{\sqrt{2\pi}}\sqrt{\frac{n}{k(n-k)}}\left(\frac{np}{k}\right)^k\left(\frac{nq}{n-k}\right)^{n-k}(1+o(1)). \tag{4}$$

Also, if

$$x = x_k = \frac{k-n\hat{T}(\Omega)p}{\sqrt{nr}},$$

then

$$1+\frac{\sqrt{r}}{\sqrt{n}\hat{T}(\Omega)p}x = 1+\frac{\sqrt{r}}{\sqrt{n}\hat{T}(\Omega)p}\frac{k-n\hat{T}(\Omega)p}{\sqrt{nr}}$$

$$= 1+\frac{k-n\hat{T}(\Omega)p}{n\hat{T}(\Omega)p}$$

$$= \frac{k}{n\hat{T}(\Omega)p},$$

$$1-\frac{\sqrt{r}}{\sqrt{n}\hat{T}(\Omega)q}x = 1-\frac{\sqrt{r}}{\sqrt{n}\hat{T}(\Omega)q}\frac{k-n\hat{T}(\Omega)p}{\sqrt{nr}}$$

$$= 1-\frac{k-n\hat{T}(\Omega)p}{n\hat{T}(\Omega)q}$$

$$= \frac{n\hat{T}(\Omega)q-k+n\hat{T}(\Omega)p}{n\hat{T}(\Omega)q}$$

$$= \frac{n-k}{n\hat{T}(\Omega)q}.$$

Therefore

$$(5) \qquad p_n(k) = \frac{1}{\hat{T}(\Omega)\sqrt{2\pi}}\sqrt{\frac{n}{k(n-k)}}\left(\frac{n\hat{T}(\Omega)p}{k}\right)^k\left(\frac{n\hat{T}(\Omega)q}{n-k}\right)^{n-k}(1+o(1)).$$

Now, using that

$$\log(1+\pm y) = \pm y - \frac{y^2}{2} + O(y^3) \qquad \text{as} \qquad y \longrightarrow 0,$$

we have

$$\log\left(\frac{np\hat{T}(\Omega)}{k}\right)^k = -k\log\left(\frac{k}{np\hat{T}(\Omega)}\right)$$

$$= -k\log\left(1+\frac{\sqrt{r}}{\sqrt{n}\hat{T}(\Omega)p}x\right)$$

$$= -k\left(\frac{\sqrt{r}}{\sqrt{n}\hat{T}(\Omega)p}x - \frac{r}{2n\hat{T}^2(\Omega)p^2}x^2\right) + O(n^{-\frac{1}{2}})$$

$$= -\left(n\hat{T}(\Omega)p + x\sqrt{nr}\right)\left(\frac{\sqrt{r}}{\sqrt{n}\hat{T}(\Omega)p}x - \frac{r}{2n\hat{T}^2(\Omega)p^2}x^2\right) + O(n^{-\frac{1}{2}})$$

$$= -\left(\sqrt{nr}x - \frac{r}{2\hat{T}(\Omega)p}x^2 + \frac{r}{\hat{T}(\Omega)p}x^2 - \frac{r^{\frac{3}{2}}}{2\sqrt{n}\hat{T}^2(\Omega)p^2}x^3\right) + O(n^{-\frac{1}{2}})$$

$$= -\left(\sqrt{nr}x + \frac{r}{2\hat{T}(\Omega)p}x^2 - \frac{r^{\frac{3}{2}}}{2\sqrt{n}\hat{T}^2(\Omega)p^2}x^3\right) + O(n^{-\frac{1}{2}}),$$

i.e.,

$$(6) \qquad \log\left(\frac{np\hat{T}(\Omega)}{k}\right)^k = -\left(\sqrt{nr}x + \frac{r}{2\hat{T}(\Omega)p}x^2 - \frac{r^{\frac{3}{2}}}{2\sqrt{n}\hat{T}^2(\Omega)p^2}x^3\right) + O(n^{-\frac{1}{2}}),$$

$$\log\left(\frac{nq\hat{T}(\Omega)}{n-k}\right)^{n-k} = -(n-k)\log\frac{n-k}{nq\hat{T}(\Omega)}$$

$$= -\left(n\hat{T}(\Omega)q - x\sqrt{nr}\right)\log\left(1-\frac{\sqrt{r}}{\sqrt{n}\hat{T}(\Omega)q}x\right)$$

$$= -\left(n\hat{T}(\Omega)q - x\sqrt{nr}\right)\left(-\frac{\sqrt{r}}{\sqrt{n}\hat{T}(\Omega)q}x - \frac{r}{2n\hat{T}^2(\Omega)q^2}x^2\right) + O(n^{-\frac{1}{2}})$$

$$= \left(n\hat{T}(\Omega)q - x\sqrt{nr}\right)\left(\frac{\sqrt{r}}{\sqrt{n}\hat{T}(\Omega)q}x + \frac{r}{2n\hat{T}^2(\Omega)q^2}x^2\right) + O(n^{-\frac{1}{2}})$$

$$= \sqrt{nr}x + \frac{r}{2\hat{T}(\Omega)q}x^2 - \frac{r}{\hat{T}(\Omega)q}x^2 - \frac{r^{\frac{3}{2}}}{2\sqrt{n}\hat{T}^2(\Omega)q^2}x^3 + O(n^{-\frac{1}{2}})$$

$$= \sqrt{nr}x - \frac{r}{2\hat{T}(\Omega)q}x^2 - \frac{r^{\frac{3}{2}}}{2\sqrt{n}\hat{T}^2(\Omega)q^2}x^3 + O(n^{-\frac{1}{2}}),$$

i.e.,

$$\log\left(\frac{nq\hat{T}(\Omega)}{n-k}\right)^{n-k} = \sqrt{nr}x - \frac{r}{2\hat{T}(\Omega)q}x^2 - \frac{r^{\frac{3}{2}}}{2\sqrt{n}\hat{T}^2(\Omega)q^2}x^3 + O(n^{-\frac{1}{2}}).$$

From the last inequality and from (6) we get

$$\log\left(\frac{np\hat{T}(\Omega)}{k}\right)^{k} + \log\left(\frac{nq\hat{T}(\Omega)}{n-k}\right)^{n-k} = -\left(\sqrt{nr}x + \frac{r}{2\hat{T}(\Omega)p}x^2 - \frac{r^{\frac{3}{2}}}{2\sqrt{n}\hat{T}^2(\Omega)p^2}x^3\right) + O(n^{-\frac{1}{2}})$$

$$+\sqrt{nr}x - \frac{r}{2\hat{T}(\Omega)q}x^2 - \frac{r^{\frac{3}{2}}}{2\sqrt{n}\hat{T}^2(\Omega)q^2}x^3 + O(n^{-\frac{1}{2}})$$

$$= -\frac{r(p+q)}{2\hat{T}(\Omega)pq}x^2 + O(n^{-\frac{1}{2}})$$

$$\longrightarrow_{n\longrightarrow\infty} -\frac{r}{2pq\hat{T}^2(\Omega)}x^2.$$

Consequently,

$$\lim_{n\longrightarrow\infty,\ \frac{k-np}{\sqrt{nr}}\longrightarrow x}\left(\frac{np\hat{T}(\Omega)}{k}\right)^{k}\left(\frac{nq\hat{T}(\Omega)}{n-k}\right)^{n-k} = e^{-\frac{r}{2pq\hat{T}^2(\Omega)}x^2}$$

and

$$\sqrt{\left(\frac{n}{k(n-k)}\right)} = \frac{\sqrt{n}}{\sqrt{n\hat{T}(\Omega)p - x\sqrt{np}}\sqrt{n\hat{T}(\Omega)q - x\sqrt{nr}}}$$

$$= \frac{\sqrt{n}}{\sqrt{n^2\hat{T}^2(\Omega)pq - xn^{\frac{3}{2}}\hat{T}(\Omega)pr^{\frac{1}{2}} - xn^{\frac{3}{2}}\hat{T}(\Omega)p^{\frac{1}{2}}q + x^2npr}}$$

$$= \frac{1}{\sqrt{n\hat{T}^2(\Omega)pq - xn^{\frac{1}{2}}\hat{T}(\Omega)pr^{\frac{1}{2}} - xn^{\frac{1}{2}}\hat{T}(\Omega)p^{\frac{1}{2}}q + x^2pr}}$$

$$= \frac{1}{\sqrt{n\hat{T}^2(\Omega)pq}}(1+o(1))$$

$$= \sqrt{\frac{r}{\hat{T}^2(\Omega)pq}}\frac{1}{\sqrt{nr}}(1+o(1))$$

$$= \sqrt{\frac{r}{\hat{T}^2(\Omega)pq}}h(1+o(1)).$$

Then

$$\hat{P}(a \le S_n^* \le b) = \Sigma_{a \le x_k \le b, x_k = \frac{k-np}{\sqrt{nr}}} p_n(k)$$

$$= \frac{1}{\sqrt{2\pi}\hat{T}(\Omega)} \Sigma_{a \le x_k \le b, x_k = \frac{k-np}{\sqrt{nr}}} \left(\frac{np\hat{T}(\Omega)}{k}\right)^k \left(\frac{nq\hat{T}(\Omega)}{(n-k)}\right)^{n-k} \sqrt{\frac{n}{k(n-k)}}$$

$$\longrightarrow_{n \longrightarrow \infty} \sqrt{\frac{r}{pq}} \frac{1}{\hat{T}^2(\Omega)\sqrt{2\pi}} \int_a^b e^{-\frac{r}{2pq\hat{T}^2(\Omega)}x^2} dx.$$

□

1.5 Conditional Iso-Expectation

Definition 1.5.1. *Let $\hat{X}$, $\hat{Y}$ be random iso-variables. Then the iso-expected value of the random iso-variable $\hat{X}$, given the random iso-variable $\hat{Y}$, denoted by $\hat{E}(\hat{X}|\hat{Y})$, is defined by*

$$\hat{\int}_A \hat{X} \hat{\times} \hat{d}\hat{P} = \hat{\int}_A \hat{E}(\hat{X}|\hat{Y}) \hat{\times} \hat{d}\hat{P}$$

for all $A \in U(\hat{Y})$.

Definition 1.5.2. *Suppose that $\mathcal{V}$ is a σ-algebra so that $\mathcal{V} \subset U$. If $\hat{X} : \Omega \longrightarrow \mathbb{R}^n$ is an iso-integrable random iso-variable we define*

$$\hat{E}(\hat{X}|\mathcal{V})$$

to be any random iso-variable on Ω such that

i) *$\hat{E}(\hat{X}|\mathcal{V})$ is $\mathcal{V}$-iso-measurable;*

ii)

$$\hat{\int}_A \hat{X} \hat{\times} \hat{d}\hat{P} = \hat{\int}_A \hat{E}(\hat{X}|\mathcal{V}) \hat{\times} \hat{d}\hat{P}$$

for all $A \in \mathcal{V}$.

Theorem 1.5.3. *We have*

$$\hat{E}(\hat{E}(\hat{X}|\mathcal{V})) = \hat{E}(\hat{X}).$$

Proof. We note that $\hat{E}(\hat{X}|\mathcal{V})$ is $\mathcal{V}$-iso-measurable. Then

$$\hat{E}(\hat{E}(\hat{X}|\mathcal{V})) = \hat{\int}_{\Omega}\hat{E}(\hat{X}|\mathcal{V})\hat{\times}\hat{d}\hat{P}$$

$$= \hat{\int}_{\Omega}\hat{X}\hat{\times}\hat{d}\hat{P}$$

$$= \hat{E}(\hat{X}).$$

□

Theorem 1.5.4. *We have*

$$\hat{E}(\hat{X}) = \hat{E}(\hat{X}|W),$$

where $W = \{\emptyset, \Omega\}$ *is the trivial* σ*-algebra.*

Proof. Really, since W is trivial σ-algebra, we have

$$\hat{E}(\hat{X}|W) = \hat{\int}_{\Omega}\hat{X}\hat{\times}\hat{d}\hat{P}$$

$$= \hat{E}(\hat{X}).$$

□

Remark 1.5.5. *For every iso-integrable random iso-variable* $\hat{X}$ *and for every* σ*-algebra* $\mathcal{V} \subset U$*, the conditional iso-expectation* $\hat{E}(\hat{X}|\mathcal{V})$ *exists and is unique up to* $\mathcal{V}$*-measurable sets of iso-probability zero.*

Exercise 1.5.6. *If* $\hat{X}$ *is* $\mathcal{V}$*-iso-measurable, prove that*

$$\hat{X} = \hat{E}(\hat{X}|\mathcal{V}) \qquad \text{i} - \text{a.s..}$$

Theorem 1.5.7. *Let a and b be real constants. Then*

$$\hat{E}(a\hat{X} + b\hat{Y}|\mathcal{V}) = a\hat{E}(\hat{X}|\mathcal{V}) + b\hat{E}(\hat{Y}|\mathcal{V}) \qquad \text{i} - \text{a.s..}$$

Proof. For every $A \in \mathcal{V}$ we have

$$\hat{\int}_{A}\hat{E}(a\hat{X} + b\hat{Y}|\mathcal{V})\hat{\times}\hat{d}\hat{P} = \hat{\int}_{A}\left(a\hat{X} + b\hat{Y}\right)\hat{\times}\hat{d}\hat{P}$$

$$= \hat{\int}_{A}a\hat{X}\hat{\times}\hat{d}\hat{P} + \hat{\int}_{A}b\hat{Y}\hat{\times}\hat{d}\hat{P}$$

$$= a\hat{\int}_{A}\hat{X}\hat{\times}\hat{d}\hat{P} + b\hat{\int}_{A}\hat{Y}\hat{\times}\hat{d}\hat{P}$$

$$= a\hat{\int}_{A}\hat{E}(\hat{X}|\mathcal{V}\hat{\times}\hat{d}\hat{P} + b\hat{\int}_{A}\hat{E}(\hat{Y}|\mathcal{V})\hat{\times}\hat{d}\hat{P}$$

$$= \hat{\int}_{A}a\hat{E}(\hat{X}|\mathcal{V})\hat{d}\hat{P} + \hat{\int}_{A}b\hat{E}(\hat{Y}|\mathcal{V}\hat{\times}\hat{d}\hat{P}$$

$$= \hat{\int}_{A}\left(a\hat{E}(\hat{X}|\mathcal{V}) + b\hat{E}(\hat{Y}|\mathcal{V})\right)\hat{\times}\hat{d}\hat{P}.$$

Because $A \in \mathcal{V}$ was arbitrarily chosen we conclude that

$$\hat{E}(a\hat{X}+b\hat{Y}|\mathcal{V}) = a\hat{E}(\hat{X}|\mathcal{V})+b\hat{E}(\hat{Y}|\mathcal{V}).$$

□

Theorem 1.5.8. *Let $\hat{X}$ be $\mathcal{V}$-iso-measurable and $\hat{X}\hat{Y}$ be integrable. Then*

$$\hat{E}(\hat{X}\hat{Y}|\mathcal{V}) = \hat{X}\hat{E}(\hat{Y}|\mathcal{V}) \qquad \mathrm{i-a.s.}$$

Proof. Let

$$\hat{X} = \sum_{i=1}^{m} b_i \chi_{B_i}, \qquad B_i \in \mathcal{V}.$$

Let also, $A \in \mathcal{V}$ be arbitrarily chosen. Then we have

$$\hat{\int}_A \hat{E}(\hat{X}\hat{Y}|\mathcal{V})\hat{\times}\hat{d}\hat{P} = \hat{\int}_A \hat{X}\hat{Y}\hat{\times}\hat{d}\hat{P}$$

$$= \hat{\int}_A \sum_{i=1}^{m} b_i \chi_{B_i}\hat{Y}\hat{\times}\hat{d}\hat{P}$$

$$= \sum_{i=1}^{m} b_i \hat{\int}_A \chi_{B_i}\hat{Y}\hat{\times}\hat{d}\hat{P}$$

$$= \sum_{i=1}^{m} b_i \hat{\int}_{A\cap B_i} \hat{Y}\hat{\times}\hat{d}\hat{P},$$

now we use that $A\cap B_i \in \mathcal{V}$ and

$$\hat{\int}_{A\cap B_i} \hat{Y}\hat{\times}\hat{d}\hat{P} = \hat{\int}_{A\cap B_i} \hat{E}(\hat{Y}|\mathcal{V})\hat{\times}\hat{d}\hat{P},$$

$$\hat{\int}_A \hat{E}(\hat{X}\hat{Y}|\mathcal{V})\hat{\times}\hat{d}\hat{P} = \sum_{i=1}^{m} b_i \hat{\int}_{A\cap B_i} \hat{E}(\hat{Y}|\mathcal{V}\hat{\times}\hat{d}\hat{P}$$

$$= \sum_{i=1}^{m} b_i \hat{\int}_A \chi_{B_i}\hat{E}(\hat{Y}|\mathcal{V})\hat{\times}\hat{d}\hat{P}$$

$$= \hat{\int}_A \sum_{i=1}^{m} b_i \hat{\chi}_{B_i}\hat{E}(\hat{Y}|\mathcal{V})\hat{\times}\hat{d}\hat{P}$$

$$= \hat{\int}_A \hat{X}\hat{E}(\hat{Y}|\mathcal{V})\hat{\times}\hat{d}\hat{P},$$

i.e.,

$$\hat{\int}_A \hat{E}(\hat{X}\hat{Y}|\mathcal{V})\hat{\times}\hat{d}\hat{P} = \hat{\int}_A \hat{X}\hat{E}(\hat{Y}|\mathcal{V})\hat{\times}\hat{d}\hat{P}.$$

Because $A \in \mathcal{V}$ was arbitrarily chosen we conclude that

$$\hat{E}(\hat{X}\hat{Y}|\mathcal{V}) = \hat{X}\hat{E}(\hat{Y}|\mathcal{V}) \qquad \mathrm{i-a.s.}$$

□

Theorem 1.5.9. *Let $\hat{X}$ be iso-independent of $\mathcal{V}$. Then*

$$\hat{E}\left(\frac{\hat{X}}{\hat{T}}\middle|\mathcal{V}\right) = \frac{1}{\hat{T}^2}\hat{E}\left(\frac{\hat{X}}{\hat{T}}\right) \qquad \mathrm{i-a.s.}$$

Proof. For every $A \in \mathcal{V}$ we have

$$\hat{\int}_A \hat{E}\left(\frac{\hat{X}}{\hat{T}}\Big|\mathcal{V}\right) \hat{\times} \hat{d}\hat{P} = \hat{\int}_A \frac{\hat{X}}{\hat{T}} \hat{\times} \hat{d}\hat{P}$$

$$= \hat{\int}_\Omega \chi_A \frac{\hat{X}}{\hat{T}} \hat{\times} \hat{d}\hat{P}$$

$$= \hat{E}\left(\frac{\chi_A \hat{X}}{\hat{T}}\right).$$

Since $\hat{X}$ is iso-independent of $\mathcal{V}$ we have that χ_A and $\hat{X}$ are iso-independent. Therefore

$$\hat{E}\left(\frac{\chi_A \hat{X}}{\hat{T}}\right) = \hat{E}\left(\frac{\chi_A}{\hat{T}}\right) \hat{E}\left(\frac{\hat{X}}{\hat{T}}\right).$$

Hence,

$$\hat{E}\left(\frac{\chi_A \hat{X}}{\hat{T}}\right) = \hat{E}\left(\frac{\hat{X}}{\hat{T}}\right) \hat{\int}_\Omega \frac{\chi_A}{\hat{T}} \hat{\times} \hat{d}\hat{P}$$

$$= \hat{E}\left(\frac{\hat{X}}{\hat{T}}\right) \hat{\int}_A \frac{1}{\hat{T}} \hat{\times} \hat{d}\hat{P}$$

$$= \hat{E}\left(\frac{\hat{X}}{\hat{T}}\right) \int_A \frac{1}{\hat{T}} \frac{1}{\hat{T}} \hat{T} \hat{T} d\hat{P}$$

$$= \hat{E}\left(\frac{\hat{X}}{\hat{T}}\right) \int_A d\hat{P}$$

$$= \int_A \hat{E}\left(\frac{\hat{X}}{\hat{T}}\right) d\hat{P}$$

$$= \int_A \frac{1}{\hat{T}} \hat{E}\left(\frac{\hat{X}}{\hat{T}}\right) \frac{1}{\hat{T}^2} \hat{T} \hat{T} d\hat{P}$$

$$= \hat{\int}_A \frac{1}{\hat{T}^2} \hat{E}\left(\frac{\hat{X}}{\hat{T}}\right) \hat{\times} \hat{d}\hat{P}.$$

Consequently,

$$\hat{\int}_A \hat{E}\left(\frac{\hat{X}}{\hat{T}}\Big|\mathcal{V}\right) \hat{\times} \hat{d}\hat{P} = \hat{\int}_A \frac{1}{\hat{T}^2} \hat{E}\left(\frac{\hat{X}}{\hat{T}}\right) \hat{\times} \hat{d}\hat{P},$$

whereupon

$$\hat{E}\left(\frac{\hat{X}}{\hat{T}}\Big|\mathcal{V}\right) = \frac{1}{\hat{T}^2} \hat{E}\left(\frac{\hat{X}}{\hat{T}}\right) \qquad \text{i} - \text{a.s.}$$

□

Theorem 1.5.10. *Let $W \subset \mathcal{V}$. Then*

$$\hat{E}(\hat{X}|W) = \hat{E}(\hat{E}(\hat{X}|\mathcal{V})|W) \qquad \text{i} - \text{a.s.}$$

Proof. Let $A \in W$ be arbitrarily chosen. Then

(7) $\quad \hat{\int} \hat{E}(\hat{X}|W) \hat{\times} \hat{d}\hat{P} = \hat{\int}_A \hat{X} \hat{\times} \hat{d}\hat{P}.$

On the other hand

$$(8) \qquad \hat{\int}_A \hat{E}(\hat{E}(\hat{X}|V)|W)\hat{\times}\hat{d}\hat{P} = \hat{\int}_A \hat{E}(\hat{X}|\mathcal{V})\hat{\times}\hat{d}\hat{P}.$$

Because $A \in W$ and $W \subset \mathcal{V}$ we have that $A \in \mathcal{V}$. From here

$$\hat{\int}_A \hat{E}(\hat{X}|\mathcal{V})\hat{\times}\hat{d}\hat{P} = \hat{\int}_A \hat{A}\hat{\times}\hat{d}\hat{P}.$$

From the last equality and from (8) it follows that

$$\hat{\int}_A \hat{E}(\hat{E}(\hat{X}|\mathcal{V})|W)\hat{\times}\hat{d}\hat{P} = \hat{\int}_A \hat{X}\hat{\times}\hat{d}\hat{P},$$

from where, using (7), we get

$$\hat{\int}_A \hat{E}(\hat{X}|W)\hat{\times}\hat{d}\hat{P} = \hat{\int}_A \hat{E}(\hat{E}(\hat{X}|\mathcal{V})|W)\hat{\times}\hat{d}\hat{P}.$$

Because $A \in W$ was arbitrarily chosen, we conclude that

$$\hat{E}(\hat{X}|W) = \hat{E}(\hat{E}(\hat{X}|\mathcal{V})|W) \qquad \mathrm{i-a.s.}$$

□

Theorem 1.5.11. *Let $W \subset \mathcal{V}$. Then*

$$\hat{E}(\hat{X}|W) = \hat{E}(\hat{E}(\hat{X}|W)|\mathcal{V}) \qquad \mathrm{i-a.s.}$$

Proof. We note that $\hat{E}(\hat{X}|W)$ is W-iso-measurable and since $W \subset \mathcal{V}$ we have that $\hat{E}(\hat{X}|W)$ is $\mathcal{V}$-iso-measurable. Therefore

$$\hat{E}(\hat{E}(\hat{X}|W)|\mathcal{V}) = \hat{E}(\hat{X}|W).$$

□

Remark 1.5.12. *From the last two theorems it follows that if $W \subset \mathcal{V}$, then*

$$\hat{E}(\hat{X}|W) = \hat{E}(\hat{E}(\hat{X}|\mathcal{V})|W) = \hat{E}(\hat{E}(\hat{X}|W)|V) \qquad \mathrm{i-a.s..}$$

Theorem 1.5.13. *Let $\hat{X} \leq \hat{Y}$ i-a.s. Then*

$$\hat{E}(\hat{X}|\mathcal{V}) \leq \hat{E}(\hat{Y}|\mathcal{V}) \qquad \mathrm{i-a.s.}$$

Proof. Let $A \in \mathcal{V}$ be arbitrarily chosen. Then

$$\begin{aligned} &\hat{\int}_A \hat{E}(\hat{Y} - \hat{X}|\mathcal{V})\hat{\times}\hat{d}\hat{P} \\ &= \int_A \tfrac{1}{\hat{T}}(\hat{Y} - \hat{X})\hat{T}\hat{T}d\hat{P} \\ &\geq 0. \end{aligned}$$

Hence,

$$\int_A \frac{1}{\hat{T}}\hat{E}(\hat{Y}-\hat{X}|\mathcal{V})\hat{T}\hat{T}d\hat{P} \geq 0.$$

Because $A \in \mathcal{V}$ was arbitrarily chosen and $\hat{T} > 0$ on U and $\mathcal{V}$, we conclude that

$$\hat{E}(\hat{Y}-\hat{X}|\mathcal{V}) \geq 0.$$

□

Theorem 1.5.14. *(iso-conditional Jensen's inequality) Suppose that* $\Phi : \mathbb{R} \longrightarrow \mathbb{R}$ *is convex with* $\hat{E}(|\Phi(\hat{X})|) < \infty$. *Then*

$$\Phi(\hat{E}(\hat{X}|\mathcal{V})) \leq \hat{E}(\Phi(\hat{X})|\mathcal{V}).$$

Proof. We suppose that $\Phi(0) = 0$. Otherwise, we will consider $\tilde{\Phi} = \Phi - \Phi(0)$. Let $A \in \mathcal{V}$ be arbitrarily chosen. Then, using that $\Phi : \mathbb{R} \longrightarrow \mathbb{R}$ is convex,

$$\hat{\int}_A \Phi(\hat{E}(\hat{X}|\mathcal{V}))\hat{\times}\hat{d}\hat{P} = \hat{\int}_\Omega \Phi(\chi_A \hat{E}(\hat{X}|\mathcal{V}))\hat{\times}\hat{d}\hat{P}$$

now we use that Φ is convex

$$\leq \hat{\int}_\Omega \chi_A \Phi(\hat{E}(\hat{X}|\mathcal{V}))\hat{\times}\hat{d}\hat{P}$$

$$= \hat{\int}\Phi(\hat{X})\hat{\times}\hat{d}\hat{P}$$

$$= \hat{\int}_A \hat{E}(\Phi(\hat{X})|\mathcal{V})\hat{\times}\hat{d}\hat{P}.$$

From the last inequality, because $A \in \mathcal{V}$ was arbitrarily chosen, we get

$$\Phi(\hat{E}(\hat{X}|\mathcal{V})) \leq \hat{E}(\Phi(\hat{X})|\mathcal{V}).$$

□

1.6 Iso-Martingales

Suppose that

$$\hat{T}\hat{Y}_1, \hat{T}\hat{Y}_2, \ldots, \hat{T}\hat{Y}_n, \ldots$$

are iso-independent real-valued random iso-variables with

(9) $\quad \hat{E}(\hat{Y}_i) = 0 \qquad i = 1, 2, \ldots.$

Define the sum

$$S_n := \hat{Y}_1 + \hat{Y}_2 + \cdots + \hat{Y}_n.$$

Then, using (9),

$$\hat{E}(S_{n+k}|S_1,S_2,\ldots,S_n) = \hat{E}(\hat{Y}_1+\hat{Y}_2+\cdots+\hat{Y}_n+\hat{Y}_{n+1}+\cdots+\hat{Y}_{n+k}|S_1,S_2,\ldots,S_n)$$

$$= \hat{E}(\hat{Y}_1+\hat{Y}_2+\cdots+\hat{Y}_n|S_1,S_2,\ldots,S_n)+\hat{E}(\hat{Y}_{n+1},\hat{Y}_{n+2}+\cdots+\hat{Y}_{n+k}|S_1,S_2,\ldots,S_n)$$

$$= \hat{Y}_1+\hat{Y}_2+\cdots+\hat{Y}_n+\hat{E}\left(\tfrac{\hat{T}\hat{Y}_{n+1}+\hat{T}\hat{Y}_{n+2}+\cdots+\hat{T}\hat{Y}_{n+k}}{\hat{T}}|S_1,S_2,\ldots,S_n\right)$$

$$= \hat{Y}_1+\hat{Y}_2+\cdots+\hat{Y}_n+\tfrac{1}{\hat{T}^2}\hat{E}\left(\tfrac{\hat{T}\hat{Y}_{n+1}+\hat{T}\hat{Y}_{n+2}+\cdots+\hat{T}\hat{Y}_{n+k}}{\hat{T}}\right)$$

$$= \hat{Y}_1+\hat{Y}_2+\cdots+\hat{Y}_n+\tfrac{1}{\hat{T}^2}\hat{E}(\hat{Y}_{n+1}+\hat{Y}_{n+2}+\cdots+\hat{Y}_{n+k})$$

$$= \hat{Y}_1+\hat{Y}_2+\cdots+\hat{Y}_n+\tfrac{1}{\hat{T}}\left(\hat{E}(\hat{Y}_{n+1})+\hat{E}(\hat{Y}_{n+2})+\cdots+\hat{E}(\hat{Y}_{n+k})\right)$$

$$= \hat{Y}_1+\hat{Y}_2+\cdots+\hat{Y}_n.$$

Definition 1.6.1. *Let $\hat{X}_1$, $\hat{X}_2$, ..., $\hat{X}_n$, ... be a sequence of real-valued random iso-variables with $\hat{E}(|\hat{X}_i|) < \infty$, $i = 1,2,\ldots$. If*

$$\hat{X}_k = \hat{E}(\hat{X}_j|\hat{X}_1,\hat{X}_2,\ldots\hat{X}_k) \qquad \text{for} \qquad \text{all} \qquad j \geq k,$$

then we call $\{\hat{X}_n\}_{n=1}^{\infty}$ an iso-discrete iso-martingal (or iso-martingal).

Definition 1.6.2. *Let $\hat{X}(\cdot)$ be a real-valued iso-stochastic process. Then*

$$U(t) := U(\hat{X}(s) : 0 \leq s \leq t)$$

is called the history of the iso-process until time $t \geq 0$.

Definition 1.6.3. *Let $\hat{X}(\cdot)$ be an iso-stochastic process, such that $\hat{E}(|\hat{X}(t)|) < \infty$ for all $t \geq 0$.*

(i) *If*

$$\hat{X}(s) = \hat{E}(\hat{X}(t)|U(s)) \qquad \mathrm{i-a.s.}$$

for all $t \geq s \geq 0$, then $\hat{X}(\cdot)$ is called an iso-martingal.

(ii) *If*

$$\hat{X}(s) \leq \hat{E}(\hat{X}(t)|U(s)) \qquad \mathrm{i-a.s.}$$

for all $t \geq s$, then $\hat{X}(\cdot)$ is called an iso-submartingal.

Theorem 1.6.4. *Suppose $\hat{X}(\cdot)$ is a real-valued martingale and $\Phi : \mathbb{R} \longrightarrow \mathbb{R}$ is convex. Then, if*

$$\hat{E}(|\Phi(\hat{X}(t))|) < \infty \qquad \text{for} \qquad \forall t \geq 0,$$

$$\Phi(\hat{X}(\cdot)) \qquad \text{is} \qquad \text{an} \qquad \text{iso}-\text{martingal}.$$

Proof. Let $t \geq s$. Since $\hat{X}(\cdot)$ is an iso-martingal, then

$$\hat{X}(s) = \hat{E}(\hat{X}(t)|\mathcal{V})$$

for every $t \geq s$. From here

$$\Phi\left(\hat{X}(s)\right) = \Phi\left(\hat{E}(\hat{X}(t)|\mathcal{V})\right),$$

now we use the iso-conditional Jensen's inequality

$$\Phi(\hat{X}(s)) \leq \hat{E}(\Phi(\hat{X}(t))|\mathcal{V})$$

for all $t \geq s$. Therefore $\Phi(\hat{X}(\cdot))$ is an iso-submartingal. □

Theorem 1.6.5. *Let* $\left\{\frac{\hat{X}_n}{\hat{T}}\right\}_{n=1}^{\infty}$ *is an iso-submartingal. Then*

$$\hat{P}\left(\max_{1\leq k\leq n} \hat{X}_k \geq \lambda\right) \leq \frac{1}{\lambda}\hat{E}\left(\frac{\hat{X}_n^+}{\hat{T}}\right)$$

for all $n = 1, 2, \ldots$ *and* $\lambda > 0$. *Here* $\hat{X}_n^+ = \max\{\hat{X}_n, 0\}$.

Proof. We define

$$A_k := \cap_{j=1}^{k-1}\{\hat{X}_j \leq \lambda\} \cap \{\hat{X}_k > \lambda\}, \qquad k = 1, 2, \ldots, n,$$

$$A := \left\{\max_{1\leq k\leq n}\hat{X}_k > \lambda\right\}.$$

We have

$$A = \bigcup_{k=1}^{n} A_k.$$

Really, let $\omega \in A$ be arbitrarily chosen. Then

$$\max_{1\leq k\leq n} \hat{X}_k(\omega) > \lambda.$$

We choose the smallest $k_1 \in \{1, 2, \ldots, n\}$ so that

$$\hat{X}_{k_1}(\omega) > \lambda.$$

We have the following cases.

1.case. $k_1 = 1$. Then $\omega \in A_{k_1}$.

2.case. $k_1 > 1$. Then

$$\hat{X}_1 \leq \lambda, \qquad \ldots, \qquad \hat{X}_{k_1-1} \leq \lambda, \qquad \hat{X}_{k_1} > \lambda,$$

i.e., $\omega \in A_{k_1}$.

Therefore

$$\omega \in \bigcup_{k=1}^{n} A_k$$

and since $\omega \in A$ was arbitrarily chosen, then

(10) $\quad A \subset \bigcup_{k=1}^{n} A_k.$

Let now $\omega \in \bigcup_{k=1}^{n} A_k$. Then there exists $k_2 \in \{1,2,\ldots,n\}$ such that

$$\omega \in A_{k_2} = \bigcap_{i=1}^{k_2-1} \{\hat{X}_i \leq \lambda\} \bigcap \{\hat{X}_{k_2} > \lambda\}.$$

Because

$$\hat{X}_{k_2}(\omega) > \lambda \qquad \text{and} \qquad k_2 \in \{1,2,\ldots,n\}$$

we conclude that

$$\max_{1\leq k\leq n} \hat{X}_k(\omega) > \lambda.$$

Consequently, $\omega \in A$. Since $\omega \in \bigcup_{k=1}^{n} A_k$ was arbitrarily chosen, we obtain that

$$\bigcup_{k=1}^{n} A_k \subset A.$$

From the last relation and from (10) we conclude that

$$A = \bigcup_{k=1}^{n} A_k.$$

Now we suppose that there are $l, m \in \{1,2,\ldots,n\}$ so that $l < m$ and $A_l \bigcap A_m \neq \emptyset$. Then there is $\omega \in A_l \bigcap A_m$. We have

$$\bigcap_{j=1}^{l-1} \{X_j \leq \lambda\} \bigcap \{X_l > \lambda\},$$

$$\bigcap_{j=1}^{m_1} \{X_j \leq \lambda\} \bigcap \{X_m > \lambda\}.$$

From $\omega \in A_l$ it follows that

$$X_j(\omega) \leq \lambda, \qquad j = 1,2,\ldots,l-1,$$

(11) $\quad X_l(\omega) > \lambda.$

From $\omega \in A_m$ it follows that

$$X_j(\omega) \leq \lambda, \qquad j = 1,2,\ldots,m-1,$$

$$X_m(\omega) > \lambda,$$

in particular, because $l < m$, we get $X_l(\omega) \leq \lambda$ which contradicts with (11).

Consequently,

$$A_l \bigcap A_m = \emptyset, \qquad l,m \in \{1,2,\ldots,n\}.$$

Also,

$$\lambda \hat{P}(A_k) = \lambda \int_{A_k} d\hat{P}$$

$$= \int_{A_k} \lambda d\hat{P}$$

$$\leq \int_{A_k} \hat{X}_k d\hat{P}$$

$$= \int_\Omega \chi_{A_k} \hat{X}_k d\hat{P}$$

$$= \int_\Omega \frac{1}{\hat{T}} \hat{T} \chi_{A_k} \hat{X}_k \frac{1}{\hat{T}^2} \hat{T} \hat{T} d\hat{P}$$

$$= \hat{\int}_\Omega \hat{T} \chi_{A_k} \hat{X}_k \frac{1}{\hat{T}^2} \hat{\times} \hat{d}\hat{P}$$

$$= \hat{\int}_\Omega \chi_{A_k} \frac{\hat{X}_k}{\hat{T}} \hat{\times} \hat{d}\hat{P}$$

$$= \hat{E}\left(\chi_{A_k} \frac{\hat{X}_k}{\hat{T}}\right),$$

and

$$\lambda \hat{P}(A) = \lambda \hat{P}\left(\bigcup_{k=1}^n A_k\right) \qquad (A_l \bigcap A_m = \emptyset, \qquad l,m \in \{1,2,\ldots,n\})$$

$$= \lambda \sum_{k=1}^n \hat{P}(A_k)$$

$$\leq \sum_{k=1}^n \hat{E}\left(\chi_{A_k} \frac{\hat{X}_k}{\hat{T}}\right),$$

and

$$\hat{E}\left(\frac{\hat{X}_n^+}{\hat{T}}\right) = \hat{\int}_\Omega \frac{\hat{X}_n^+}{\hat{T}} \hat{\times} \hat{d}\hat{P}$$

$$\geq \hat{\int}_A \frac{\hat{X}_n^+}{\hat{T}} \hat{\times} \hat{d}\hat{P}$$

$$= \hat{\int}_{\bigcup_{k=1}^n A_k} \frac{\hat{X}_n^+}{\hat{T}} \hat{\times} \hat{d}\hat{P}$$

$$= \sum_{k=1}^n \hat{\int}_{A_k} \frac{\hat{X}_n^+}{\hat{T}} \hat{\times} \hat{d}\hat{P}$$

$$= \sum_{k=1}^n \hat{\int}_\Omega \chi_{A_k} \frac{\hat{X}_n^+}{\hat{T}} \hat{\times} \hat{d}\hat{P}$$

$$= \sum_{k=1}^n \hat{E}\left(\chi_{A_k} \frac{\hat{X}_n^+}{\hat{T}}\right)$$

$$= \Sigma_{k=1}^{n} \hat{E}\left(\hat{E}\left(\chi_{A_k}\frac{\hat{X}_n^+}{\hat{T}}|\frac{\hat{X}_1}{\hat{T}},\frac{\hat{X}_2}{\hat{T}},\ldots,\frac{\hat{X}_k}{\hat{T}}\right)\right)$$

$$= \Sigma_{k=1}^{n} \hat{E}\left(\chi_{A_k}\hat{E}\left(\frac{\hat{X}_n^+}{\hat{T}}|\frac{\hat{X}_1}{\hat{T}},\frac{\hat{X}_2}{\hat{T}},\ldots,\frac{\hat{X}_k}{\hat{T}}\right)\right)$$

$$\geq \Sigma_{k=1}^{n} \hat{E}\left(\chi_{A_k}\hat{E}\left(\frac{\hat{X}_n}{\hat{T}}|\frac{\hat{X}_1}{\hat{T}},\frac{\hat{X}_2}{\hat{T}},\ldots,\frac{\hat{X}_k}{\hat{T}}\right)\right)$$

$$\geq \Sigma_{k=1}^{n} \hat{E}\left(\chi_{A_k}\frac{\hat{X}_k}{\hat{T}}\right)$$

$$\geq \lambda\hat{P}(A),$$

i.e.,

$$\lambda\hat{P}\left(\max_{1\leq k\leq n}\hat{X}_k > \lambda\right) \leq \hat{E}\left(\frac{\hat{X}_n^+}{\hat{T}}\right),$$

or

$$\hat{P}\left(\max_{1\leq k\leq n}\hat{X}_k > \lambda\right) \leq \frac{1}{\lambda}\hat{E}\left(\frac{\hat{X}_n^+}{\hat{T}}\right).$$

□

Remark 1.6.6. We note that

$$F(\hat{X}\leq\lambda) = \hat{P}(\hat{X}\leq\lambda)$$

and

$$F(\hat{X}\leq\lambda) = \hat{\int}_{-\infty}^{\lambda}\hat{f}(t)\hat{\times}\hat{d}\hat{t},$$

$$F(\hat{X}\leq\infty) = \hat{P}(\hat{X}\leq\infty)$$

$$= \frac{1}{\hat{T}(\Omega)}$$

$$= \int_{-\infty}^{\infty}\hat{f}(t)\hat{\times}\hat{d}\hat{t}.$$

Then

$$\hat{P}(\hat{X}>\lambda) = \frac{1}{\hat{T}(\Omega)} - \hat{P}(\hat{X}\leq\lambda)$$

$$= \hat{\int}_{-\infty}^{\infty}\hat{f}(t)\hat{\times}\hat{d}\hat{t} - \hat{\int}_{-\infty}^{\lambda}\hat{f}(t)\hat{\times}\hat{d}\hat{t}$$

$$= \hat{\int}_{\lambda}^{\infty}\hat{f}(t)\hat{\times}\hat{d}\hat{t}$$

$$= \int_{\lambda}^{\infty}\frac{1}{\hat{T}(t)}\hat{f}(t)\hat{T}(t)\hat{T}(t)d\hat{t}$$

$$= \int_{\lambda}^{\infty}\hat{f}(t)\hat{T}(t)d\hat{t}$$

$$= \int_{\lambda}^{\infty}\hat{f}(t)\left(1-\frac{\hat{T}'(t)}{\hat{T}(t)}\right)dt.$$

Hence,

$$\frac{d}{d\lambda}\hat{P}(\hat{X} > \lambda) = -\hat{f}(\lambda)\left(1 - \frac{\hat{T}'(\lambda)}{\hat{T}(\lambda)}\right).$$

Therefore

$$\hat{E}\left(\tfrac{|\hat{X}|^p}{\hat{T}}\right) = \hat{\int}_{-\infty}^{\infty}\lambda^p\hat{f}(\lambda)\hat{\times}\hat{d\lambda}$$

$$= \int_{-\infty}^{\infty}\tfrac{1}{\hat{T}(\lambda)}\lambda^p\hat{f}(\lambda)\hat{T}(\lambda)\hat{T}(\lambda)d\hat{\lambda}$$

$$= \int_{-\infty}^{\infty}\lambda^p\hat{f}(\lambda)\left(1 - \tfrac{\hat{T}'(\lambda)}{\hat{T}(\lambda)}\right)d\lambda$$

$$= -\int_{-\infty}^{\infty}\lambda^p d\hat{P}(\hat{X} > \lambda)$$

$$= -\int_{0}^{\infty}\lambda^p d\hat{P}(\hat{X} > \lambda) - \int_{-\infty}^{0}\lambda^p d\hat{P}(\hat{X} > \lambda).$$

From here, when $\hat{X} > 0$, we have

$$\hat{E}\left(\frac{|\hat{X}|^p}{\hat{T}}\right) = -\int_0^{\infty}\lambda^p d\hat{P}(\hat{X} > \lambda).$$

Theorem 1.6.7. *If $\{\hat{X}_n\}_{n=1}^{\infty}$ is an iso-martingal and $1 < p < 1$, then*

$$\hat{E}\left(\frac{\max_{1\leq k\leq n}|\hat{X}_k|^p}{\hat{T}}\right) \leq \left(\frac{p}{p-1}\right)^p\hat{\int}_{\Omega}\frac{|\hat{X}_n|^p}{\hat{T}}\hat{\times}\hat{d}\hat{P}.$$

Remark 1.6.8. *This assertion we can rewrite as follows.*

$$\hat{E}\left(\frac{\max_{1\leq k\leq n}|\hat{X}_k|^p}{\hat{T}}\right) \leq \left(\frac{p}{p-1}\right)^p\hat{E}\left(\frac{|\hat{X}_n|^p}{\hat{T}}\right).$$

Proof. Let

$$\hat{P}(\max_{1\leq k\leq n}\hat{X}_k > \lambda) > 0.$$

Then

$$\hat{P}(\max_{1\leq k\leq n}\hat{X}_k \leq \lambda) = 0.$$

From the previous theorem we have that

$$\lambda\hat{P}(\max{}_{1\leq k\leq n}\hat{X}_k > \lambda) \leq \hat{E}\left(\tfrac{\hat{X}_n^+}{\hat{T}}\right)$$

$$= \hat{\int}_{\Omega}\tfrac{\hat{X}_n^+}{\hat{T}}\hat{\times}\hat{d}\hat{P}$$

$$= \hat{\int}_{\max_{1\leq k\leq n}\hat{X}_k > \lambda}\tfrac{\hat{X}_n^+}{\hat{T}}\hat{\times}\hat{d}\hat{P} + \hat{\int}_{\max_{1\leq k\leq n}\hat{X}_k \leq \lambda}\tfrac{\hat{X}_n^+}{\hat{T}}\hat{\times}\hat{d}\hat{P}$$

$$= \hat{\int}_{\max_{1\leq k\leq n}\hat{X}_k > \lambda}\tfrac{\hat{X}_n^+}{\hat{T}}\hat{\times}\hat{d}\hat{P}.$$

Now we apply the last equality for $|\hat{X}_k|$ and we get

$$\lambda\hat{P}(\max_{1\le k\le n}|\hat{X}_k|>\lambda)\le \hat{\int}_{\max_{1\le k\le n}\hat{X}_k>\lambda}\frac{|\hat{X}_n|}{\hat{T}}\hat{\times}\hat{d}\hat{P}.$$

Let

$$\hat{X}:=\max_{1\le k\le n}|\hat{X}_k|,\qquad \hat{Y}:=|\hat{X}_n|.$$

Then

$$\begin{aligned}\lambda\hat{P}(\hat{X}>\lambda) &\le \hat{\int}_{\hat{X}>\lambda}\tfrac{\hat{Y}}{\hat{T}}\hat{\times}\hat{d}\hat{P}\\ &= \int_{\hat{X}>\lambda}\tfrac{1}{\hat{T}}\tfrac{\hat{Y}}{\hat{T}}\hat{T}\hat{T}d\hat{P} \qquad (12)\\ &= \int_{\hat{X}>\lambda}\hat{Y}d\hat{P}.\end{aligned}$$

Let also,

$$\hat{P}(\lambda)=\hat{P}(\hat{X}>\lambda).$$

Then, applying (12), we have

$$\begin{aligned}\hat{E}\left(\tfrac{\hat{X}^p}{\hat{T}}\right) &= -\int_0^\infty \lambda^p d\hat{P}(\lambda)\\ &= -\lambda^p\hat{P}(\lambda)\Big|_{\lambda=0}^{\lambda=\infty}+p\int_0^\infty\lambda^{p-1}\hat{P}(\lambda)d\lambda\\ &= p\int_0^\infty\lambda^{p-1}\hat{P}(\lambda)d\lambda\\ &\le p\int_0^\infty\lambda^{p-1}\left(\tfrac{1}{\lambda}\int_{\hat{X}>\lambda}\hat{Y}d\hat{P}\right)d\lambda\\ &= p\int_\Omega\hat{Y}\int_0^{\hat{X}}\lambda^{p-2}d\lambda d\hat{P}\\ &= \tfrac{p}{p-1}\int_\Omega\hat{Y}\hat{X}^{p-1}d\hat{P}\\ &\le \tfrac{p}{p-1}\left(\int_\Omega\hat{Y}^pd\hat{P}\right)^{\frac{1}{p}}\left(\hat{X}^pd\hat{P}\right)^{\frac{p-1}{p}}\\ &= \tfrac{p}{p-1}\left(\hat{\int}\hat{Y}^pd\hat{P}\right)^{\frac{1}{p}}\left(\int_\Omega\tfrac{1}{\hat{T}}\tfrac{\hat{X}^p}{\hat{T}}\hat{T}\hat{T}d\hat{P}\right)^{\frac{p-1}{p}}\\ &= \tfrac{p}{p-1}\left(\hat{\int}\hat{Y}^pd\hat{P}\right)^{\frac{1}{p}}\left(\hat{\int}\tfrac{\hat{X}^p}{\hat{T}}\hat{\times}\hat{d}\hat{P}\right)^{\frac{p-1}{p}}\\ &= \tfrac{p}{p-1}\left(\hat{\int}\hat{Y}^pd\hat{P}\right)^{\frac{1}{p}}\hat{E}\left(\tfrac{\hat{X}^p}{\hat{T}}\right)^{\frac{p-1}{p}},\end{aligned}$$

i.e.,

$$\hat{E}\left(\frac{\hat{X}^p}{\hat{T}}\right)\le\frac{p}{p-1}\left(\hat{\int}\hat{Y}^pd\hat{P}\right)^{\frac{1}{p}}\hat{E}\left(\frac{\hat{X}^p}{\hat{T}}\right)^{\frac{p-1}{p}},$$

whereupon

$$\hat{E}\left(\frac{\hat{X}^p}{\hat{T}}\right)^{\frac{1}{p}} \leq \frac{p}{p-1}\left(\hat{\int}\hat{Y}^p d\hat{P}\right)^{\frac{1}{p}}$$

or

$$\hat{E}\left(\frac{\hat{X}^p}{\hat{T}}\right) \leq \left(\frac{p}{p-1}\right)^p \int_\Omega \hat{Y}^p d\hat{P}$$

$$= \left(\frac{p}{p-1}\right)^p \int_\Omega \frac{1}{\hat{T}}\frac{\hat{Y}^p}{\hat{T}}\hat{T}\hat{T}d\hat{P}$$

$$= \left(\frac{p}{p-1}\right)^p \hat{\int}_\Omega \frac{\hat{Y}^p}{\hat{T}}\hat{\times}d\hat{P}$$

$$= \left(\frac{p}{p-1}\right)^p \hat{E}\left(\frac{\hat{Y}^p}{\hat{T}}\right).$$

Consequently,

$$\hat{E}\left(\frac{\max_{1\leq k\leq n}|\hat{X}_k|^p}{\hat{T}}\right) \leq \left(\frac{p}{p-1}\right)^p \hat{E}\left(\frac{|\hat{X}_n|^p}{\hat{T}}\right).$$

□

Remark 1.6.9. *The last two results we can generalize as follows.*

Theorem 1.6.10. *Let $\hat{X}(\cdot)$ be an iso-stochastic process with continuous sample paths i-a.s.*

(i) *If $\hat{X}(\cdot)$ is an iso-submarginal, then*

$$(13)\qquad \hat{P}\left(\max_{0\leq s\leq t}\hat{X}(s) \geq \lambda\right) \leq \frac{1}{\lambda}\hat{E}\left(|\hat{X}(t)|^+\right)$$

for all $\lambda > 0$ and $t \geq 0$.

(ii) *If $\hat{X}(\cdot)$ is an iso-martingale and $1 < p < \infty$, then*

$$(14)\qquad \hat{E}\left(\max_{0\leq s\leq t}|\hat{X}(s)|^p\right) \leq \left(\frac{p}{p-1}\right)^p \hat{E}\left(|\hat{X}(t)|^p\right).$$

Outline of Proof. *We choose $\lambda > 0$, $t > 0$ and select*

$$0 = t_0 < t_2 \ldots < t_n = t.$$

We check that $\{\hat{X}(t_i)\}_{i=1}^n$ is an iso-submartingale/iso-martingale and apply the above inequalities. Next choose a finer and finer partition of $[0,t]$ and pass to limits.

Exercise 1.6.11. *Prove* (13).

Exercise 1.6.12. *Prove* (14).

1.7 Iso-Brownian Motion and Iso-White Noise

Below we will suppose that

$$0 \neq 0 \left| \int_{-\infty}^{\infty} s \frac{\tilde{T}(s) - s\tilde{T}'(s)}{\tilde{T}(s)} ds \right| < \infty.$$

Firstly, we will note that if $\hat{X}$ is an $N^1(m, \sigma^2)$ random iso-variable, then it has iso-density

$$\hat{f}(x) = \frac{1}{\sqrt{2\pi\sigma^2}} e^{-\frac{(x-m)^2}{2\sigma^2}} \frac{1}{\tilde{T}(x)}.$$

Hence,

$$\begin{aligned} F(x) &= \hat{\int}_{-\infty}^{x} \hat{f}(s) \hat{\times} d\hat{s} \\ &= \frac{1}{\sqrt{2\pi\sigma^2}} \hat{\int}_{-\infty}^{x} e^{-\frac{(s-m)^2}{2\sigma^2}} \frac{1}{\tilde{T}(s)} \hat{\times} d\hat{s} \\ &= \frac{1}{\sqrt{2\pi\sigma^2}} \int_{-\infty}^{x} \frac{1}{\tilde{T}(s)} e^{-\frac{(s-m)^2}{2\sigma^2}} \frac{1}{\tilde{T}(s)} \tilde{T}(s)\tilde{T}(s) d\frac{s}{\tilde{T}(s)} \\ &= \frac{1}{\sqrt{2\pi\sigma^2}} \int_{-\infty}^{x} e^{-\frac{(s-m)^2}{2\sigma^2}} \frac{\tilde{T}(s) - s\tilde{T}'(s)}{\tilde{T}^2(s)} ds. \end{aligned}$$

In particular, when $m = 0$ and $\sigma^2 = t$, we have

$$\hat{f}(x) = \frac{1}{\sqrt{2\pi t}} e^{-\frac{x^2}{2t}} \frac{1}{\tilde{T}(t)},$$

$$F(x) = \frac{1}{\sqrt{2\pi t}} \int_{-\infty}^{x} e^{-\frac{s^2}{t}} \frac{\tilde{T}(s) - s\tilde{T}'(s)}{\tilde{T}^2(s)} ds.$$

Let

$$\tilde{f}(x) = \hat{f}(x) - C,$$

where C is a constant which is chosen so that

$$\hat{\int}_{-\infty}^{\infty} s\tilde{f}(s) \hat{\times} d\hat{s} = 0,$$

or

$$\begin{aligned} 0 &= \hat{\int}_{-\infty}^{\infty} s(\hat{f}(s) - C) \hat{\times} d\hat{s} \\ &= \int_{-\infty}^{\infty} \frac{1}{\tilde{T}(s)} s \left(\frac{1}{\sqrt{2\pi t}} e^{-\frac{s^2}{2t}} \frac{1}{\tilde{T}(s)} - C \right) \tilde{T}(s)\tilde{T}(s) d\hat{s} \\ &= \frac{1}{\sqrt{2\pi t}} \int_{-\infty}^{\infty} s e^{-\frac{s^2}{2t}} \frac{1}{\tilde{T}^2(s)} \tilde{T}^2(s) d\hat{s} - C \int_{-\infty}^{\infty} s \frac{1}{\tilde{T}(s)} \tilde{T}(s)\tilde{T}(s) d\hat{s} \\ &= \frac{1}{\sqrt{2\pi t}} \int_{-\infty}^{\infty} s e^{-\frac{s^2}{2t}} d\frac{s}{\tilde{T}(s)} - C \int_{-\infty}^{\infty} s\tilde{T}(s) d\frac{s}{\tilde{T}(s)} \\ &= \frac{1}{\sqrt{2\pi t}} s e^{-\frac{s^2}{2t}} \frac{\tilde{T}(s) - s\tilde{T}'(s)}{\tilde{T}^2(s)} ds - C \int_{-\infty}^{\infty} s\tilde{T}(s) \frac{\tilde{T}(s) - s\tilde{T}'(s)}{\tilde{T}^2(s)} ds \\ &= \frac{1}{\sqrt{2\pi t}} \int_{-\infty}^{\infty} s e^{-\frac{s^2}{2t}} \frac{\tilde{T}(s) - s\tilde{T}'(s)}{\tilde{T}^2(s)} ds - C \int_{-\infty}^{\infty} s \frac{\tilde{T}(s) - s\tilde{T}'(s)}{\tilde{T}(s)} ds, \end{aligned}$$

whereupon,

$$(15) \qquad C=\frac{\frac{1}{\sqrt{2\pi t}}\int_{-\infty}^{\infty} s e^{-\frac{s^2}{2t}}\frac{\tilde{T}(s)-s\tilde{T}'(s)}{\tilde{T}^2(s)}ds}{\int_{-\infty}^{\infty} s\frac{\tilde{T}(s)-s\tilde{T}'(s)}{\tilde{T}(s)}ds}.$$

Definition 1.7.1. *We say that $\hat{X}$ is an $\hat{N}^1(0,t)$ -random iso-variable if it has iso-density*

$$\frac{1}{\sqrt{2\pi t}}e^{-\frac{x^2}{2t}}\frac{1}{\tilde{T}(x)}-C,$$

where C is a constant which is defined by (15).

Let now $\hat{X}$ is $N^2(m,\sigma^2)$-random iso-variable. Then it has iso-density

$$\hat{f}(x)=\frac{1}{\sqrt{2\pi\sigma^2}}e^{-\frac{(x\tilde{T}(x)-m)^2}{2\sigma^2}}\frac{1}{\tilde{T}(x)}.$$

From here,

$$\begin{aligned}
F(x) &= \hat{\int}_{-\infty}^{x}\hat{f}(s)\hat{\times}\hat{d}\hat{s}\\
&= \hat{\int}_{-\infty}^{x}\frac{1}{\sqrt{2\pi\sigma^2}}e^{-\frac{(x\tilde{T}(x)-m)^2}{2\sigma^2}}\frac{1}{\tilde{T}(s)}\hat{\times}\hat{d}\hat{s}\\
&= \frac{1}{\sqrt{2\pi\sigma^2}}\int_{-\infty}^{x}\frac{1}{\tilde{T}(s)}e^{-\frac{(s\tilde{T}(s)-m)^2}{2\sigma^2}}\frac{1}{\tilde{T}(s)}\tilde{T}(s)\tilde{T}(s)d\hat{s}\\
&= \frac{1}{\sqrt{2\pi\sigma^2}}\int_{-\infty}^{x}e^{-\frac{(s\tilde{T}(s)-m)^2}{2\sigma^2}}d\frac{s}{\tilde{T}(s)}\\
&= \frac{1}{\sqrt{2\pi\sigma^2}}\int_{-\infty}^{x}e^{-\frac{(s\tilde{T}(s)-m)^2}{2\sigma^2}}\frac{\tilde{T}(s)-s\tilde{T}'(s)}{\tilde{T}^2(s)s}ds.
\end{aligned}$$

In particular, when $m=0$, $\sigma^2=t$, we have

$$\hat{f}(x)=\frac{1}{\sqrt{2\pi t}}e^{-\frac{x^2\tilde{T}^2(x)}{2t}}\frac{1}{\tilde{T}(x)},$$

$$F(x)=\frac{1}{\sqrt{2\pi t}}\int_{-\infty}^{x}e^{-\frac{s^2\tilde{T}^2(s)}{2t}}\frac{\tilde{T}(s)-s\tilde{T}'(s)}{\tilde{T}^2(s)}ds.$$

Let

$$\tilde{f}(x)=\hat{f}(x)-C,$$

where C is a constant which is chosen so that

$$\hat{\int}_{-\infty}^{\infty} s\tilde{f}(s)\hat{\times}\hat{d}\hat{s}=0,$$

i.e.,

$$\hat{\int}_{-\infty}^{\infty} s\left(\hat{f}(s)-C\right)\hat{\times}\hat{d}\hat{s}=0,$$

or

$$0=\int_{-\infty}^{\infty}\frac{1}{\tilde{T}(s)}s\left(\frac{1}{\sqrt{2\pi t}}e^{-\frac{s^2\tilde{T}^2(s)}{2t}}\frac{1}{\tilde{T}(s)}-C\right)\tilde{T}(s)\tilde{T}(s)d\hat{s}$$

$$=\frac{1}{\sqrt{2\pi t}}\int_{-\infty}^{\infty}se^{-\frac{s^2\tilde{T}^2(s)}{2t}}\frac{1}{\tilde{T}^2(s)}\tilde{T}^2(s)d\hat{s}-C\int_{-\infty}^{\infty}s\frac{1}{\tilde{T}(s)}\tilde{T}(s)\tilde{T}(s)d\hat{s}$$

$$=\frac{1}{\sqrt{2\pi t}}\int_{-\infty}^{\infty}se^{-\frac{s^2\tilde{T}^2(s)}{2t}}d\hat{s}-C\int_{-\infty}^{\infty}s\tilde{T}(s)d\hat{s}$$

$$=\frac{1}{\sqrt{2\pi t}}\int_{-\infty}^{\infty}se^{-\frac{s^2\tilde{T}^2(s)}{2t}}d\frac{s}{\tilde{T}(s)}-C\int_{-\infty}^{\infty}s\tilde{T}(s)d\frac{s}{\tilde{T}(s)}$$

$$=\frac{1}{\sqrt{2\pi t}}\int_{-\infty}^{\infty}se^{-\frac{s^2\tilde{T}^2(s)}{2t}}\frac{\tilde{T}(s)-s\tilde{T}'(s)}{\tilde{T}^2(s)}ds-C\int_{-\infty}^{\infty}s\tilde{T}(s)\frac{\tilde{T}(s)-s\tilde{T}'(s)}{\tilde{T}^2(s)}ds$$

$$=\frac{1}{\sqrt{2\pi t}}\int_{-\infty}^{\infty}se^{-\frac{s^2\tilde{T}^2(s)}{2t}}\frac{\tilde{T}(s)-s\tilde{T}'(s)}{\tilde{T}^2(s)}ds-C\int_{-\infty}^{\infty}s\frac{\tilde{T}(s)-s\tilde{T}'(s)}{\tilde{T}(s)}ds,$$

from where

$$(16)\qquad C=\frac{\frac{1}{\sqrt{2\pi t}}\int_{-\infty}^{\infty}se^{-\frac{s^2\tilde{T}^2(s)}{2t}}\frac{\tilde{T}(s)-s\tilde{T}'(s)}{\tilde{T}^2(s)}ds}{\int_{-\infty}^{\infty}s\frac{\tilde{T}(s)-s\tilde{T}'(s)}{\tilde{T}(s)}ds}.$$

Definition 1.7.2. *We say that $\hat{X}$ is $\hat{N}^2(0,t)$-random iso-variable if it has iso-density*

$$\frac{1}{\sqrt{2\pi t}}e^{-\frac{x^2\tilde{T}^2(x)}{2t}}\frac{1}{\tilde{T}(x)}-C,$$

where C is defined by (16).

Now we suppose that $\hat{X}$ is $N^3(m,\sigma^2)$-random iso-variable. Then it has iso-density

$$\hat{f}(x)=\frac{1}{\sqrt{2\pi\sigma^2}}e^{-\frac{\left(\frac{x}{\tilde{T}(x)}-m\right)^2}{2\sigma^2}}\frac{1}{\tilde{T}(x)}.$$

Therefore

$$F(x)=\hat{\int}_{-\infty}^{x}\hat{f}(s)\hat{\times}\hat{d}\hat{s}$$

$$=\hat{\int}_{-\infty}^{x}\frac{1}{\sqrt{2\pi\sigma^2}}e^{-\frac{\left(\frac{s}{\tilde{T}(s)}-m\right)^2}{2\sigma^2}}\frac{1}{\tilde{T}(s)}\hat{\times}\hat{d}\hat{s}$$

$$=\frac{1}{\sqrt{2\pi\sigma^2}}\int_{-\infty}^{x}\frac{1}{\tilde{T}(s)}e^{-\frac{\left(\frac{s}{\tilde{T}(s)}-m\right)^2}{2\sigma^2}}\frac{1}{\tilde{T}(s)}\tilde{T}(s)\tilde{T}(s)d\hat{s}$$

$$=\frac{1}{\sqrt{2\pi\sigma^2}}\int_{-\infty}^{x}e^{-\frac{\left(\frac{s}{\tilde{T}(s)}-m\right)^2}{2\sigma^2}}d\frac{s}{\tilde{T}(s)}$$

$$=\frac{1}{\sqrt{2\pi\sigma^2}}\int_{-\infty}^{x}e^{-\frac{\left(\frac{s}{\tilde{T}(s)}-m\right)^2}{2\sigma^2}}\frac{\tilde{T}(s)-s\tilde{T}'(s)}{\tilde{T}^2(s)}ds.$$

When $m = 0$, $\sigma^2 = t$ we have

$$\hat{f}(x) = \frac{1}{\sqrt{2\pi t}} e^{-\frac{x^2}{2t\tilde{T}^2(x)}} \frac{1}{\tilde{T}(x)},$$

$$F(x) = \frac{1}{\sqrt{2\pi t}} \int_{-\infty}^{x} e^{-\frac{s^2}{2t\tilde{T}^2(s)}} \frac{\tilde{T}(s) - s\tilde{T}'(s)}{\tilde{T}^2(s)} ds.$$

Let

$$\tilde{f}(x) = \hat{f}(x) - C,$$

where C is a constant which is chosen so that

$$\hat{\int}_{-\infty}^{\infty} s\tilde{f}(s)\hat{\times}\hat{d}\hat{s} = 0,$$

i.e.,

$$\hat{\int}_{-\infty}^{\infty} s\left(\hat{f}(s) - C\right)\hat{\times}\hat{d}\hat{s} = 0,$$

or

$$0 = \hat{\int}_{-\infty}^{\infty} s\left(\frac{1}{\sqrt{2\pi t}} e^{-\frac{s^2}{2t\tilde{T}^2(s)}} \frac{1}{\tilde{T}(s)} - C\right)\hat{\times}\hat{d}\hat{s}$$

$$= \int_{-\infty}^{\infty} \frac{1}{\tilde{T}(s)} s\left(\frac{1}{\sqrt{2\pi t}} e^{-\frac{s^2}{2t\tilde{T}^2(s)}} \frac{1}{\tilde{T}(s)} - C\right)\tilde{T}(s)\tilde{T}(s)d\hat{s}$$

$$= \frac{1}{\sqrt{2\pi t}} \int_{-\infty}^{\infty} se^{-\frac{s^2}{2t\tilde{T}^2(s)}} \frac{1}{\tilde{T}^2(s)}\tilde{T}(s)\tilde{T}(s)d\hat{s} - C\int_{-\infty}^{\infty} \frac{1}{\tilde{T}(s)} s\tilde{T}(s)\tilde{T}(s)d\hat{s}$$

$$- \frac{1}{\sqrt{2\pi t}} \int_{-\infty}^{\infty} se^{-\frac{s^2}{2t\tilde{T}^2(s)}} d\hat{s} - C\int_{-\infty}^{\infty} s\tilde{T}(s)d\hat{s}$$

$$= \frac{1}{\sqrt{2\pi t}} \int_{-\infty}^{\infty} se^{-\frac{s^2}{2t\tilde{T}^2(s)}} d\frac{s}{\tilde{T}(s)} - C\int_{-\infty}^{\infty} s\tilde{T}(s)d\frac{s}{\tilde{T}(s)}$$

$$= \frac{1}{\sqrt{2\pi t}} \int_{-\infty}^{\infty} se^{-\frac{s^2}{2t\tilde{T}^2(s)}} \frac{\tilde{T}(s) - s\tilde{T}'(s)}{\tilde{T}^2(s)} ds - C\int_{-\infty}^{\infty} s\tilde{T}(s)\frac{\tilde{T}(s) - s\tilde{T}'(s)}{\tilde{T}^2(s)} ds$$

$$= \frac{1}{\sqrt{2\pi t}} \int_{-\infty}^{\infty} se^{-\frac{s^2}{2t\tilde{T}^2(s)}} \frac{\tilde{T}(s) - s\tilde{T}'(s)}{\tilde{T}^2(s)} ds - C\int_{-\infty}^{\infty} s\frac{\tilde{T}(s) - s\tilde{T}'(s)}{\tilde{T}(s)} ds,$$

whereupon

$$(17) \qquad C = \frac{\frac{1}{\sqrt{2\pi t}} \int_{-\infty}^{\infty} se^{-\frac{s^2}{2t\tilde{T}^2(s)}} \frac{\tilde{T}(s) - s\tilde{T}'(s)}{\tilde{T}^2(s)} ds}{\int_{-\infty}^{\infty} s\frac{\tilde{T}(s) - s\tilde{T}'(s)}{\tilde{T}(s)} ds}.$$

Definition 1.7.3. *We say that $\hat{X}$ is $\hat{N}^3(0,t)$-random iso-variable if it has iso-density*

$$\frac{1}{\sqrt{2\pi t}} e^{-\frac{x^2}{2t\tilde{T}^2(x)}} \frac{1}{\tilde{T}(x)} - C,$$

where C is defined by (17).

We suppose that $\hat{X}$ is $N^4(m,\sigma^2)$-random iso-variable. Then it has iso-density

$$\hat{f}(x) = \frac{1}{\sqrt{2\pi\sigma^2}} e^{-\frac{(x\tilde{T}(x)-m)^2}{2\sigma^2}}.$$

Then

$$F(x) = \hat{\int}_{-\infty}^{x} \hat{f}(s) \hat{\times} \hat{d}\hat{s}$$

$$= \hat{\int}_{-\infty}^{x} \frac{1}{\sqrt{2\pi\sigma^2}} e^{-\frac{(x\tilde{T}(x)-m)^2}{2\sigma^2}} \hat{\times} \hat{d}\hat{s}$$

$$= \frac{1}{\sqrt{2\pi\sigma^2}} \int_{-\infty}^{x} \frac{1}{\tilde{T}(s)} e^{-\frac{(x\tilde{T}(x)-m)^2}{2\sigma^2}} \tilde{T}(s)\tilde{T}(s) d\hat{s}$$

$$= \frac{1}{\sqrt{2\pi\sigma^2}} \int_{-\infty}^{x} e^{-\frac{(x\tilde{T}(x)-m)^2}{2\sigma^2}} \tilde{T}(s) d\frac{s}{\tilde{T}(s)}$$

$$= \frac{1}{\sqrt{2\pi\sigma^2}} \int_{-\infty}^{x} e^{-\frac{(x\tilde{T}(x)-m)^2}{2\sigma^2}} \frac{\tilde{T}(s)-s\tilde{T}'(s)}{\tilde{T}(s)} ds.$$

Let $m = 0$, $\sigma^2 = t$. In this case we have

$$\hat{f}(x) = \frac{1}{\sqrt{2\pi t}} e^{-\frac{x^2\tilde{T}^2(x)}{2t}},$$

$$F(x) = \frac{1}{\sqrt{2\pi t}} \int_{-\infty}^{x} e^{-\frac{s^2\tilde{T}^2(s)}{2t}} ds.$$

Let also,

$$\tilde{f}(s) = \hat{f}(s) - C,$$

where C is a constant which is chosen so that

$$\hat{\int}_{-\infty}^{\infty} s\tilde{f}(s) \hat{\times} \hat{d}\hat{s} = 0,$$

i.e.,

$$\hat{\int}_{-\infty}^{\infty} s\left(\hat{f}(s) - C\right) \hat{\times} \hat{d}\hat{s} = 0,$$

or

$$0 = \hat{\int}_{-\infty}^{\infty} s\left(\frac{1}{\sqrt{2\pi t}} e^{-\frac{s^2\tilde{T}^2(s)}{2t}} - C\right) \hat{\times} \hat{d}\hat{s}$$

$$= \int_{-\infty}^{\infty} s\frac{1}{\tilde{T}(s)} \left(\frac{1}{\sqrt{2\pi t}} e^{-\frac{s^2\tilde{T}^2(s)}{2t}} - C\right) \tilde{T}(s)\tilde{T}(s) \hat{\times} \hat{d}\hat{s}$$

$$= \frac{1}{\sqrt{2\pi t}} \int_{-\infty}^{\infty} s e^{-\frac{s^2\tilde{T}^2(s)}{2t}} \tilde{T}(s) d\hat{s} - C\int_{-\infty}^{\infty} s\tilde{T}(s) d\hat{s}$$

$$= \frac{1}{\sqrt{2\pi t}} \int_{-\infty}^{\infty} s e^{-\frac{s^2\tilde{T}^2(s)}{2t}} \tilde{T}(s) d\frac{s}{\tilde{T}(s)} - C\int_{-\infty}^{\infty} s\tilde{T}(s) d\frac{s}{\tilde{T}(s)}$$

$$= \frac{1}{\sqrt{2\pi t}} \int_{-\infty}^{\infty} s e^{-\frac{s^2\tilde{T}^2(s)}{2t}} \frac{\tilde{T}(s)-s\tilde{T}'(s)}{\tilde{T}(s)} ds - C\int_{-\infty}^{\infty} s\frac{\tilde{T}(s)-s\tilde{T}'(s)}{\tilde{T}(s)} ds,$$

from where

$$(18)\qquad C=\frac{\frac{1}{\sqrt{2\pi t}}\int_{-\infty}^{\infty}se^{-\frac{s^2\tilde{T}^2(s)}{2t}}\frac{\tilde{T}(s)-s\tilde{T}'(s)}{\tilde{T}(s)}ds}{\int_{-\infty}^{\infty}s\frac{\tilde{T}(s)-s\tilde{T}'(s)}{\tilde{T}(s)}ds}.$$

Definition 1.7.4. *We will say that* $\hat{X}$ *is* $\hat{N}^4(0,t)$*-random-iso-variable if its iso-density is*

$$\frac{1}{\sqrt{2\pi t}}e^{-\frac{x^2\tilde{T}^2(x)}{2t}}-C,$$

where C *is defined by* (18).

Let now $\hat{X}$ is $N^5(m,\sigma^2)$ - random iso-variable. Then

$$\hat{f}(x)=\frac{1}{\sqrt{2\pi\sigma^2}}e^{-\frac{\left(\frac{x}{\tilde{T}(x)}-m\right)^2}{2\sigma^2}}$$

and

$$F(x)=\hat{\int}_{-\infty}^{x}\hat{f}(s)\hat{\times}\hat{d}s$$

$$=\hat{\int}_{-\infty}^{x}\frac{1}{\sqrt{2\pi\sigma^2}}e^{-\frac{\left(\frac{s}{\tilde{T}(s)}-m\right)^2}{2\sigma^2}}\hat{\times}\hat{d}\hat{s}$$

$$=\frac{1}{\sqrt{2\pi\sigma^2}}\int_{-\infty}^{x}\frac{1}{\tilde{T}(s)}e^{-\frac{\left(\frac{s}{\tilde{T}(s)}-m\right)^2}{2\sigma^2}}\tilde{T}(s)\tilde{T}(s)d\hat{s}$$

$$=\frac{1}{\sqrt{2\pi\sigma^2}}\int_{-\infty}^{x}e^{-\frac{\left(\frac{s}{\tilde{T}(s)}-m\right)^2}{2\sigma^2}}\tilde{T}(s)d\frac{s}{\tilde{T}(s)}$$

$$=\frac{1}{\sqrt{2\pi\sigma^2}}\int_{-\infty}^{x}e^{-\frac{\left(\frac{s}{\tilde{T}(s)}-m\right)^2}{2\sigma^2}}\frac{\tilde{T}(s)-s\tilde{T}'(s)}{\tilde{T}(s)}ds.$$

When $m=0$, $\sigma^2=t$ we have

$$\hat{f}(x)=\frac{1}{\sqrt{2t\pi}}e^{-\frac{x^2}{2t\tilde{T}^2(x)}},$$

$$F(x)=\int_{-\infty}^{x}e^{-\frac{s^2}{2t\tilde{T}^2(s)}}\frac{\tilde{T}(s)-s\tilde{T}'(s)}{\tilde{T}(s)}ds.$$

Let also,

$$\tilde{f}(x)=\hat{f}(x)-C,$$

where C is a constant which is chosen so that

$$\hat{\int}_{-\infty}^{\infty}s\tilde{f}(s)\hat{\times}\hat{d}\hat{s}=0,$$

i.e.,

$$\hat{\int}_{-\infty}^{\infty} s\left(\hat{f}(s)-C\right)\hat{\times}\hat{d}\hat{s}=0,$$

or

$$0=\hat{\int}_{-\infty}^{\infty} s\left(\frac{1}{\sqrt{2\pi t}}e^{-\frac{s^2}{2t\tilde{T}^2(s)}}-C\right)\hat{\times}\hat{d}\hat{s}$$

$$=\int_{-\infty}^{\infty} s\frac{1}{\tilde{T}(s)}\left(\frac{1}{\sqrt{2\pi t}}e^{-\frac{s^2}{2t\tilde{T}^2(s)}}-C\right)\tilde{T}(s)\tilde{T}(s)d\hat{s}$$

$$=\frac{1}{\sqrt{2t\pi}}\int_{-\infty}^{\infty} se^{-\frac{s^2}{2t\tilde{T}^2(s)}}\tilde{T}(s)d\hat{s}-C\int_{-\infty}^{\infty} s\tilde{T}(s)d\hat{s}$$

$$=\frac{1}{\sqrt{2t\pi}}\int_{-\infty}^{\infty} se^{-\frac{s^2}{2t\tilde{T}^2(s)}}\tilde{T}(s)d\frac{s}{\tilde{T}(s)}-C\int_{-\infty}^{\infty} s\tilde{T}(s)d\frac{s}{\tilde{T}(s)}$$

$$=\frac{1}{\sqrt{2t\pi}}\int_{-\infty}^{\infty} se^{-\frac{s^2}{2t\tilde{T}^2(s)}}\frac{\tilde{T}(s)-s\tilde{T}'(s)}{\tilde{T}(s)}ds-C\int_{-\infty}^{\infty} s\frac{\tilde{T}(s)-s\tilde{T}'(s)}{\tilde{T}(s)}ds,$$

whereupon

$$(19)\qquad C=\frac{\frac{1}{\sqrt{2t\pi}}\int_{-\infty}^{\infty} se^{-\frac{s^2}{2t\tilde{T}^2(s)}}\frac{\tilde{T}(s)-s\tilde{T}'(s)}{\tilde{T}(s)}ds}{\int_{-\infty}^{\infty} s\frac{\tilde{T}(s)-s\tilde{T}'(s)}{\tilde{T}(s)}ds}.$$

Definition 1.7.5. *We will say that $\hat{X}$ is $\hat{N}^5(0,t)$-random iso-variable if its iso-density is*

$$\frac{1}{\sqrt{2t\pi}}e^{-\frac{x^2}{2t\tilde{T}^2(x)}}-C,$$

where C is defined by (19).

Definition 1.7.6. *A real-valued iso-stochastic process $\hat{W}(\cdot)$ is called an iso-Brownian motion of the ith kind, $i=1,2,3,4,5$, or iso-Wiener process of the ith kind if*

(i) $\hat{W}(0)=0$ *i-a.s,*

(ii) $\hat{W}(t)-\hat{W}(s)$ *is* $\hat{N}^i(0,t)$*-random iso-variable for all* $t\geq s\geq 0$,

(iii) *For all times*

$$0<t_1<t_2\ldots<t_n$$

the random iso-variables

$$\hat{W}(t_1),\quad \hat{W}(t_2)-\hat{W}(t_1),\quad \ldots,\quad \hat{W}(t_n)-\hat{W}(t_{n_1})$$

are independent.

Because $\hat{W}(0)=0$ we have that

$$\hat{W}(t)-\hat{W}(0)=\hat{W}(t)\qquad\text{for}\qquad \forall t\geq 0.$$

therefore $\hat{W}$ is $\hat{N}^i(0,t)$-random iso-variable. Then

$$\hat{E}\left(\frac{\hat{W}(t)}{\hat{T}}\right) = 0.$$

For convenience, below we will use iso-Brownian motion or iso-Wiener process instead iso-Brownian motion of ith kind or iso-Wiener process of ith kind.

Theorem 1.7.7. *We have*

$$\hat{E}\left(\frac{\hat{W}(t_{i-1})\left(\hat{W}(t_i)-\hat{W}(t_{i-1})\right)}{\hat{T}}\right) = 0.$$

Proof. Really, since $\hat{W}(t_{i-1})$ and $\hat{W}(t_i)-\hat{W}(t_{i-1})$ are independent, then

$$\hat{E}\left(\frac{\hat{W}(t_{i-1})\left(\hat{W}(t_i)-\hat{W}(t_{i-1})\right)}{\hat{T}}\right) = \hat{E}\left(\frac{\hat{W}(t_{i-1})}{\hat{T}}\right)\hat{E}\left(\frac{\hat{W}(t_i)-\hat{W}(t_{i-1})}{\hat{T}}\right),$$

now we use that

$$\hat{E}\left(\frac{\hat{W}(t_{i-1})}{\hat{T}}\right) = 0,$$

whereupon

$$\hat{E}\left(\frac{\hat{W}(t_{i-1})\left(\hat{W}(t_i)-\hat{W}(t_{i-1})\right)}{\hat{T}}\right) = 0.$$

□

Corollary 1.7.8. *We have*

$$\hat{E}\left(\frac{\hat{W}(t_{i-1})\hat{W}(t_i)}{\hat{T}}\right) = \hat{E}\left(\frac{\hat{W}^2(t_{i-1})}{\hat{T}}\right).$$

We can rewrite this corollary as follows

$$\hat{E}\left(\frac{\hat{W}(t)\hat{W}(s)}{\hat{T}}\right) = \hat{E}\left(\frac{\hat{W}^2(t\wedge s)}{\hat{T}}\right),$$

where

$$t \wedge s = \min\{t,s\}.$$

Definition 1.7.9. *1-dimensional iso-white noise is called*

$$\hat{\xi}(t) := \hat{W}^{\circledast}(t).$$

Using the definition for first-order iso-derivative we get

$$\hat{W}^{\circledast}(t) = \hat{d}\hat{W}(t) \nearrow \hat{d}\hat{t}$$

$$= \frac{1}{\tilde{T}(t)} \frac{\hat{d}\hat{W}(t)}{\hat{d}\hat{t}}$$

$$= \frac{1}{\tilde{T}(t)} \frac{\tilde{T}(t)d\hat{W}(t)}{\tilde{T}(t)d\hat{t}}$$

$$= \frac{1}{\tilde{T}(t)} \frac{d\hat{W}(t)}{d\hat{t}}$$

$$= \frac{1}{\tilde{T}(t)} \frac{d\hat{W}(t)}{d\frac{t}{\tilde{T}(t)}}$$

$$= \frac{1}{\tilde{T}(t)} \frac{d\hat{W}(t)}{\frac{\tilde{T}(t)-t\tilde{T}'(t)}{\tilde{T}^2(t)}dt}$$

$$= \frac{\tilde{T}(t)}{\tilde{T}(t)-t\tilde{T}'(t)} \frac{d\hat{W}(t)}{dt},$$

i.e.,

$$\hat{\xi}(t) = \frac{\tilde{T}(t)}{\tilde{T}(t)-t\tilde{T}'(t)} \frac{d\hat{W}(t)}{dt}.$$

Now we suppose that $\hat{X}$ is $N^1(0,t)$-random iso-variable. Then its iso-density is

$$\hat{f}(x) = \frac{1}{\sqrt{2\pi t}} e^{-\frac{x^2}{2t}} \frac{1}{\tilde{T}(x)}.$$

We define

$$\hat{G}^1(\hat{X}) := \frac{\tilde{T}^2(\hat{X})}{\tilde{T}(\hat{X}) - \hat{X}\tilde{T}'(\hat{X})}.$$

Then

$$\hat{E}\left(\frac{\hat{X}\hat{G}^1(\hat{X})}{\hat{T}}\right) = \hat{\int}_{-\infty}^{\infty} x\hat{G}^1(x)\hat{f}(x)\hat{\times}\hat{d}\hat{x}$$

$$= \hat{\int}_{-\infty}^{\infty} x\frac{\tilde{T}^2(x)}{\tilde{T}(x)-x\tilde{T}'(x)} \frac{1}{\sqrt{2\pi t}} e^{-\frac{x^2}{2t}} \frac{1}{\tilde{T}(x)} \hat{\times}\hat{d}\hat{x}$$

$$= \frac{1}{\sqrt{2\pi t}} \hat{\int}_{-\infty}^{\infty} x\frac{\tilde{T}(x)}{\tilde{T}(x)-x\tilde{T}'(x)} e^{-\frac{x^2}{2t}} \hat{\times}\hat{d}\hat{x}$$

$$= \frac{1}{\sqrt{2\pi t}} \int_{-\infty}^{\infty} \frac{1}{\tilde{T}(x)} x\frac{\tilde{T}(x)}{\tilde{T}(x)-x\tilde{T}'(x)} e^{-\frac{x^2}{2t}} \tilde{T}(x)\tilde{T}(x)d\hat{x}$$

$$= \frac{1}{\sqrt{2\pi t}} \int_{-\infty}^{\infty} x \frac{\tilde{T}^2(x)}{\tilde{T}(x)-x\tilde{T}'(x)} e^{-\frac{x^2}{2t}} d\hat{x}$$

$$= \frac{1}{\sqrt{2\pi t}} \int_{-\infty}^{\infty} x \frac{\tilde{T}^2(x)}{\tilde{T}(x)-x\tilde{T}'(x)} e^{-\frac{x^2}{2t}} d\frac{x}{\tilde{T}(x)}$$

$$= \frac{1}{\sqrt{2\pi t}} \int_{-\infty}^{\infty} x \frac{\tilde{T}^2(x)}{\tilde{T}(x)-x\tilde{T}'(x)} e^{-\frac{x^2}{2t}} \frac{\tilde{T}(x)-x\tilde{T}'(x)}{\tilde{T}^2(x)} dx$$

$$= \frac{1}{\sqrt{2\pi t}} \int_{-\infty}^{\infty} x e^{-\frac{x^2}{2t}} dx$$

$$= \frac{1}{\sqrt{2\pi t}} \int_{-\infty}^{\infty} e^{-\frac{x^2}{2t}} d\frac{x^2}{2}$$

$$= -\frac{t}{\sqrt{2\pi t}} \int_{-\infty}^{\infty} e^{-\frac{x^2}{2t}} d\left(-\frac{x^2}{2t}\right)$$

$$= -\sqrt{\frac{t}{2\pi}} e^{-\frac{x^2}{2t}} \Big|_{x=-\infty}^{x=\infty}$$

$$= 0.$$

Also, as in above,

$$\hat{E}\left(\frac{\hat{G}^1(\hat{X})}{\hat{T}}\right) = \hat{\int}_{-\infty}^{\infty} \hat{G}^1(x)\hat{f}(x) \hat{\times} d\hat{x}$$

$$= \frac{1}{\sqrt{2\pi t}} \int_{-\infty}^{\infty} e^{-\frac{x^2}{2t}} dx$$

$$= \frac{1}{\sqrt{\pi}} \int_{-\infty}^{\infty} e^{-\frac{x^2}{2t}} d\frac{x}{\sqrt{2t}}$$

$$= \frac{1}{\sqrt{\pi}} \sqrt{\pi}$$

$$= 1.$$

Below we will suppose that $\tilde{T} : \mathbb{R} \longrightarrow (0,\infty)$ is a bounded function.

Now we suppose that $\hat{X}$ is $N^2(0,t)$-random iso-variable. Then its iso-density is

$$\hat{f}(x) = \frac{1}{\sqrt{2\pi t}} e^{-\frac{x^2 \tilde{T}^2(x)}{2t}} \frac{1}{\tilde{T}(x)}.$$

We define

$$\hat{G}^2(\hat{X}) := \frac{\tilde{T}^2(\hat{X})}{\tilde{T}(\hat{X}) - \hat{X}\tilde{T}'(\hat{X})} \left(\tilde{T}(\hat{X}) + \hat{X}\tilde{T}(\hat{X})\right).$$

Then

$$\hat{E}\left(\frac{\hat{G}^2(\hat{X})}{\hat{T}}\right) = \hat{\int}_{-\infty}^{\infty} \hat{G}^2(x)\hat{f}(x) \hat{\times} d\hat{x}$$

$$= \hat{\int}_{-\infty}^{\infty} \frac{\tilde{T}^2(x)}{\tilde{T}(x)-x\tilde{T}'(x)} \left(\tilde{T}(x) + x\tilde{T}'(x)\right) \frac{1}{\sqrt{2\pi t}} e^{-\frac{x^2 \tilde{T}^2(x)}{2t}} \frac{1}{\tilde{T}(x)} \hat{\times} d\hat{x}$$

$$= \frac{1}{\sqrt{2\pi t}} \hat{\int}_{-\infty}^{\infty} \frac{\tilde{T}(x)}{\tilde{T}(x)-x\tilde{T}'(x)} \left(\tilde{T}(x) + x\tilde{T}'(x)\right) e^{-\frac{x^2 \tilde{T}^2(x)}{2t}} \hat{\times} d\hat{x}$$

$$
\begin{aligned}
&= \tfrac{1}{\sqrt{2\pi t}} \int_{-\infty}^{\infty} \tfrac{1}{\tilde{T}(x)} \tfrac{\tilde{T}(x)}{\tilde{T}(x)-x\tilde{T}'(x)} \left(\tilde{T}(x)+x\tilde{T}'(x)\right) e^{-\frac{x^2\tilde{T}^2(x)}{2t}} \tilde{T}(x)\tilde{T}(x) d\hat{x} \\
&= \tfrac{1}{\sqrt{2\pi t}} \int_{-\infty}^{\infty} \tfrac{\tilde{T}^2(x)}{\tilde{T}(x)-x\tilde{T}'(x)} \left(\tilde{T}(x)+x\tilde{T}'(x)\right) e^{-\frac{x^2\tilde{T}^2(x)}{2t}} d\hat{x} \\
&= \tfrac{1}{\sqrt{2\pi t}} \int_{-\infty}^{\infty} \tfrac{\tilde{T}^2(x)}{\tilde{T}(x)-x\tilde{T}'(x)} \left(\tilde{T}(x)+x\tilde{T}'(x)\right) e^{-\frac{x^2\tilde{T}^2(x)}{2t}} d\tfrac{x}{\tilde{T}(x)} \\
&= \tfrac{1}{\sqrt{2\pi t}} \int_{-\infty}^{\infty} \tfrac{\tilde{T}^2(x)}{\tilde{T}(x)-x\tilde{T}'(x)} \left(\tilde{T}(x)+x\tilde{T}'(x)\right) e^{-\frac{x^2\tilde{T}^2(x)}{2t}} \tfrac{\tilde{T}(x)-x\tilde{T}'(x)}{\tilde{T}^2(x)} dx \\
&= \tfrac{1}{\sqrt{2\pi t}} \int_{-\infty}^{\infty} \left(\tilde{T}(x)+x\tilde{T}'(x)\right) e^{-\frac{x^2\tilde{T}^2(x)}{2t}} dx \\
&= \tfrac{1}{\sqrt{2\pi t}} \int_{-\infty}^{\infty} e^{-\frac{x^2\tilde{T}^2(x)}{2t}} d\left(x\tilde{T}(x)\right) \\
&= \tfrac{1}{\sqrt{\pi}} \int_{-\infty}^{\infty} e^{-\frac{x^2\tilde{T}^2(x)}{2t}} d\tfrac{x\tilde{T}(x)}{\sqrt{2t}} \\
&= \tfrac{1}{\sqrt{\pi}} \sqrt{\pi} \\
&= 1.
\end{aligned}
$$

Also, as in above,

$$
\begin{aligned}
\hat{E}\left(\tfrac{\hat{X}\tilde{T}(\hat{X})\hat{G}^2(\hat{X})}{\tilde{T}}\right) &= \hat{\int}_{-\infty}^{\infty} x\tilde{T}(x)\hat{G}^2(x)\hat{f}(x)\hat{\times} d\hat{x} \\
&= \tfrac{1}{\sqrt{2\pi t}} \int_{-\infty}^{\infty} x\tilde{T}(x) e^{-\frac{x^2\tilde{T}^2(x)}{2t}} d\left(x\tilde{T}(x)\right) \\
&= \tfrac{1}{\sqrt{2\pi t}} \int_{-\infty}^{\infty} e^{-\frac{x^2\tilde{T}^2(x)}{2t}} d\tfrac{x^2\tilde{T}^2(x)}{2} \\
&= -\tfrac{t}{\sqrt{2\pi t}} \int_{-\infty}^{\infty} e^{-\frac{x^2\tilde{T}^2(x)}{2t}} d\left(-\tfrac{x^2\tilde{T}^2(x)}{2t}\right) \\
&= -\sqrt{\tfrac{t}{2\pi}} e^{-\frac{x^2\tilde{T}^2(x)}{2t}} \Big|_{x=-\infty}^{x=\infty} \\
&= 0.
\end{aligned}
$$

Now we suppose that $\hat{X}$ is $N^3(0,t)$-random iso-variable. Then its iso-density is

$$
\hat{f}(x) = \frac{1}{\sqrt{2\pi t}} e^{-\frac{x^2}{2t\tilde{T}^2(x)}} \frac{1}{\tilde{T}(x)}.
$$

We define the function

$$
\hat{G}^3(\hat{X}) := \frac{\hat{X}}{\tilde{T}(\hat{X})}.
$$

Then

$$\hat{E}\left(\frac{\hat{G}^3(\hat{X})}{\hat{T}}\right) = \hat{\int}_{-\infty}^{\infty} \hat{G}^3(x)\hat{f}(x)\hat{\times}\hat{d}\hat{x}$$

$$= \hat{\int}_{-\infty}^{\infty} \frac{x}{\tilde{T}(x)} \frac{1}{\sqrt{2\pi t}} e^{-\frac{x^2}{2t\tilde{T}^2(x)}} \frac{1}{\tilde{T}(x)} \hat{\times}\hat{d}\hat{x}$$

$$= \frac{1}{\sqrt{2\pi t}} \hat{\int}_{-\infty}^{\infty} \frac{x}{\tilde{T}^2(x)} e^{-\frac{x^2}{2t\tilde{T}^2(x)}} \hat{\times}\hat{d}\hat{x}$$

$$= \frac{1}{\sqrt{2\pi t}} \int_{-\infty}^{\infty} \frac{1}{\tilde{T}(x)} \frac{x}{\tilde{T}^2(x)} e^{-\frac{x^2}{2t\tilde{T}^2(x)}} \tilde{T}(x)\tilde{T}(x)d\hat{x}$$

$$= \frac{1}{\sqrt{2\pi t}} \int_{-\infty}^{\infty} \frac{x}{\tilde{T}(x)} e^{-\frac{x^2}{2t\tilde{T}^2(x)}} d\hat{x}$$

$$= \frac{1}{\sqrt{2\pi t}} \int_{-\infty}^{\infty} \frac{x}{\tilde{T}(x)} e^{-\frac{x^2}{2t\tilde{T}^2(x)}} d\frac{x}{\tilde{T}(x)}$$

$$= \frac{1}{\sqrt{2\pi t}} \int_{-\infty}^{\infty} e^{-\frac{x^2}{2t\tilde{T}^2(x)}} d\frac{x^2}{2\tilde{T}^2(x)}$$

$$= -\frac{t}{\sqrt{2\pi t}} \int_{-\infty}^{\infty} e^{-\frac{x^2}{2t\tilde{T}^2(x)}} d\left(-\frac{x^2}{2t\tilde{T}^2(x)}\right)$$

$$= \sqrt{\frac{t}{2\pi}} e^{-\frac{x^2}{2t\tilde{T}^2(x)}} \Big|_{x=-\infty}^{x=\infty}$$

$$= 0.$$

Also, as in above,

$$\hat{E}\left(\frac{\tilde{T}(\hat{X})}{\hat{X}} \frac{\hat{G}^3(\hat{X})}{\hat{T}}\right) = \hat{\int}_{-\infty}^{\infty} \frac{\tilde{T}(x)}{x} \hat{G}^3(x)\hat{f}(x)\hat{\times}\hat{d}\hat{x}$$

$$= \frac{1}{\sqrt{2\pi t}} \int_{-\infty}^{\infty} e^{-\frac{x^2}{2t\tilde{T}^2(x)}} d\frac{x}{\tilde{T}(x)}$$

$$= \frac{1}{\sqrt{\pi}} \int_{-\infty}^{\infty} e^{-\frac{x^2}{2t\tilde{T}^2(x)}} d\frac{x}{\sqrt{2t}\tilde{T}(x)}$$

$$= \frac{1}{\sqrt{\pi}}\sqrt{\pi}$$

$$= 1.$$

Now we suppose that $\hat{X}$ is $N^4(0,t)$-random iso-variable. Then its iso-density is

$$\hat{f}(x) = \frac{1}{\sqrt{2\pi t}} e^{-\frac{x^2\tilde{T}^2(x)}{2t}}.$$

We define the function

$$\hat{G}^4(\hat{X}) := \frac{\tilde{T}(\hat{X})}{\tilde{T}(\hat{X}) - \hat{X}\tilde{T}'(\hat{X})}(\tilde{T}(\hat{X}) + \hat{X}\tilde{T}'(\hat{X})).$$

Then

$$\hat{E}\left(\frac{\hat{G}^4(\hat{X})}{\hat{T}}\right) = \hat{\int}_{-\infty}^{\infty} \hat{G}^4(x)\hat{f}(x)\hat{\times}\hat{d}\hat{x}$$

$$= \hat{\int}_{-\infty}^{\infty} \frac{\tilde{T}(x)}{\tilde{T}(x)-x\tilde{T}'(x)}(\tilde{T}(x)+x\tilde{T}'(x))\frac{1}{\sqrt{2\pi t}}e^{-\frac{x^2\tilde{T}^2(x)}{2t}}\hat{\times}\hat{d}\hat{x}$$

$$= \frac{1}{\sqrt{2\pi t}}\hat{\int}_{-\infty}^{\infty} \frac{\tilde{T}(x)}{\tilde{T}(x)-x\tilde{T}'(x)}(\tilde{T}(x)+x\tilde{T}'(x))e^{-\frac{x^2\tilde{T}^2(x)}{2t}}\hat{\times}\hat{d}\hat{x}$$

$$= \frac{1}{\sqrt{2\pi t}}\int_{-\infty}^{\infty} \frac{1}{\tilde{T}(x)}\frac{\tilde{T}(x)}{\tilde{T}(x)-x\tilde{T}'(x)}(\tilde{T}(x)+x\tilde{T}'(x))e^{-\frac{x^2\tilde{T}^2(x)}{2t}}\tilde{T}(x)\tilde{T}(x)d\hat{x}$$

$$= \frac{1}{\sqrt{2\pi t}}\int_{-\infty}^{\infty} \frac{\tilde{T}^2(x)}{\tilde{T}(x)-x\tilde{T}'(x)}(\tilde{T}(x)+x\tilde{T}'(x))e^{-\frac{x^2\tilde{T}^2(x)}{2t}}d\hat{x}$$

$$= \frac{1}{\sqrt{2\pi t}}\int_{-\infty}^{\infty} \frac{\tilde{T}^2(x)}{\tilde{T}(x)-x\tilde{T}'(x)}(\tilde{T}(x)+x\tilde{T}'(x))e^{-\frac{x^2\tilde{T}^2(x)}{2t}}d\frac{x}{\tilde{T}(x)}$$

$$= \frac{1}{\sqrt{2\pi t}}\int_{-\infty}^{\infty} \frac{\tilde{T}^2(x)}{\tilde{T}(x)-x\tilde{T}'(x)}(\tilde{T}(x)+x\tilde{T}'(x))e^{-\frac{x^2\tilde{T}^2(x)}{2t}}\frac{\tilde{T}(x)-x\tilde{T}'(x)}{\tilde{T}^2(x)}dx$$

$$= \frac{1}{\sqrt{2\pi t}}\int_{-\infty}^{\infty}(\tilde{T}(x)+x\tilde{T}'(x))e^{-\frac{x^2\tilde{T}^2(x)}{2t}}dx$$

$$= \frac{1}{\sqrt{2\pi t}}\int_{-\infty}^{\infty} e^{-\frac{x^2\tilde{T}^2(x)}{2t}}d(x\tilde{T}(x))$$

$$= \frac{1}{\sqrt{\pi}}\int_{-\infty}^{\infty} e^{-\frac{x^2\tilde{T}^2(x)}{2t}}d\frac{x\tilde{T}(x)}{\sqrt{2t}}$$

$$= \frac{1}{\sqrt{\pi}}\sqrt{\pi}$$

$$= 1.$$

Also, as in above, we have

$$\hat{E}\left(\frac{\hat{G}^4(\hat{X})\hat{X}\tilde{T}(\hat{X})}{\hat{T}}\right) = \hat{\int}_{-\infty}^{\infty} \hat{G}^4(x)x\tilde{T}(x)\hat{f}(x)\hat{\times}\hat{d}\hat{x}$$

$$= \frac{1}{\sqrt{2\pi t}}\int_{-\infty}^{\infty} x\tilde{T}(x)e^{-\frac{x^2\tilde{T}^2(x)}{2t}}d(x\tilde{T}(x))$$

$$= \frac{1}{\sqrt{2\pi t}}\int_{-\infty}^{\infty} e^{-\frac{x^2\tilde{T}^2(x)}{2t}}d\frac{x^2\tilde{T}^2(x)}{2}$$

$$= -\frac{t}{\sqrt{2\pi t}}\int_{-\infty}^{\infty} e^{-\frac{x^2\tilde{T}^2(x)}{2t}}d\left(-\frac{x^2\tilde{T}^2(x)}{2t}\right)$$

$$= -\sqrt{\frac{t}{2\pi}}\int_{-\infty}^{\infty} e^{-\frac{x^2\tilde{T}^2(x)}{2t}}d\left(-\frac{x^2\tilde{T}^2(x)}{2t}\right)$$

$$= -\sqrt{\frac{t}{2\pi}} e^{-\frac{x^2 \tilde{T}^2(x)}{2t}} \Big|_{x=-\infty}^{x=\infty}$$

$$= 0.$$

Let now $\hat{X}$ is $N^5(0,t)$-random iso-variable. Then its iso-density is

$$\hat{f}(x) = \frac{1}{\sqrt{2\pi t}} e^{-\frac{x^2}{2t\tilde{T}^2(x)}}.$$

We define the function

$$\hat{G}^5(\hat{X}) := \frac{1}{\tilde{T}(\hat{X})}.$$

Hence,

$$\hat{E}\left(\frac{\hat{G}^5(\hat{X})}{\hat{T}}\right) = \hat{\int}_{-\infty}^{\infty} \hat{G}^5(x)\hat{f}(x)\hat{\times}\hat{d}\hat{x}$$

$$= \hat{\int}_{-\infty}^{\infty} \frac{1}{\tilde{T}(x)} \frac{1}{\sqrt{2\pi t}} e^{-\frac{x^2}{2t\tilde{T}^2(x)}} \hat{\times}\hat{d}\hat{x}$$

$$= \frac{1}{\sqrt{2\pi t}} \hat{\int}_{-\infty}^{\infty} \frac{1}{\tilde{T}(x)} e^{-\frac{x^2}{2t\tilde{T}^2(x)}} \hat{\times}\hat{d}\hat{x}$$

$$= \frac{1}{\sqrt{2\pi t}} \int_{-\infty}^{\infty} \frac{1}{\tilde{T}(x)} \frac{1}{\tilde{T}(x)} e^{-\frac{x^2}{2t\tilde{T}^2(x)}} \tilde{T}(x)\tilde{T}(x) d\hat{x}$$

$$= \frac{1}{\sqrt{2\pi t}} \int_{-\infty}^{\infty} e^{-\frac{x^2}{2t\tilde{T}^2(x)}} d\hat{x}$$

$$= \frac{1}{\sqrt{2\pi t}} \int_{-\infty}^{\infty} e^{-\frac{x^2}{2t\tilde{T}^2(x)}} d\frac{x}{\tilde{T}(x)}$$

$$= \frac{1}{\sqrt{\pi}} \int_{-\infty}^{\infty} e^{-\frac{x^2}{2t\tilde{T}^2(x)}} d\frac{x}{\sqrt{2t}\tilde{T}(x)}$$

$$= \frac{1}{\sqrt{\pi}} \sqrt{\pi}$$

$$= 1.$$

Also, as in above, we obtain

$$\hat{E}\left(\frac{\hat{X}\hat{G}^5(\hat{X})}{\tilde{T}(\hat{X})\hat{T}}\right) = \hat{\int}_{-\infty}^{\infty} \frac{x\hat{G}^5(x)}{\tilde{T}(x)} \hat{f}(x) \hat{\times} \hat{d}\hat{x}$$

$$= \frac{1}{\sqrt{2\pi t}} \int_{-\infty}^{\infty} \frac{x}{\tilde{T}(x)} e^{-\frac{x^2}{2t\tilde{T}^2(x)}} d\frac{x}{\tilde{T}(x)}$$

$$= \frac{1}{\sqrt{2\pi t}} \int_{-\infty}^{\infty} e^{-\frac{x^2}{2t\tilde{T}^2(x)}} d\frac{x^2}{2\tilde{T}^2(x)}$$

$$= \frac{t}{\sqrt{2\pi t}} \int_{-\infty}^{\infty} e^{-\frac{x^2}{2t\tilde{T}^2(x)}} d\left(-\frac{x^2}{2t\tilde{T}^2(x)}\right)$$

$$= -\sqrt{\frac{t}{2\pi}} \int_{-\infty}^{\infty} e^{-\frac{x^2}{2t\tilde{T}^2(x)}} d\left(-\frac{x^2}{2t\tilde{T}^2(x)}\right)$$

$$= -\sqrt{\frac{t}{2\pi}} e^{-\frac{x^2}{2t\tilde{T}^2(x)}} \Big|_{x=-\infty}^{x=\infty}$$

$$= 0.$$

Definition 1.7.10. *We will say that the random iso-variable $\hat{X}$ is $\tilde{N}^1(m, \sigma^2)$-random iso-variable if its iso-density is*

$$\tilde{f}(x) = \frac{1}{\sqrt{2\pi\sigma^2}} e^{-\frac{(x-m)^2}{2\sigma^2}} \frac{\tilde{T}(x)}{\tilde{T}(x) - x\tilde{T}'(x)}.$$

In particular, when $m = 0$, $\sigma^2 = t$, we have

$$\tilde{f}(x) = \frac{1}{\sqrt{2\pi t}} e^{-\frac{x^2}{2t}} \frac{\tilde{T}(x)}{\tilde{T}(x) - x\tilde{T}'(x)}.$$

We suppose that $\hat{X}$ is $\tilde{N}^1(0, t)$-random iso-variable. Then

$$\hat{E}\left(\frac{\hat{X}}{\hat{T}}\right) = \hat{\int}_{-\infty}^{\infty} x\tilde{f}(x) \hat{\times} \hat{d}\hat{x}$$

$$= \hat{\int}_{-\infty}^{\infty} x \frac{1}{\sqrt{2\pi t}} e^{-\frac{x^2}{2t}} \frac{\tilde{T}(x)}{\tilde{T}(x) - x\tilde{T}'(x)} \hat{\times} \hat{d}\hat{x}$$

$$= \frac{1}{\sqrt{2\pi t}} \hat{\int}_{-\infty}^{\infty} x e^{-\frac{x^2}{2t}} \frac{\tilde{T}(x)}{\tilde{T}(x) - x\tilde{T}'(x)} \hat{\times} \hat{d}\hat{x}$$

$$= \frac{1}{\sqrt{2\pi t}} \int_{-\infty}^{\infty} \frac{1}{\tilde{T}(x)} x e^{-\frac{x^2}{2t}} \frac{\tilde{T}(x)}{\tilde{T}(x) - x\tilde{T}'(x)} \tilde{T}(x)\tilde{T}(x) d\hat{x}$$

$$= \frac{1}{\sqrt{2\pi t}} \int_{-\infty}^{\infty} x e^{-\frac{x^2}{2t}} \frac{\tilde{T}^2(x)}{\tilde{T}(x) - x\tilde{T}'(x)} d\hat{x}$$

$$= \frac{1}{\sqrt{2\pi t}} \int_{-\infty}^{\infty} x e^{-\frac{x^2}{2t}} \frac{\tilde{T}^2(x)}{\tilde{T}(x) - x\tilde{T}'(x)} d\frac{x}{\tilde{T}(x)}$$

$$= \frac{1}{\sqrt{2\pi t}} \int_{-\infty}^{\infty} x e^{-\frac{x^2}{2t}} \frac{\tilde{T}^2(x)}{\tilde{T}(x) - x\tilde{T}'(x)} \frac{\tilde{T}(x) - x\tilde{T}'(x)}{\tilde{T}^2(x)} dx$$

$$= \frac{1}{\sqrt{2\pi t}} \int_{-\infty}^{\infty} x e^{-\frac{x^2}{2t}} dx$$

$$= \frac{1}{\sqrt{2\pi t}} \int_{-\infty}^{\infty} e^{-\frac{x^2}{2t}} d\frac{x^2}{2}$$

$$= -\frac{t}{\sqrt{2\pi t}} \int_{-\infty}^{\infty} e^{-\frac{x^2}{2t}} d\left(-\frac{x^2}{2t}\right)$$

$$= -\sqrt{\frac{t}{2\pi}} e^{-\frac{x^2}{2t}} \Big|_{x=-\infty}^{x=\infty}$$

$$= 0.$$

As in above, we have

$$\hat{E}\left(\frac{\hat{X}^2}{\hat{T}}\right) = \hat{\int}_{-\infty}^{\infty} x^2 \tilde{f}(x) \hat{\times} \hat{d}\hat{x}$$

$$= \frac{1}{\sqrt{2\pi t}} \int_{-\infty}^{\infty} x^2 e^{-\frac{x^2}{2t}} dx$$

$$= \frac{1}{\sqrt{2\pi t}} \int_{-\infty}^{\infty} x e^{-\frac{x^2}{2t}} d\left(\frac{x^2}{2}\right)$$

$$- -\frac{t}{\sqrt{2\pi t}} \int_{-\infty}^{\infty} x e^{-\frac{x^2}{2t}} d\left(-\frac{x^2}{2t}\right)$$

$$= \sqrt{\frac{t}{2\pi}} \int_{-\infty}^{\infty} x de^{-\frac{x^2}{2t}}$$

$$= -\sqrt{\frac{t}{2\pi}} x e^{-\frac{x^2}{2t}} \Big|_{x=-\infty}^{x=\infty} + \sqrt{\frac{t}{2\pi}} \int_{-\infty}^{\infty} e^{-\frac{x^2}{2t}} dx$$

$$= \frac{t}{\sqrt{\pi}} \int_{-\infty}^{\infty} e^{-\frac{x^2}{2t}} d\frac{x}{\sqrt{2t}}$$

$$= \frac{t}{\sqrt{\pi}} \sqrt{\pi}$$

$$= t.$$

Definition 1.7.11. *Let*

$$0!! = 1, \qquad 1!! = 1.$$

We define

$$k!! = k(k-2)\ldots 1 \quad \text{for} \quad k \ \text{odd},$$

$$k!! = k(k-2)\ldots 2 \quad \text{for} \quad k \ \text{even}.$$

From the definition it follows that

$$k!! = k(k-2)!!.$$

Example 1.7.12.

$$2!! = 2 \cdot 0!! = 2 \cdot 1 = 2,$$

$$3!! = 3 \cdot 1!! = 3 \cdot 1 = 3,$$

$$5!! = 5 \cdot 3!! = 5 \cdot 3 = 15,$$

$$\begin{aligned} 10!! &= 10 \cdot 8!! \\ &= 10 \cdot 8 \cdot 6!! \\ &= 80 \cdot 6 \cdot 4!! \\ &= 480 \cdot 4 \cdot 2!! \\ &= 1920 \cdot 2 \\ &= 3840. \end{aligned}$$

Lemma 1.7.13. *Let $a > 0$. Then*

$$I = \int_{-\infty}^{\infty} x^k e^{-ax^2} dx = \begin{cases} 0 & \text{when } k \text{ is odd} \\ \frac{(k-1)!!}{2^{\frac{k}{2}} a^{\frac{k}{2}}} \sqrt{\frac{\pi}{a}} & \text{when } k \text{ is even.} \end{cases}$$

Proof. 1. k is odd.

(a) $k = 1$. Then

$$\begin{aligned} I &= \int_{-\infty}^{\infty} x e^{-ax^2} dx \\ &= \frac{1}{2} \int_{-\infty}^{\infty} e^{-ax^2} dx^2 \\ &= -\frac{1}{2a} \int_{-\infty}^{\infty} e^{-ax^2} d(-ax^2) \\ &= -\frac{1}{2a} e^{-ax^2} \Big|_{x=-\infty}^{x=\infty} \\ &= 0. \end{aligned}$$

(b) We suppose that

$$\int_{-\infty}^{\infty} x^k e^{-ax^2} dx = 0$$

for some $k \in \mathbb{N}$, k is odd.

(c) We will prove that

$$\int_{-\infty}^{\infty} x^{k+2} e^{-ax^2} dx = 0.$$

Really,

$$\int_{-\infty}^{\infty} x^{k+2} e^{-ax^2} dx = \frac{1}{2}\int_{-\infty}^{\infty} x^{k+1} e^{-ax^2} dx^2$$

$$= -\frac{1}{2a}\int_{-\infty}^{\infty} x^{k+1} de^{-ax^2}$$

$$= -\frac{1}{2a} x^{k+1} e^{-ax^2}\Big|_{x=-\infty}^{x=\infty} + \frac{k+1}{2a}\int_{-\infty}^{\infty} x^k e^{-ax^2} dx$$

$$= 0.$$

2. k is even.

   (a) $k = 2$. Then

$$I = \int_{-\infty}^{\infty} x^2 e^{-ax^2} dx$$

$$= \frac{1}{2}\int_{-\infty}^{\infty} x e^{-ax^2} dx^2$$

$$= -\frac{1}{2a}\int_{-\infty}^{\infty} x e^{-ax^2} d(-ax^2)$$

$$= -\frac{1}{2a}\int_{-\infty}^{\infty} x de^{-ax^2}$$

$$= -\frac{1}{2a} x e^{-ax^2}\Big|_{x=-\infty}^{x=\infty} + \frac{1}{2a}\int_{-\infty}^{\infty} e^{-ax^2} dx$$

$$= \frac{1}{2a}\int_{-\infty}^{\infty} e^{-ax^2} dx$$

$$= \frac{1}{2a}\int_{-\infty}^{\infty} e^{-(\sqrt{a}x)^2} dx$$

$$= \frac{1}{2a\sqrt{a}}\int_{-\infty}^{\infty} e^{-(\sqrt{a}x)^2} d(\sqrt{a}x)$$

$$= \frac{1}{2a\sqrt{a}}\sqrt{\pi}$$

$$= \frac{1}{2a}\sqrt{\frac{\pi}{a}}.$$

   (b) We suppose that

$$\int_{-\infty}^{\infty} x^k e^{-ax^2} dx = \frac{(k-1)!!}{2^{\frac{k}{2}} a^{\frac{k}{2}}}\sqrt{\frac{\pi}{a}}$$

   for some $k \in \mathbb{N}$, k is even.

   (c) We will prove that

$$\int_{-\infty}^{\infty} x^{k+2} e^{-ax^2} dx = \frac{(k+1)!!}{2^{\frac{k}{2}+1} a^{\frac{k}{2}+1}}\sqrt{\frac{\pi}{a}}.$$

Indeed,

$$\int_{-\infty}^{\infty} x^{k+2} e^{-ax^2} dx = \frac{1}{2}\int_{-\infty}^{\infty} x^{k+1} e^{-ax^2} d(x^2)$$

$$= -\frac{1}{2a}\int_{-\infty}^{\infty} x^{k+1} e^{-ax^2} d(-ax^2)$$

$$= -\frac{1}{2a}\int_{-\infty}^{\infty} x^{k+1} de^{-ax^2}$$

$$= -\frac{1}{2a} x^{k+1} e^{-ax^2}\Big|_{x=-\infty}^{x=\infty} + \frac{k+1}{2a}\int_{-\infty}^{\infty} x^k e^{-ax^2} dx$$

$$= \frac{k+1}{2a}\int_{-\infty}^{\infty} x^k e^{-ax^2} dx$$

$$= \frac{k+1}{2a}\frac{(k-1)!!}{2^{\frac{k}{2}} a^{\frac{k}{2}}}\sqrt{\frac{\pi}{a}}$$

$$= \frac{(k+1)!!}{2^{\frac{k}{2}+1} a^{\frac{k}{2}+1}}\sqrt{\frac{\pi}{a}}.$$

□

Corollary 1.7.14. *Let $t > 0$. Then*

$$\frac{1}{\sqrt{2\pi t}}\int_{-\infty}^{\infty} x^k e^{-\frac{x^2}{2t}} dx = \begin{cases} 0 & \text{when } k \text{ is } odd \\ (k-1)!!t^{\frac{k}{2}} & \text{when } k \text{ is even.} \end{cases}$$

Proof. Let k is even. Then

$$\frac{1}{\sqrt{2\pi t}}\int_{-\infty}^{\infty} x^k e^{-\frac{x^2}{2t}} dx = \frac{1}{\sqrt{2\pi t}}\frac{(k-1)!!}{2^{\frac{k}{2}}\left(\frac{1}{2t}\right)^{\frac{k}{2}}}\sqrt{\frac{\pi}{\frac{1}{2t}}}$$

$$= \frac{1}{\sqrt{2\pi t}}\frac{(k-1)!!}{2^{\frac{k}{2}}\frac{1}{2^{\frac{k}{2}} t^{\frac{k}{2}}}}\sqrt{2\pi t}$$

$$= (k-1)!!t^{\frac{k}{2}}.$$

□

Exercise 1.7.15. *Let $\hat{X}$ be $\tilde{N}^1(0,t)$-random iso-variable. Prove that*

$$\hat{E}\left(\frac{\hat{X}^k}{\hat{T}}\right) = \frac{1}{\sqrt{2\pi t}}\int_{-\infty}^{\infty} x^k e^{-\frac{x^2}{t}} dx$$

$$= \begin{cases} 0 & \text{when } k \text{ is } odd \\ (k-1)!!t^{\frac{k}{2}} & \text{when } k \text{ is even.} \end{cases}$$

Definition 1.7.16. *We will say the random iso-variable $\hat{X}$ is $\tilde{N}^2(m,\sigma^2)$-random iso-variable if its iso-density is*

$$\tilde{f}(x) = \frac{1}{\sqrt{2\pi\sigma^2}} e^{-\frac{(x\tilde{T}(x)-m)^2}{2\sigma^2}} \frac{\tilde{T}(x)}{\tilde{T}(x) - x\tilde{T}'(x)}(\tilde{T}(x) + x\tilde{T}'(x)).$$

In particular, when $m = 0$, $\sigma^2 = t$, we have

$$\tilde{f}(x) = \frac{1}{\sqrt{2\pi t}} e^{-\frac{x^2\tilde{T}^2(x)}{2t}} \frac{\tilde{T}(x)}{\tilde{T}(x) - x\tilde{T}'(x)} (\tilde{T}(x) + x\tilde{T}'(x)).$$

We suppose that $\hat{X}$ is $\tilde{N}^2(0,t)$-random iso-variable. Then

$$\hat{E}\left(\frac{\hat{X}\tilde{T}(\hat{X})}{\hat{T}}\right) = \hat{\int}_{-\infty}^{\infty} x\tilde{T}(x)\tilde{f}(x)\hat{\times}\hat{d}\hat{x}$$

$$= \frac{1}{\sqrt{2\pi t}} \hat{\int}_{-\infty}^{\infty} x\tilde{T}(x) \frac{1}{\sqrt{2\pi t}} e^{-\frac{x^2\tilde{T}^2(x)}{2t}} \frac{\tilde{T}(x)}{\tilde{T}(x) - x\tilde{T}'(x)} (\tilde{T}(x) + x\tilde{T}'(x)) \hat{\times}\hat{d}\hat{x}$$

$$= \frac{1}{\sqrt{2\pi t}} \hat{\int}_{-\infty}^{\infty} x e^{-\frac{x^2\tilde{T}^2(x)}{2t}} \frac{\tilde{T}^2(x)}{\tilde{T}(x) - x\tilde{T}'(x)} (\tilde{T}(x) + x\tilde{T}'(x)) \hat{\times}\hat{d}\hat{x}$$

$$= \frac{1}{\sqrt{2\pi t}} \int_{-\infty}^{\infty} \frac{1}{\tilde{T}(x)} x e^{-\frac{x^2\tilde{T}^2(x)}{2t}} \frac{\tilde{T}^2(x)}{\tilde{T}(x) - x\tilde{T}'(x)} (\tilde{T}(x) + x\tilde{T}'(x)) \tilde{T}(x)\tilde{T}(x) d\hat{x}$$

$$= \frac{1}{\sqrt{2\pi t}} \int_{-\infty}^{\infty} x e^{-\frac{x^2\tilde{T}^2(x)}{2t}} \frac{\tilde{T}^3(x)}{\tilde{T}(x) - x\tilde{T}'(x)} (\tilde{T}(x) + x\tilde{T}'(x)) d\frac{x}{\tilde{T}(x)}$$

$$= \frac{1}{\sqrt{2\pi t}} \int_{-\infty}^{\infty} x e^{-\frac{x^2\tilde{T}^2(x)}{2t}} \frac{\tilde{T}^3(x)}{\tilde{T}(x) - x\tilde{T}'(x)} (\tilde{T}(x) + x\tilde{T}'(x)) \frac{\tilde{T}(x) - x\tilde{T}'(x)}{\tilde{T}^2(x)} dx$$

$$= \frac{1}{\sqrt{2\pi t}} \int_{-\infty}^{\infty} x\tilde{T}(x) e^{-\frac{x^2\tilde{T}^2(x)}{2t}} (\tilde{T}(x) + x\tilde{T}'(x)) dx$$

$$= \frac{1}{\sqrt{2\pi t}} \int_{-\infty}^{\infty} x\tilde{T}(x) e^{-\frac{x^2\tilde{T}^2(x)}{2t}} d(x\tilde{T}(x)) \qquad (x\tilde{T}(x) = y)$$

$$= \frac{1}{\sqrt{2\pi t}} \int_{-\infty}^{\infty} e e^{-\frac{y^2}{2t}} dy$$

$$= 0.$$

Also, as in above,

$$\hat{E}\left(\frac{\hat{X}^2\tilde{T}^2(\hat{X})}{\hat{T}}\right) = \hat{\int}_{-\infty}^{\infty} x^2\tilde{T}^2(x)\tilde{f}(x)\hat{\times}\hat{d}\hat{x}$$

$$= \frac{1}{\sqrt{2\pi t}} \int_{-\infty}^{\infty} x^2\tilde{T}^2(x) e^{-\frac{x^2\tilde{T}^2(x)}{2t}} d(x\tilde{T}(x)) \qquad (x\tilde{T}(x) = y)$$

$$= \frac{1}{\sqrt{2\pi t}} \int_{-\infty}^{\infty} y^2 e^{-\frac{y^2}{2t}} dy$$

$$= t.$$

Exercise 1.7.17. *Let $\hat{X}$ be $\tilde{N}^2(0,t)$-random iso-variable. Prove that*

$$\hat{E}\left(\frac{\hat{X}^k\tilde{T}^k(\hat{X})}{\hat{T}}\right) = \frac{1}{\sqrt{2\pi t}}\int_{-\infty}^{\infty} x^k\tilde{T}^k(x)e^{-\frac{x^2\tilde{T}^2(x)}{t}}d(x\tilde{T}(x))$$

$$= \begin{cases} 0 & \text{when} \quad k \quad \text{is} \quad odd \\ (k-1)!!t^{\frac{k}{2}} & \text{when} \quad k \quad \text{is} \quad \text{even.} \end{cases}$$

Definition 1.7.18. *We define*

$$\tilde{N}^3(m,\sigma^2) := \hat{N}^3(m,\sigma^2).$$

We suppose that $\hat{X} \in \tilde{N}^3(0,t)$. Then

$$\hat{E}\left(\frac{\hat{X}}{\tilde{T}(\hat{X})\hat{T}}\right) = 0, \qquad \hat{E}\left(\frac{\hat{X}^2}{\tilde{T}^2(\hat{X})\hat{T}}\right) = t.$$

Exercise 1.7.19. *Let $\hat{X}$ be $\tilde{N}^3(0,t)$-random iso-variable. Prove that*

$$\hat{E}\left(\frac{\frac{\hat{X}^k}{\tilde{T}^k(\hat{X})}}{\hat{T}}\right) = \frac{1}{\sqrt{2\pi t}}\int_{-\infty}^{\infty}\frac{x^k}{\tilde{T}^k(x)}e^{-\frac{x^2}{2t\tilde{T}^2(x)}}d\frac{x}{\tilde{T}(x)}$$

$$= \begin{cases} 0 & \text{when} \quad k \quad \text{is} \quad odd \\ (k-1)!!t^{\frac{k}{2}} & \text{when} \quad k \quad \text{is} \quad \text{even.} \end{cases}$$

Definition 1.7.20. *A real-valued iso-stochastic process $\tilde{W}^1(\cdot)$ is called an iso-Brownian iso-motion of the first kind or iso-Wiener iso-process of the first kind if*

(i) $\tilde{W}^1(0) = 0$ *i-a.s.,*

(ii) $\tilde{W}^1(t) - \tilde{W}^1(s)$ *is* $\tilde{N}^1(0,t-s)$ *for all* $t \geq s \geq 0$,

(iii) *for all times*

$$0 < t_1 < t_2 < \cdots < t_n,$$

the iso-random variables

$$\tilde{W}^1(t_1), \qquad \tilde{W}^2(t_2) - \tilde{W}(t_1), \ldots, \tilde{W}^1(t_n) - \tilde{W}(t_{n-1})$$

are independent.

We note that

$$\hat{E}\left(\frac{\tilde{W}^1(t)}{\hat{T}}\right) = 0, \qquad \hat{E}\left(\frac{(\tilde{W}^1(t))^2}{\hat{T}}\right) = t,$$

$$\hat{E}\left(\frac{(\tilde{W}^1(t))^k}{\hat{T}}\right) = \begin{cases} 0 & \text{when} \quad k \quad \text{is} \quad odd \\ (k-1)!!t^{\frac{k}{2}} & \text{when} \quad k \quad \text{is} \quad \text{even.} \end{cases}$$

for all $t \geq 0$.

Theorem 1.7.21. *Let $\tilde{W}^1(\cdot)$ be an iso-Wiener iso-process of the first kind. Then*

$$\hat{E}\left(\frac{\tilde{W}^1(t)\tilde{W}^1(s)}{\hat{T}}\right) = t \wedge s = \min\{t,s\}$$

for all $t,s \geq 0$.

Proof. Let $t \geq s \geq 0$. Then

$$\hat{E}\left(\frac{\tilde{W}^1(t)\tilde{W}^1(s)}{\hat{T}}\right) = \hat{E}\left(\frac{1}{\hat{T}}(\tilde{W}^1(t) - \tilde{W}^1(s) + \tilde{W}^1(s))\tilde{W}^1(s)\right)$$

$$\hat{E}\left(\frac{1}{\hat{T}}(\tilde{W}^1(s))^2 + \frac{1}{\hat{T}}(\tilde{W}^1(t) - \tilde{W}^1(s))\tilde{W}^1(s)\right)$$

$$= \hat{E}\left(\frac{1}{\hat{T}}((\tilde{W}^1(s))^2\right) + \hat{E}\left(\frac{1}{\hat{T}}((\tilde{W}^1(t) - \tilde{W}^1(s))\tilde{W}^1(s))\right)$$

$$= s + \hat{E}\left(\frac{1}{\hat{T}}(\tilde{W}^1(t) - \tilde{W}^1(s))\right)\hat{E}\left(\frac{1}{\hat{T}}(\tilde{W}^1(s))\right)$$

$$= s + 0$$

$$= s$$

$$= t \wedge s.$$

□

Definition 1.7.22. 1*-dimensional iso-white iso-noise of the first kind we define as follows*

$$\tilde{\xi}^1(t) := \left(\tilde{W}^1\right)^{\circledast}(t).$$

We can rewrite $\tilde{\xi}^1(t)$ as follows.

$$\tilde{\xi}^1(t) = \hat{d}\tilde{W}^1(t) \nearrow \hat{d}\hat{t}$$

$$= \frac{1}{\tilde{T}(t)} \frac{\hat{d}\tilde{W}^1(t)}{\hat{d}\hat{t}}$$

$$= \frac{1}{\tilde{T}(t)} \frac{\tilde{T}(t)d\tilde{W}^1(t)}{\tilde{T}(t)d\hat{t}}$$

$$= \frac{1}{\tilde{T}(t)} \frac{d\tilde{W}^1(t)}{d\frac{t}{\tilde{T}(t)}}$$

$$= \frac{1}{\tilde{T}(t)} \frac{d\tilde{W}^1(t)}{\frac{\tilde{T}(t)-t\tilde{T}'(t)}{\tilde{T}^2(t)}dt}$$

$$= \frac{\tilde{T}(t)}{\tilde{T}(t)-t\tilde{T}'(t)} \frac{d\tilde{W}^1(t)}{dt}.$$

Definition 1.7.23. *A real-valued iso-stochastic process* $\tilde{W}^2(\cdot)$ *is called an iso-Brownian iso-motion of the second kind or iso-Wiener iso-process of the second kind if*

(i) $\tilde{W}^2(0) = 0$ *i-a.s.,*

(ii) $\tilde{W}^2(t)\tilde{T}(\tilde{W}^2(t)) - \tilde{W}^2(s)\tilde{T}(\tilde{W}^2(s))$ *is* $\tilde{N}^2(0, t-s)$ *for all* $t \geq s \geq 0$*,*

(iii) *for all times*

$$0 < t_1 < t_2 < \cdots < t_n,$$

the iso-random variables

$$\tilde{W}^2(t_1)\tilde{T}(\tilde{W}^2(t_1)), \quad \tilde{W}^2(t_2)\tilde{T}(\tilde{W}^2(t_2)) - \tilde{W}^2(t_1)\tilde{T}(\tilde{W}^2(t_1)),$$

$$\cdots, \tilde{W}^2(t_n)\tilde{T}(\tilde{W}^2(t_n)) - \tilde{W}^2(t_{n-1})\tilde{T}(\tilde{W}^2(t_{n-1}))$$

are independent.

We have that

$$\hat{E}\left(\frac{\tilde{W}^2(t)\tilde{T}(\tilde{W}^2(t))}{\hat{T}}\right) = 0, \qquad \hat{E}\left(\frac{\left(\tilde{W}^2(t)\tilde{T}(\tilde{W}^2(t))\right)^2}{\hat{T}}\right) = t^2,$$

$$\hat{E}\left(\frac{(\tilde{W}^2(t)\tilde{T}(\tilde{W}^2(t)))^k}{\hat{T}}\right) = \begin{cases} 0 & \text{when} \quad k \quad \text{is} \quad odd \\ (k-1)!!t^{\frac{k}{2}} & \text{when} \quad k \quad \text{is} \quad \text{even.} \end{cases}$$

for all $t \geq 0$.

Theorem 1.7.24. *Let* $\tilde{W}^2(\cdot)$ *be an iso-Wiener iso-process of the second kind. Then*

$$\hat{E}\left(\tilde{W}^2(t)\tilde{T}(\tilde{W}^2(t))\tilde{W}^2(s)\tilde{T}(\tilde{W}^2(s))\right) = t \wedge s$$

for all $t, s \geq 0$.

Proof. Let $t \geq s$. Then

$$\begin{aligned}
&\hat{E}\left(\tilde{W}^2(t)\tilde{T}(\tilde{W}^2(t))\tilde{W}^2(s)\tilde{T}(\tilde{W}^2(s))\right) \\
&= \hat{E}\left(\left(\tilde{W}^2(t)\tilde{T}(\tilde{W}^2(t)) - \tilde{W}^2(s)\tilde{T}(\tilde{W}^2(s)) + \tilde{W}^2(s)\tilde{T}(\tilde{W}^2(s))\right)\tilde{W}^2(s)\tilde{T}(\tilde{W}^2(s))\right) \\
&= \hat{E}\left(\left(\tilde{W}^2(t)\tilde{T}(\tilde{W}^2(t)) - \tilde{W}^2(s)\tilde{T}(\tilde{W}^2(s))\right)\tilde{W}^2(s)\tilde{T}(\tilde{W}^2(s)) + \left(\tilde{W}^2(s)\tilde{T}(\tilde{W}^2(s))\right)^2\right) \\
&= \hat{E}\left(\left(\tilde{W}^2(t)\tilde{T}(\tilde{W}^2(t)) - \tilde{W}^2(s)\tilde{T}(\tilde{W}^2(s))\right)\tilde{W}^2(s)\tilde{T}(\tilde{W}^2(s))\right) + \hat{E}\left(\left(\tilde{W}^2(s)\tilde{T}(\tilde{W}^2(s))\right)^2\right) \\
&= \hat{E}\left(\left(\tilde{W}^2(t)\tilde{T}(\tilde{W}^2(t)) - \tilde{W}^2(s)\tilde{T}(\tilde{W}^2(s))\right)\right)\hat{E}\left(\tilde{W}^2(s)\tilde{T}(\tilde{W}^2(s))\right) + s \\
&= 0 + s \\
&= s \\
&= t \wedge s.
\end{aligned}$$

□

Definition 1.7.25. *1-dimensional iso-white iso-noise of the second kind we define as follows*

$$\tilde{\xi}^2(t) := \left(\tilde{W}^2\right)^{\circledast}(t).$$

We can rewrite $\tilde{\xi}^2(t)$ as follows

$$\tilde{\xi}^2(t) = \frac{\tilde{T}(t)}{\tilde{T}(t) - t\tilde{T}'(t)} \frac{d\tilde{W}^2(t)}{dt}.$$

Definition 1.7.26. *A real-valued iso-stochastic process $\tilde{W}^3(\cdot)$ is called an iso-Brownian iso-motion of the third kind or iso-Wiener iso-process of the third kind if*

(i) $\tilde{W}^3(0) = 0$ *i-a.s.,*

(ii) $\frac{\tilde{W}^3(t)}{\tilde{T}(\tilde{W}^3(t))} - \frac{\tilde{W}^3(s)}{\tilde{T}(\tilde{W}^3(s))}$ *is $\tilde{N}^3(0, t-s)$ for all $t \geq s \geq 0$,*

(iii) *for all times*

$$0 < t_1 < t_2 < \cdots < t_n,$$

the iso-random variables

$$\frac{\tilde{W}^3(t_1)}{\tilde{T}(\tilde{W}^3(t_1))}, \quad \frac{\tilde{W}^3(t_2)}{\tilde{T}(\tilde{W}^3(t_2))} - \frac{\tilde{W}^3(t_1)}{\tilde{T}(\tilde{W}^3(t_1))},$$

$$\ldots, \frac{\tilde{W}^3(t_n)}{\tilde{T}(\tilde{W}^3(t_n))} - \frac{\tilde{W}^3(t_{n-1})}{\tilde{T}(\tilde{W}^3(t_{n-1}))}$$

are independent.

We have that

$$\hat{E}\left(\frac{\frac{\tilde{W}^3(t)}{\tilde{T}(\tilde{W}^3(t))}}{\hat{T}}\right) = 0, \qquad \hat{E}\left(\frac{\left(\frac{\tilde{W}^3(t)}{\tilde{T}(\tilde{W}^3(t))}\right)^2}{\hat{T}}\right) = t^2,$$

$$\hat{E}\left(\frac{\left(\frac{\tilde{W}^3(t)}{\tilde{T}(\tilde{W}^3(t))}\right)^k}{\hat{T}}\right) = \begin{cases} 0 & \text{when} \quad k \quad \text{is} \quad odd \\ (k-1)!!t^{\frac{k}{2}} & \text{when} \quad k \quad \text{is} \quad \text{even}. \end{cases}$$

for all $t \geq 0$.

Theorem 1.7.27. *Let $\tilde{W}^2(\cdot)$ be an iso-Wiener iso-process of the second kind. Then*

$$\hat{E}\left(\frac{\tilde{W}^2(t)}{\tilde{T}(\tilde{W}^2(t))} \frac{\tilde{W}^2(s)}{\tilde{T}(\tilde{W}^2(s))}\right) = t \wedge s$$

for all $t, s \geq 0$.

Proof. Let $t \geq s$. Then

$$\begin{aligned}
&\hat{E}\left(\frac{\tilde{W}^3(t)}{\tilde{T}(\tilde{W}^3(t))}\frac{\tilde{W}^3(s)}{\tilde{T}(\tilde{W}^3(s))}\right)\\
&=\hat{E}\left(\left(\frac{\tilde{W}^3(t)}{\tilde{T}(\tilde{W}^3(t))}-\frac{\tilde{W}^3(s)}{\tilde{T}(\tilde{W}^3(s))}+\frac{\tilde{W}^3(s)}{\tilde{T}(\tilde{W}^3(s))}\right)\frac{\tilde{W}^3(s)}{\tilde{T}(\tilde{W}^3(s))}\right)\\
&=\hat{E}\left(\left(\frac{\tilde{W}^3(t)}{\tilde{T}(\tilde{W}^3(t))}-\frac{\tilde{W}^3(s)}{\tilde{T}(\tilde{W}^3(s))}\right)\frac{\tilde{W}^3(s)}{\tilde{T}(\tilde{W}^3(s))}+\left(\frac{\tilde{W}^3(s)}{\tilde{T}(\tilde{W}^3(s))}\right)^2\right)\\
&=\hat{E}\left(\left(\frac{\tilde{W}^3(t)}{\tilde{T}(\tilde{W}^3(t))}-\frac{\tilde{W}^3(s)}{\tilde{T}(\tilde{W}^3(s))}\right)\frac{\tilde{W}^3(s)}{\tilde{T}(\tilde{W}^3(s))}\right)+\hat{E}\left(\left(\frac{\tilde{W}^3(s)}{\tilde{T}(\tilde{W}^3(s))}\right)^2\right)\\
&=\hat{E}\left(\left(\frac{\tilde{W}^3(t)}{\tilde{T}(\tilde{W}^3(t))}-\frac{\tilde{W}^3(s)}{\tilde{T}(\tilde{W}^3(s))}\right)\right)\hat{E}\left(\frac{\tilde{W}^3(s)}{\tilde{T}(\tilde{W}^3(s))}\right)+s\\
&=0+s\\
&=s\\
&=t\wedge s.
\end{aligned}$$

□

Definition 1.7.28. 1*-dimensional iso-white iso-noise of the third kind we define as follows*

$$\tilde{\xi}^3(t) := \left(\tilde{W}^3\right)^{\circledast}(t).$$

We can rewrite $\tilde{\xi}^3(t)$ as follows

$$\tilde{\xi}^3(t) = \frac{\tilde{T}(t)}{\tilde{T}(t)-t\tilde{T}'(t)}\frac{d\tilde{W}^3(t)}{dt}.$$

1.8 Paley-Wiener-Zygmund-Iso-integral

Suppose that $g : [0,1] \longrightarrow \mathbb{R}$ is continuously-differentiable, with $g(0) = g(1) = 0$.

Note that g is an ordinary, deterministic function, which can be also an iso-function of the first, second, third, fourth or fifth kind and not an iso-stochastic process.

Definition 1.8.1. *We define the Paley-Wiener-Zygmund-iso-integral as follows*

$$\hat{\int}_0^1 g\hat{d}\frac{\hat{W}}{\hat{T}} := -\hat{\int} g^{\circledast}(t)\hat{\times}\frac{\hat{W}(t)}{\hat{T}}\hat{\times}\hat{d}\hat{t}.$$

We remember that

$$g^{\circledast}(t) = \hat{d}g(t) \nearrow \hat{d}\hat{t}$$

$$= \frac{1}{\tilde{T}(t)} \frac{\hat{d}g(t)}{\hat{d}\hat{t}}$$

$$= \frac{1}{\tilde{T}(t)} \frac{\tilde{T}(t)dg(t)}{\tilde{T}(t)d\hat{t}}$$

$$= \frac{1}{\tilde{T}(t)} \frac{dg(t)}{d\hat{t}}$$

$$= \frac{1}{\tilde{T}(t)} \frac{dg(t)}{d\frac{t}{\tilde{T}(t)}}$$

$$= \frac{1}{\tilde{T}(t)} \frac{dg(t)}{\frac{\tilde{T}(t)-t\tilde{T}'(t)}{\tilde{T}^2(t)}dt}$$

$$= \frac{\tilde{T}(t)}{\tilde{T}(t)-t\tilde{T}'(t)} \frac{dg(t)}{dt}$$

$$= \frac{\tilde{T}(t)}{\tilde{T}(t)-t\tilde{T}'(t)} g'(t),$$

hence,

$$\hat{\int}_0^1 g\hat{d}\frac{\hat{W}(t)}{\hat{T}} = -\int_0^1 \frac{1}{\tilde{T}(t)} \frac{\tilde{T}(t)}{\tilde{T}(t)-t\tilde{T}'(t)} g'(t)\tilde{T}(t)\frac{\hat{W}(t)}{\hat{T}}\tilde{T}(t)\tilde{T}(t)d\hat{t}$$

$$= -\int_0^1 \frac{\tilde{T}^3(t)}{\tilde{T}(t)-t\tilde{T}'(t)} g'(t)\frac{\hat{W}(t)}{\hat{T}} d\frac{t}{\tilde{T}(t)}$$

$$= -\int_0^1 \frac{\tilde{T}^3(t)}{\tilde{T}(t)-t\tilde{T}'(t)} g'(t)\frac{\hat{W}(t)}{\hat{T}} \frac{\tilde{T}(t)-t\tilde{T}'(t)}{\tilde{T}^2} dt$$

$$= -\int_0^1 \tilde{T}(t)g'(t)\frac{\hat{W}(t)}{\hat{T}} dt.$$

Theorem 1.8.2. *We have*

$$\hat{E}\left(\hat{\int}_0^1 g\hat{d}\frac{\hat{W}}{\hat{T}}\right) = 0.$$

Proof. We note that

(20) $\hat{E}\left(\frac{\hat{W}}{\hat{T}}\right) = 0.$

Also,

$$\hat{E}\left(\hat{\int}_0^1 g\hat{d}\frac{\hat{W}}{\hat{T}}\right) = \hat{E}\left(-\int_0^1 \tilde{T}(t)g'(t)\frac{\hat{W}(t)}{\hat{T}}dt\right)$$

$$= -\int_0^1 \hat{E}\left(\tilde{T}(t)g'(t)\frac{\hat{W}(t)}{\hat{T}}\right)dt$$

$$= -\int_0^1 \tilde{T}(t)g'(t)\hat{E}\left(\frac{\hat{W}(t)}{\hat{T}}\right)dt,$$

i.e.,

$$\hat{E}\left(\hat{\int}_0^1 g\hat{d}\frac{\hat{W}}{\hat{T}}\right) = -\int_0^1 \tilde{T}(t)g'(t)\hat{E}\left(\frac{\hat{W}(t)}{\hat{T}}\right)dt.$$

Hence, using (20), we get

$$\hat{E}\left(\hat{\int}_0^1 g\hat{d}\frac{\hat{W}}{\hat{T}}\right) = 0.$$

□

Definition 1.8.3. *We define the Paley-Wiener-Zygmund iso-integral of the first kind as follows*

$$\hat{\int}_0^1 gd\frac{\tilde{W}^1}{\hat{T}} := -\hat{\int}_0^1 g^{\circledast}(t)\hat{\times}\frac{\tilde{W}^1(t)}{\hat{T}}\hat{d}\hat{t}.$$

We can rewrite the Paley-Wiener-Zygmund iso-integral of the first kind as follows.

$$\hat{\int}_0^1 gd\frac{\tilde{W}^1}{\hat{T}} = -\int_0^1 \frac{1}{\tilde{T}(t)}\frac{\tilde{T}(t)}{\tilde{T}(t)-t\tilde{T}'(t)}g'(t)\tilde{T}(t)\frac{\tilde{W}^1(t)}{\hat{T}}\tilde{T}(t)d\hat{t}$$

$$= -\int_0^1 \frac{\tilde{T}^2(t)}{\tilde{T}(t)-t\tilde{T}'(t)}g'(t)\frac{\tilde{W}^1(t)}{\hat{T}}d\hat{t}$$

$$-\int_0^1 \frac{\tilde{T}^2(t)}{\tilde{T}(t)-t\tilde{T}'(t)}g'(t)\frac{\tilde{W}^1(t)}{\hat{T}}d\frac{t}{\tilde{T}(t)}$$

$$= -\int_0^1 \frac{\tilde{T}^2(t)}{\tilde{T}(t)-t\tilde{T}'(t)}g'(t)\frac{\tilde{W}^1(t)}{\hat{T}}\frac{\tilde{T}(t)-t\tilde{T}'(t)}{\tilde{T}^2(t)}dt$$

$$= -\int_0^1 g'(t)\frac{\tilde{W}^1(t)}{\hat{T}}dt.$$

Theorem 1.8.4. *We have*

$$\hat{E}\left(\hat{\int}_0^1 gd\frac{\tilde{W}^1}{\hat{T}}\right) = 0.$$

Proof. We note that

(21) $\quad \hat{E}\left(\frac{\tilde{W}^1}{\hat{T}}\right) = 0.$

Then

$$\hat{E}\left(\hat{\int}_0^1 gd\frac{\tilde{W}^1}{\hat{T}}\right) = \hat{E}\left(-\int_0^1 g'(t)\frac{\tilde{W}^1(t)}{\hat{T}}dt\right)$$

$$= -\int_0^1 \hat{E}\left(g'(t)\frac{\tilde{W}^1(t)}{\hat{T}}\right)dt$$

$$= -\int_0^1 g'(t)\hat{E}\left(\frac{\tilde{W}^1(t)}{\hat{T}}\right)dt \qquad ((21))$$

$$= 0.$$

□

Definition 1.8.5. *For a random iso-variable $\hat{X}$ we define iso-square as follows*

$$\hat{X}^{\hat{2}} = \hat{X}\hat{T}\hat{X}.$$

Also, we can define

$$\hat{X}^{\hat{3}} := \hat{X}\hat{T}\hat{X}^{\hat{2}},$$

$$\dots$$

$$\hat{X}^{\hat{n}} := \hat{X}\hat{T}\hat{X}^{\widehat{n-1}}, \qquad n \in \mathbb{N}.$$

Using the above definition we have

$$\hat{X}^{\hat{3}} = \hat{X}\hat{T}\hat{X}^{\hat{2}}$$

$$= \hat{X}\hat{T}\hat{X}\hat{T}\hat{X}$$

$$= \hat{X}^3\hat{T}^2,$$

$$\hat{X}^{\hat{4}} = \hat{X}\hat{T}\hat{X}^{\hat{3}}$$

$$= \hat{X}\hat{T}\hat{X}^3\hat{T}^2$$

$$= \hat{X}^4\hat{T}^3.$$

We suppose that

$$\hat{X}^{\hat{n}} = \hat{X}^n\hat{T}^{n-1}$$

for some $n \in \mathbb{N}$.

We will prove that

$$\hat{X}^{\widehat{n+1}} = \hat{X}^{n+1}\hat{T}^n.$$

Indeed,

$$\hat{X}^{\widehat{n+1}} = \hat{X}\hat{T}\hat{X}^{\hat{n}}$$

$$= \hat{X}\hat{T}\hat{X}^n\hat{T}^{n-1}$$

$$= \hat{X}^{n+1}\hat{T}^n.$$

Exercise 1.8.6. *Let $\hat{X}$, $\hat{Y}$ be random iso-variables. Find*

$$(\hat{X}+\hat{Y})^{\hat{3}}, \qquad (\hat{X}-\hat{Y})^{\hat{4}}.$$

Answer.

$$(\hat{X}+\hat{Y})^3\hat{T}^2, \qquad (\hat{X}-\hat{Y})^4\hat{T}^3.$$

Theorem 1.8.7. *We have*

$$\hat{E}\left(\left(\hat{\int}_0^1 gd\frac{\tilde{W}^1}{\hat{T}}\right)^{\hat{2}}\right)=\int_0^1 g^2(t)dt.$$

Proof. We note that

$$(22) \qquad \hat{E}\left(\frac{\tilde{W}^1(t)\tilde{W}^1(s)}{\hat{T}}\right)=t\wedge s$$

for all $t,s\geq 0$.

Then

$$\hat{E}\left(\left(\hat{\int}_0^1 gd\frac{\tilde{W}^1}{\hat{T}}\right)^{\hat{2}}\right)=\hat{E}\left(\left(-\int_0^1 g'(t)\frac{\tilde{W}^1(t)}{\hat{T}}dt\right)^{\hat{2}}\right)$$

$$=\hat{E}\left(\left(\int_0^1 g'(t)\frac{\tilde{W}^1(t)}{\hat{T}}dt\right)^{\hat{2}}\right)$$

$$=\hat{E}\left(\int_0^1 g'(t)\frac{\tilde{W}^1(t)}{\hat{T}}dt\hat{T}\int_0^1 g'(s)\frac{\tilde{W}^1(s)}{\hat{T}}ds\right)$$

$$=\hat{E}\left(\int_0^1\int_0^1 g'(t)g'(s)\frac{\tilde{W}^1(t)\tilde{W}^1(s)}{\hat{T}}dsdt\right)$$

$$=\int_0^1\int_0^1\hat{E}\left(g'(t)g'(s)\frac{\tilde{W}^1(t)\tilde{W}^1(s)}{\hat{T}}\right)dsdt$$

$$=\int_0^1\int_0^1 g'(t)g'(s)\hat{E}\left(\frac{\tilde{W}^1(t)\tilde{W}^1(s)}{\hat{T}}\right)dsdt \qquad ((22))$$

$$=\int_0^1\int_0^1 g'(t)g'(s)(t\wedge s)dsdt$$

$$=\int_0^1 g'(t)\left(\int_0^t g'(s)(t\wedge s)ds+\int_t^1 g'(s)(t\wedge s)ds\right)dt$$

$$=\int_0^1 g'(t)\left(\int_0^t g'(s)sds+t\int_t^1 g'(s)ds\right)dt$$

$$= \int_0^1 g'(t)\left(\int_0^t sdg(s) + tg(s)\Big|_{s=t}^{s=1}\right)dt$$

$$= \int_0^1 g'(t)\left(sg(s)\Big|_{s=0}^{s=t} - \int_0^t g(s)ds + tg(1) - tg(t)\right)dt$$

$$= \int_0^1 g'(t)\left(tg(t) - \int_0^t g(s)ds - t(t)\right)dt$$

$$= -\int_0^1 g'(t)\left(\int_0^t g(s)ds\right)dt$$

$$= -\int_0^1 \left(\int_0^t g(s)ds\right)dg(t)$$

$$= -\left(\int_0^t g(s)ds\right)g(t)\Big|_{t=0}^{t=1} + \int_0^1 g^2(t)dt$$

$$= \int_0^1 g^2(t)dt.$$

□

Definition 1.8.8. *We define $\hat{L}^{\hat{2}}(\Omega)$ as the space of all random iso-variables $\hat{X}$ such that*

$$\hat{E}\left(\hat{X}^{\hat{2}}\right) < \infty.$$

A norm in $\hat{L}^{\hat{2}}(\Omega)$ we define as follows

$$||\hat{X}|| := \left(\hat{E}\left(\hat{X}^{\hat{2}}\right)\right)^{\frac{1}{2}} \qquad \text{for} \qquad \hat{X} \in \hat{L}^{\hat{2}}(\Omega).$$

Exercise 1.8.9. *Check the axioms for norm.*

An inner product in $\hat{L}^{\hat{2}}(\Omega)$ we define as follows

$$(\hat{X}, \hat{Y}) = \hat{E}(\hat{X}\hat{Y}).$$

Exercise 1.8.10. *Check the axioms for inner product.*

We will say that the sequence $\{\hat{X}_n\}_{n=1}^{\infty}$ of elements of the space $\hat{L}^{\hat{2}}(\Omega)$ is convergent to the element $\hat{X} \in \hat{L}^{\hat{2}}(\Omega)$ in the sense of the space $\hat{L}^{\hat{2}}(\Omega)$ if

$$\lim_{n \longrightarrow \infty} \hat{E}\left(\left(\hat{X}_n - \hat{X}\right)^{\hat{2}}\right) = 0.$$

We will say that the sequence $\{\hat{X}_n\}_{n=1}^{\infty}$ of elements of the space $\hat{L}^{\hat{2}}(\Omega)$ is a fundamental sequence or Cauchy sequence in $\hat{L}^{\hat{2}}(\Omega)$ if

$$\lim_{m,n \longrightarrow \infty} \hat{E}\left(\left(\hat{X}_n - \hat{X}_m\right)^{\hat{2}}\right) = 0.$$

Exercise 1.8.11. *Prove that every convergent sequence in $\hat{L}^{\hat{2}}$ is a fundamental sequence in $\hat{L}^{\hat{2}}(\Omega)$.*

Exercise 1.8.12. *Prove that $\hat{L}^{\hat{2}}(\Omega)$ is a completely normed space.*

Two elements $\hat{X}$ and $\hat{Y}$ of the space $\hat{L}^2(\Omega)$ are called orthogonal elements in $\hat{L}^{\hat{2}}(\Omega)$ if

$$(\hat{X},\hat{Y}) = 0.$$

Exercise 1.8.13. *Let $\hat{X}$ and $\hat{Y}$ be orthogonal elements in $\hat{L}^{\hat{2}}(\Omega)$. Prove that*

$$||\hat{X}+\hat{Y}||^2 = 2\left(||\hat{X}||^2 + ||\hat{Y}||^2\right).$$

Remark 1.8.14. *Suppose now $g \in L^2(0,1)$. Then we can take a sequence of C^1-functions $\{g_n\}_{n=1}^{\infty}$, as above so that*

$$\int_0^1 (g_n - g)^2 dt \longrightarrow_{n\longrightarrow\infty} 0.$$

Then

$$\hat{E}\left(\left(\hat{\int}_0^1 g_m d\tilde{W}^1 - \hat{\int}_0^1 g_n d\tilde{W}^1\right)^{\hat{2}}\right) = \int_0^1 (g_m - g_n)^2 dt.$$

Therefore the sequence

$$\left\{\hat{\int}_0^1 g_n d\tilde{W}^1\right\}_{n=1}^{\infty}$$

is a Cauchy sequence in $\hat{L}^{\hat{2}}(\Omega)$.

Consequently, we can define

$$\hat{\int}_0^1 g d\tilde{W}^1 := \lim_{n\longrightarrow\infty} \hat{\int}_0^1 g_n d\tilde{W}^1.$$

Now we suppose that $a > 0$. We will construct a Riemann sum approximation corresponding to

$$\hat{\int}_0^b \frac{\tilde{W}^1}{\tilde{T}} \hat{\times} \hat{d}\frac{\tilde{W}^1}{\hat{T}}.$$

Definition 1.8.15. (i) *If $[0,b]$ is an interval, a partition D of $[0,b]$ is a finite collection of points in $[a,b]$, namely*

$$D := \{0 = t_0 < t_1 < t_2 \ldots < t_{m-1} < t_m = b\}.$$

(ii) *Let the mesh size of D be*

$$|D| := \max_{0\le k\le m-1} |t_{k+1} - t_k|.$$

(iii) *For fixed* $0 \leq \lambda \leq 1$ *and* D *a given partition of* $[0,b]$, *we set*

$$\tau_k := (1-\lambda)t_k + \lambda t_{k+1}, \qquad k = 0,1,\ldots,m-1.$$

For such a partition D *and* $0 \leq \lambda \leq 1$ *we define*

$$R = R(D,\lambda) := \sum_{k=1}^{m-1} \frac{1}{\hat{T}} \tilde{W}^1(\tau_k)\left(\tilde{W}^1(t_{k+1}) - \tilde{W}(t_k)\right).$$

This is the corresponding Riemann sum approximation of $\hat{\int}_0^b \frac{\tilde{W}^1}{\hat{T}} \hat{\times} \hat{d}\frac{\tilde{W}^1}{\hat{T}}$.

Theorem 1.8.16. *(quadratic variation) Let* $[a,b]$ *be an interval in* $[0,\infty)$, *and suppose*

$$D^m := \{a = t_0 < t_1 < t_2 < \ldots < t_{m-1} < t_m = b\}$$

are partitions of $[a,b]$ *with* $|D^m| \longrightarrow_{m\longrightarrow\infty} 0$. *Then*

$$\frac{1}{\hat{T}} \sum_{k=0}^{m-1} \left(\tilde{W}^1(t_{k+1}) - \tilde{W}^1(t_k)\right)^2 \longrightarrow_{m\longrightarrow\infty} (b-a)\frac{\hat{T}(\Omega)}{\hat{T}}$$

in $\hat{L}^{\hat{2}}(\Omega)$.

Remark 1.8.17. *This assertion says us that*

$$\frac{\left(d\tilde{W}^1\right)^2}{\hat{T}} \approx \frac{\hat{T}(\Omega)}{\hat{T}} dt$$

in the sense of $\hat{L}^{\hat{2}}(\Omega)$.

Proof. We set

$$Q_m := \sum_{k=0}^{m-1} \frac{\left(\tilde{W}^1(t_{k+1}) - \tilde{W}^1(t_k)\right)^2}{\hat{T}}.$$

Then

$$Q_m - \hat{T}(\Omega)\tfrac{b-a}{\hat{T}} = \textstyle\sum_{k=0}^{m-1} \frac{\left(\tilde{W}^1(t_{k+1}) - \tilde{W}^1(t_k)\right)^2}{\hat{T}} - \frac{b-a}{\hat{T}}\hat{T}(\Omega)$$

$$= \tfrac{1}{\hat{T}}\left(\textstyle\sum_{k=0}^{m-1}\left(\tilde{W}^1(t_{k+1}) - \tilde{W}^1(t_k)\right)^2 - \hat{T}(\Omega)(b-a)\right)$$

$$= \tfrac{1}{\hat{T}}\left(\textstyle\sum_{k=0}^{m-1}\left(\tilde{W}^1(t_{k+1}) - \tilde{W}^1(t_k)\right)^2 - \hat{T}(\Omega)\sum_{k=0}^{m-1}(t_{k+1}-t_k)\right)$$

$$= \tfrac{1}{\hat{T}}\textstyle\sum_{k=0}^{m-1}\left(\left(\tilde{W}^1(t_{k+1}) - \tilde{W}^1(t_k)\right)^2 - \hat{T}(\Omega)(t_{k+1}-t_k)\right),$$

$$\left(Q_m - \tfrac{b-a}{\hat{T}}\hat{T}(\Omega)\right)^2 = \tfrac{1}{\hat{T}}\textstyle\sum_{k=0}^{m-1}\sum_{j=0}^{m-1}\left(\left(\tilde{W}^1(t_{k+1}) - \tilde{W}^1(t_k)\right)^2 - \hat{T}(\Omega)(t_{k+1}-t_k)\right)$$

$$\times\left(\left(\tilde{W}^1(t_{j+1}) - \tilde{W}^1(t_j)\right)^2 - \hat{T}(\Omega)(t_{j+1}-t_j)\right).$$

Hence,

$$\hat{E}\left(\left(Q_m - \frac{b-a}{\hat{T}}\hat{T}(\Omega)\right)^2\right)$$

$$= \hat{E}\left(\frac{1}{\hat{T}}\sum_{k=0}^{m-1}\sum_{j=0}^{m-1}\left(\left(\tilde{W}^1(t_{k+1}) - \tilde{W}^1(t_k)\right)^2 - \hat{T}(\Omega)(t_{k+1}-t_k)\right)\right.$$

$$\left.\times\left(\left(\tilde{W}^1(t_{j+1}) - \tilde{W}^1(t_j)\right)^2 - \hat{T}(\Omega)(t_{j+1}-t_j)\right)\right)$$

$$= \sum_{k=0}^{m-1}\sum_{j=0}^{m-1}\hat{E}\left(\frac{1}{\hat{T}}\left(\left(\tilde{W}^1(t_{k+1}) - \tilde{W}^1(t_k)\right)^2 - \hat{T}(\Omega)(t_{k+1}-t_k)\right)\right.$$

$$\left.\times\left(\left(\tilde{W}^1(t_{j+1}) - \tilde{W}^1(t_j)\right)^2 - \hat{T}(\Omega)(t_{j+1}-t_j)\right)\right)$$

For $k \neq j$ we have that

$$\left(\tilde{W}^1(t_{k+1}) - \tilde{W}^1(t_k)\right)^2 - \hat{T}(\Omega)(t_{k+1}-t_k), \quad \left(\tilde{W}^1(t_{j+1}) - \tilde{W}^1(t_j)\right)^2 - \hat{T}(\Omega)(t_{j+1}-t_j)$$

are independent and because they are $\tilde{N}^1(0, t_{k+1}-t_k)$ and $\tilde{N}^1(0, t_{j+1}-t_j)$, respectively,

$$\hat{E}\left(\frac{\left(\tilde{W}^1(t_{k+1}) - \tilde{W}^1(t_k)\right)^2 - \hat{T}(\Omega)(t_{k+1}-t_k)}{\hat{T}}\right) = \hat{E}\left(\frac{\left(\tilde{W}^1(t_{k+1}) - \tilde{W}^1(t_k)\right)^2}{\hat{T}}\right) - \hat{E}\left(\frac{\hat{T}(\Omega)(t_{k+1}-t_k)}{\hat{T}}\right)$$

$$= t_{k+1} - t_k - \hat{\int}_\Omega \hat{T}(\Omega)\frac{t_{k+1}-t_k}{\hat{T}}\hat{\times}\hat{d}\hat{P}$$

$$= t_{k+1} - t_k - \hat{T}(\Omega)\int_\Omega \frac{1}{\hat{T}}\frac{t_{k+1}-t_k}{\hat{T}}\hat{T}\hat{T}d\hat{P}$$

$$= t_{k+1} - t_k - \hat{T}(\Omega)\int_\Omega (t_{k+1}-t_k)d\hat{P}$$

$$= t_{k+1} - t_k - \hat{T}(\Omega)(t_{k+1}-t_k)\int_\Omega d\hat{P}$$

$$= t_{k+1} - t_k - \hat{T}(\Omega)(t_{k+1}-t_k)\hat{P}(\Omega)$$

$$= t_{k+1} - t_k - \hat{T}(\Omega)(t_{k+1}-t_k)\frac{1}{\hat{T}(\Omega)}$$

$$= t_{k+1} - t_k - (t_{k+1}-t_k)$$

$$= 0.$$

As in above,

$$\hat{E}\left(\frac{\left(\tilde{W}^1(t_{j+1}) - \tilde{W}^1(t_j)\right)^2 - \hat{T}(\Omega)(t_{j+1}-t_j)}{\hat{T}}\right) = 0.$$

Consequently,

$$\hat{E}\left(\left(Q_m - \tfrac{b-a}{\hat{T}}\hat{T}(\Omega)\right)^2\right) = \sum_{k=0}^{m-1}\hat{E}\left(\frac{\left(\left(\tilde{W}^1(t_{k+1})-\tilde{W}^1(t_k)\right)^2-\hat{T}(\Omega)(t_{k+1}-t_k)\right)^2}{\hat{T}}\right)$$

$$= \sum_{k=0}^{m-1}(t_{k+1}-t_k)^2\hat{E}\left(\frac{\left(\frac{\left(\tilde{W}^1(t_{k+1})-\tilde{W}^1(t_k)\right)^2}{t_{k+1}-t_k}-\hat{T}(\Omega)\right)^2}{\hat{T}}\right)$$

$$\leq C\sum_{k=0}^{m-1}(t_{k+1}-t_k)^2$$

$$\leq C|D^m|\sum_{k=0}^{m-1}(t_{k+1}-t_k)$$

$$= C|D^m|(b-a)$$

$$\longrightarrow_{m\longrightarrow\infty} 0.$$

□

Theorem 1.8.18. *Let*

$$D^m := \{0 = t_0 < t_1 < t_2 \ldots < t_{m-1} < t_m = T\}$$

are partitions of the interval $[0,T]$ *with* $|D^m| \longrightarrow_{m\longrightarrow\infty} 0$, *and*

$$R_m := \sum_{k=0}^{m-1}\frac{\tilde{W}^1(\tau_k)\left(\tilde{W}^1(t_{k+1})-\tilde{W}^1(t_k)\right)}{\hat{T}}.$$

Then

$$\lim_{m\longrightarrow\infty} R_m = \frac{\left(\tilde{W}^1(T)\right)^2}{2\hat{T}} + \left(\lambda - \frac{1}{2}\right)\frac{T}{\hat{T}}\hat{T}(\Omega)$$

in the sense of $\hat{L}^{\hat{2}}(\Omega)$, *i.e.,*

$$\hat{E}\left(\left(R_m - \frac{\left(\tilde{W}^1(T)\right)^2}{2\hat{T}} - \left(\lambda-\frac{1}{2}\right)\frac{T}{\hat{T}}\hat{T}(\Omega)\right)^{\hat{2}}\right) \longrightarrow_{m\longrightarrow\infty} 0.$$

Proof. We note that

$$R_m = \sum_{k=0}^{m-1}\frac{\tilde{W}^1(\tau_k)\left(\tilde{W}^1(t_{k+1})-\tilde{W}^1(t_k)\right)}{\hat{T}}$$

$$= \frac{\left(\tilde{W}^1(T)\right)^2}{2\hat{T}} - \frac{1}{2}\frac{1}{\hat{T}}\sum_{k=0}^{m-1}\left(\left(\tilde{W}^1(t_{k+1})\right)^2 - \left(\tilde{W}^1(t_k)\right)^2\right)$$

$$+\sum_{k=0}^{m-1}\frac{1}{\hat{T}}\tilde{W}^1(\tau_k)\left(\tilde{W}^1(t_{k+1})-\tilde{W}^1(t_k)\right)$$

$$= \frac{\left(\tilde{W}^1(T)\right)^2}{2\hat{T}} - \frac{1}{2}\frac{1}{\hat{T}}\sum_{k=0}^{m-1}\left(\tilde{W}^1(t_{k+1})-\tilde{W}^1(t_k)\right)\left(\tilde{W}^1(t_{k+1})+\tilde{W}^1(t_k)\right)$$

$$+\frac{1}{\hat{T}}\sum_{k=0}^{m-1}\tilde{W}^1(\tau_k)\left(\tilde{W}^1(t_{k+1})-\tilde{W}^1(t_k)\right)$$

$$= \frac{\left(\tilde{W}^1(T)\right)^2}{2\hat{T}} - \frac{1}{2}\frac{1}{\hat{T}}\sum_{k=0}^{m-1}\left(\tilde{W}^1(t_{k+1}) - \tilde{W}^1(t_k)\right)\left(\tilde{W}^1(t_{k+1}) + \tilde{W}^1(t_k) - 2\tilde{W}^1(\tau_k)\right)$$

$$= \frac{\left(\tilde{W}^1(T)\right)^2}{2\hat{T}}$$

$$-\frac{1}{2}\frac{1}{\hat{T}}\sum_{k=0}^{m-1}\left(\tilde{W}^1(t_{k+1}) - \tilde{W}^1(t_k)\right)\left(\tilde{W}^1(t_{k+1}) + \tilde{W}^1(t_k) - \tilde{W}^1(t_k) + 2\tilde{W}^1(t_k) - 2\tilde{W}^1(\tau_k)\right)$$

$$= \frac{\left(\tilde{W}^1(T)\right)^2}{2\hat{T}} - \frac{1}{2\hat{T}}\sum_{k=0}^{m-1}\tilde{\left(\tilde{W}^1(t_{k+1}) - \tilde{W}^1(t_k)\right)}^{\tilde{2}}$$

$$-\frac{1}{\hat{T}}\sum_{k=0}^{m-1}\left(\tilde{W}^1(t_{k+1}) - \tilde{W}^1(t_k)\right)\left(\tilde{W}^1(t_k) - \tilde{W}^1(\tau_k)\right).$$

Hence,

$$R_m - \frac{\left(\tilde{W}^1(T)\right)^2}{2\hat{T}} = -\frac{1}{2\hat{T}}\sum_{k=0}^{m-1}\tilde{\left(\tilde{W}^1(t_{k+1}) - \tilde{W}^1(t_k)\right)}^{\tilde{2}}$$

$$-\frac{1}{\hat{T}}\sum_{k=0}^{m-1}\left(\tilde{W}^1(t_{k+1}) - \tilde{W}^1(t_k)\right)\left(\tilde{W}^1(t_k) - \tilde{W}^1(\tau_k)\right)$$

$$:= A + B.$$

Using the previous theorem we have that

$$A \longrightarrow_{m\longrightarrow\infty} -\frac{1}{2\hat{T}}T\hat{T}(\Omega)$$

in the sense of $\hat{L}^{\hat{2}}(\Omega)$.

Now we consider B. We have

$$B = -\frac{P1}{\hat{T}}\sum_{k=0}^{m-1}\left(\tilde{W}^1(t_{k+1}) - \tilde{W}^1(t_k)\right)\left(\tilde{W}^1(t_k) - \tilde{W}^1(\tau_k)\right)$$

$$= -\frac{1}{\hat{T}}\sum_{k=0}^{m-1}\left(\tilde{W}^1(t_{k+1}) - \tilde{W}^1(\tau_k) + \tilde{W}^1(\tau_k) - \tilde{W}^1(t_k)\right)\left(\tilde{W}^1(t_k) - \tilde{W}^1(\tau_k)\right)$$

$$= \frac{1}{\hat{T}}\sum_{k=0}^{m-1}\left(\tilde{W}^1(\tau_k) - \tilde{W}^1(t_k)\right)^2 + \frac{1}{\hat{T}}\sum_{k=0}^{\infty}\left(\tilde{W}^1(t_{k+1}) - \tilde{W}^1(\tau_k)\right)\left(\tilde{W}^1(\tau_k) - \tilde{W}^1(t_k)\right)$$

$$:= B_1 + B_2.$$

Using the previous theorem we conclude that

$$B_1 \longrightarrow_{m\longrightarrow\infty} \frac{\lambda}{\hat{T}}\hat{T}(\Omega)T.$$

For B_2 we have

$$\hat{E}\left(B_2^{\hat{2}}\right) = \hat{E}\left(B_2^2\hat{T}\right)$$

$$= \hat{E}\left(\frac{1}{\hat{T}}\sum_{k=0}^{m-1}\sum_{j=0}^{m-1}\left(\tilde{W}^1(t_{k+1}) - \tilde{W}^1(\tau_k)\right)\left(\tilde{W}^1(\tau_k) - \tilde{W}^1(t_k)\right)\right.$$

$$(23) \qquad \left.\times\left(\tilde{W}^1(t_{j+1}) - \tilde{W}^1(\tau_j)\right)\left(\tilde{W}^1(\tau_j) - \tilde{W}^1(t_j)\right)\right)$$

$$= \sum_{k=0}^{m-1}\sum_{j=0}^{m-1}\hat{E}\left(\frac{\left(\tilde{W}^1(t_{k+1})-\tilde{W}^1(\tau_k)\right)\left(\tilde{W}^1(\tau_k)-\tilde{W}^1(t_k)\right)\left(\tilde{W}^1(t_{j+1})-\tilde{W}^1(\tau_j)\right)\left(\tilde{W}^1(\tau_j)-\tilde{W}^1(t_j)\right)}{\hat{T}}\right).$$

For $k \neq j$ we have

$$\hat{E}\left(\frac{\left(\tilde{W}^1(t_{k+1})-\tilde{W}^1(\tau_k)\right)\left(\tilde{W}^1(\tau_k)-\tilde{W}^1(t_k)\right)\left(\tilde{W}^1(t_{j+1})-\tilde{W}^1(\tau_j)\right)\left(\tilde{W}^1(\tau_j)-\tilde{W}^1(t_j)\right)}{\hat{T}}\right)$$

$$= \hat{E}\left(\frac{\tilde{W}^1(t_{k+1})-\tilde{W}^1(\tau_k)}{\hat{T}}\right)\hat{E}\left(\frac{\tilde{W}^1(\tau_k)-\tilde{W}^1(t_k)}{\hat{T}}\right)\hat{E}\left(\frac{\tilde{W}^1(t_{j+1})-\tilde{W}^1(\tau_j)}{\hat{T}}\right)\hat{E}\left(\frac{\tilde{W}^1(\tau_j)-\tilde{W}^1(t_j)}{\hat{T}}\right)$$

$$= 0.$$

Hence, using (23),

$$\hat{E}\left(B_2^{\hat{2}}\right) = \sum_{k=0}^{m-1}\hat{E}\left(\frac{\left(\tilde{W}^1(t_{k+1})-\tilde{W}^1(\tau_k)\right)^2}{\hat{T}}\right)\hat{E}\left(\frac{\left(\tilde{W}^1(\tau_k)-\tilde{W}^1(t_k)\right)^2}{\hat{T}}\right)$$

$$= \sum_{k=0}^{m-1}(t_{k+1}-\tau_k)(\tau_k-t_k)$$

$$= \sum_{k=0}^{m-1}(t_{k+1}-(1-\lambda)t_k-\lambda t_{k+1})((1-\lambda)t_k+\lambda t_{k+1}-t_k)$$

$$= \sum_{k=0}^{m-1}((1-\lambda)t_{k+1}-(1-\lambda)t_k)(\lambda t_{k+1}-\lambda t_k)$$

$$= \lambda(1-\lambda)\sum_{k=0}^{m-1}(t_{k+1}-t_k)^2$$

$$\leq \lambda(1-\lambda)|D^m|\sum_{k=0}^{m-1}(t_{k+1}-t_k)$$

$$= \lambda(1-\lambda)|D^m|T$$

$$\longrightarrow_{m\longrightarrow\infty} 0.$$

Consequently,

$$B = B_1 + B_2 \longrightarrow_{m\longrightarrow\infty} \frac{\lambda}{\hat{T}}\hat{T}(\Omega)T$$

in the sense of the space $\hat{L}^{\hat{2}}(\Omega)$.

From here,

$$R_m - \frac{\left(\tilde{W}^1(T)\right)^2}{2\hat{T}} \longrightarrow_{m\longrightarrow\infty} \frac{\lambda}{\hat{T}}\hat{T}(\Omega)T - \frac{1}{2}\frac{1}{\hat{T}}\hat{T}(\Omega)T$$

$$= \left(\lambda - \frac{1}{2}\right)\frac{\hat{T}(\Omega)}{\hat{T}}T.$$

□

Definition 1.8.19. *We define the Paley-Wiener-Zygmund iso-integral of the second kind as follows*

$$\hat{\int}_0^1 gd\frac{\tilde{T}(\tilde{W}^2)\tilde{W}^2}{\hat{T}} := -\hat{\int}_0^1 g^{\circledast}(t)\hat{\times}\frac{\tilde{W}^2\tilde{T}(\tilde{W}^2)}{\hat{T}}d\hat{t}.$$

We can rewrite the Paley-Wiener-Zygmund iso-integral of the second kind as follows

$$\hat{\int}_0^1 gd\frac{\tilde{T}(\tilde{W}^2)\tilde{W}^2}{\hat{T}} = -\int_0^1 \frac{1}{\tilde{T}(t)} g'(t) \frac{\tilde{T}(t)}{\tilde{T}(t)-t\tilde{T}'(t)} \tilde{T}(t) \frac{\tilde{W}^2\tilde{T}(\tilde{W}^2)}{\hat{T}} \tilde{T} d\hat{t}$$

$$= -\int_0^1 g'(t) \frac{\tilde{T}^2(t)}{\tilde{T}(t)-t\tilde{T}'(t)} \frac{\tilde{W}^2\tilde{T}(\tilde{W}^2)}{\hat{T}} d\hat{t}$$

$$= -\int_0^1 g'(t) \frac{\tilde{T}^2(t)}{\tilde{T}(t)-t\tilde{T}'(t)} \frac{\tilde{W}^2\tilde{T}(\tilde{W}^2)}{\hat{T}} d\frac{t}{\tilde{T}(t)}$$

$$= -\int_0^1 g'(t) \frac{\tilde{T}^2(t)}{\tilde{T}(t)-t\tilde{T}'(t)} \frac{\tilde{W}^2\tilde{T}(\tilde{W}^2)}{\hat{T}} \frac{\tilde{T}(t)-t\tilde{T}'(t)}{\tilde{T}^2(t)} dt$$

$$= -\int_0^1 g'(t) \frac{\tilde{W}^2\tilde{T}(\tilde{W}^2)}{\hat{T}} dt.$$

Theorem 1.8.20. *We have*

$$\hat{E}\left(\hat{\int}_0^1 gd\frac{\tilde{T}(\tilde{W}^2)\tilde{W}^2}{\hat{T}}\right) = 0.$$

Proof. We note that

$$\hat{E}\left(\frac{\tilde{T}(\tilde{W}^2)\tilde{W}^2}{\hat{T}}\right) = 0. \tag{24}$$

Hence,

$$\hat{E}\left(\hat{\int}_0^1 gd\frac{\tilde{T}(\tilde{W}^2)\tilde{W}^2}{\hat{T}}\right)$$

$$= \hat{E}\left(-\int_0^1 g'(t)\frac{\tilde{T}(\tilde{W}^2)\tilde{W}^2}{\hat{T}} dt\right)$$

$$= -\hat{E}\left(\int_0^1 g'(t)\frac{\tilde{T}(\tilde{W}^2)\tilde{W}^2}{\hat{T}} dt\right)$$

$$= -\int_0^1 \hat{E}\left(g'(t)\frac{\tilde{T}(\tilde{W}^2)\tilde{W}^2}{\hat{T}}\right) dt$$

$$= -\int_0^1 g'(t)\hat{E}\left(\frac{\tilde{T}(\tilde{W}^2)\tilde{W}^2}{\hat{T}}\right) dt \qquad ((24))$$

$$= 0.$$

□

Theorem 1.8.21. *We have*

$$\hat{E}\left(\left(\hat{\int}_0^1 gd\frac{\tilde{W}^2\tilde{T}(\tilde{W}^2)}{\hat{T}}\right)^{\hat{2}}\right) = \int_0^1 g^2(t)dt.$$

Proof. We note that

$$\hat{E}\left(\frac{\tilde{W}^2(t)\tilde{T}(\tilde{W}^2(t))\tilde{W}^2(s)\tilde{T}(\tilde{W}^2(s))}{\hat{T}}\right) = t \wedge s. \tag{25}$$

Then

$$\hat{E}\left(\left(\hat{\int}_0^1 gd\frac{\tilde{W}^2\tilde{T}(\tilde{W}^2)}{\hat{T}}\right)^{\hat{2}}\right) = \hat{E}\left(\left(\hat{\int}_0^1 gd\frac{\tilde{W}^2\tilde{T}(\tilde{W}^2)}{\hat{T}}\right)^2 \hat{T}\right)$$

$$= \hat{E}\left(\hat{T}\left(-\int_0^1 g'(t)\frac{\tilde{W}^2(t)\tilde{T}(\tilde{W}^2(t))}{\hat{T}}dt\right)^2\right)$$

$$= \hat{E}\left(\hat{T}\left(\int_0^1 g'(t)\frac{\tilde{W}^2(t)\tilde{T}(\tilde{W}^2(t))}{\hat{T}}dt\right)^2\right)$$

$$= \hat{E}\left(\hat{T}\int_0^1 g'(t)\frac{\tilde{W}^2(t)\tilde{T}(\tilde{W}^2(t))}{\hat{T}}dt\int_0^1 g'(s)\frac{\tilde{W}^2(s)\tilde{T}(\tilde{W}^2(s))}{\hat{T}}ds\right)$$

$$= \hat{E}\left(\int_0^1\int_0^1 g'(t)g'(s)\frac{\tilde{W}^2(t)\tilde{T}(\tilde{W}^2(t))\tilde{W}^2(s)\tilde{T}(\tilde{W}^2(s))}{\hat{T}}dtds\right)$$

$$= \int_0^1\int_0^1 \hat{E}\left(g'(t)g'(s)\frac{\tilde{W}^2(t)\tilde{T}(\tilde{W}^2(t))\tilde{W}^2(s)\tilde{T}(\tilde{W}^2(s))}{\hat{T}}\right)dtds$$

$$= \int_0^1\int_0^1 g'(t)g'(s)\hat{E}\left(\frac{\tilde{W}^2(t)\tilde{T}(\tilde{W}^2(t))\tilde{W}^2(s)\tilde{T}(\tilde{W}^2(s))}{\hat{T}}\right)dtds$$

$$= \int_0^1\int_0^1 g'(t)g'(s)t\wedge s dtds$$

$$= \int_0^1 g^2(t)dt.$$

□

Definition 1.8.22. *We suppose that $g \in L^2(0,1)$. Then there exists a sequence $\{g_n\}_{n=1}^{\infty}$ of elements of $C^1(0,1)$, $g_n(0) = g_n(1) = 0$, $n = 1,2,\ldots$, such that*

$$\lim_{n\longrightarrow\infty} g_n = g$$

in $L^2(0,1)$.

Then using the last theorem we have

$$\hat{E}\left(\left(\hat{\int}_0^1 g_m d\frac{\tilde{W}^2\tilde{T}(\tilde{W}^2)}{\hat{T}} - \hat{\int}_0^1 g_n d\frac{\tilde{W}^2\tilde{T}(\tilde{W}^2)}{\hat{T}}\right)^{\hat{2}}\right) = \hat{E}\left(\left(\hat{\int}_0^1 (g_m - g_n)d\frac{\tilde{W}^2\tilde{T}(\tilde{W}^2)}{\hat{T}}\right)^{\hat{2}}\right)$$

$$= \int_0^1 (g_m - g_n)^2 dt$$

$$\longrightarrow_{m,n\longrightarrow\infty} 0.$$

Therefore the sequence

$$\left\{\hat{\int}_0^1 g_m d\frac{\tilde{W}^2\tilde{T}(\tilde{W}^2)}{\hat{T}}\right\}_{m=1}^{\infty}$$

is a fundamental sequence in $\hat{L}^{\hat{2}}(\Omega)$.

We define

$$\hat{\int}_0^1 gd\frac{\tilde{W}^2\tilde{T}(\tilde{W}^2)}{\hat{T}} = \lim_{m\longrightarrow\infty}\hat{\int}_0^1 g_m d\frac{\tilde{W}^2\tilde{T}(\tilde{W}^2)}{\hat{T}}$$

in $\hat{L}^{\hat{2}}(\Omega)$.

Definition 1.8.23. *If*

$$D := \{0 = t_0 < t_2 < \ldots < t_{m-1} < t_m\},$$

then

$$R = R(D,\lambda) := \sum_{k=1}^{m-1}\frac{1}{\hat{T}}\tilde{W}^2(\tau_k)\tilde{T}(\tilde{W}^2(\tau_k))\left(\tilde{W}^2(t_{k+1})\tilde{T}(\tilde{W}^2(t_{k+1})) - \tilde{W}^2(t_k)\tilde{T}(\tilde{W}^2(t_k))\right)$$

will be the corresponding Riemann sum approximation of

$$\hat{\int}_0^T \frac{\tilde{W}^2\tilde{T}(\tilde{W}^2)}{\tilde{T}}\hat{\times}\hat{d}\frac{\tilde{W}^2\tilde{T}(\tilde{W}^2)}{\hat{T}}.$$

Exercise 1.8.24. *Let* $[a,b]$ *be an interval in* $[0,\infty)$, *and*

$$D^m := \{0 = t_0 < t_1 < \ldots < t_{m-1} < t_m = T\}$$

be partitions of $[0,T]$ *with* $|D^m| \longrightarrow_{m\longrightarrow\infty} 0$. *Prove that*

$$\sum_{k=0}^{\infty}\frac{\left(\tilde{W}^2(t_{k+1})\tilde{T}(\tilde{W}^2(t_{k+1})) - \tilde{W}^2(t_k)\tilde{T}(\tilde{W}^2(t_k))\right)^2}{\hat{T}} \longrightarrow_{m\longrightarrow\infty} (b-a)\frac{\hat{T}(\Omega)}{\hat{T}}$$

in $\hat{L}^{\hat{2}}(\Omega)$.

Exercise 1.8.25. *Let*

$$D^m := \{0 = t_0 < t_1 < \ldots < t_{m-1} < t_m = T\}$$

be partitions of the interval $[0,T]$ *with* $|D^m| \longrightarrow_{m\longrightarrow\infty} 0$, *and*

$$R_m = \sum_{k=0}^{m-1}\frac{\left(\tilde{W}^2(t_{k+1})\tilde{T}(\tilde{W}^2(t_{k+1})) - \tilde{W}^2(t_k)\tilde{T}(\tilde{W}^2(t_k))\right)\tilde{W}^2(\tau_k)\tilde{T}(\tilde{W}^2(\tau_k))}{\hat{T}}.$$

Prove that

$$R_m \longrightarrow_{m\longrightarrow\infty} \frac{\left(\tilde{W}^2(T)\tilde{T}(\tilde{W}^2(T))\right)^2}{2\hat{T}} + \left(\lambda - \frac{1}{2}\right)\frac{T}{\hat{T}}\hat{T}(\Omega)$$

in $\hat{L}^{\hat{2}}(\Omega)$.

Definition 1.8.26. *We define the Paley-Wiener-Zygmund iso-integral of the third kind as follows*

$$\hat{\int}_0^1 gd\frac{\tilde{W}^3}{\tilde{T}(\tilde{W}^3)\hat{T}} := -\hat{\int}_0^1 g^{\circledast}(t)\hat{\times}\frac{\tilde{W}^3}{\tilde{T}(\tilde{W}^3)\hat{T}}\hat{d}\hat{t}.$$

We can rewrite the Paley-Wiener-Zygmund iso-integral of the third kind in the following manner

$$\hat{\int}_0^1 g d\frac{\tilde{W}^3}{\tilde{T}(\tilde{W}^3)\hat{T}} = -\hat{\int}_0^1 g^{\circledast}(t)\hat{\times}\frac{\tilde{W}^3}{\tilde{T}(\tilde{W}^3)\hat{T}}d\hat{t}$$

$$= -\int_0^1 \frac{1}{\tilde{T}(t)}g'(t)\frac{\tilde{T}(t)}{\tilde{T}(t)-t\tilde{T}'(t)}\tilde{T}(t)\frac{\tilde{W}^3}{\tilde{T}(\tilde{W}^3)\hat{T}}\tilde{T}(t)d\hat{t}$$

$$= -\int_0^1 g'(t)\frac{\tilde{T}^2(t)}{\tilde{T}(t)-t\tilde{T}'(t)}\frac{\tilde{W}^3}{\tilde{T}(\tilde{W}^3)\hat{T}}d\hat{t}$$

$$= -\int_0^1 g'(t)\frac{\tilde{T}^2(t)}{\tilde{T}(t)-t\tilde{T}'(t)}\frac{\tilde{W}^3}{\tilde{T}(\tilde{W}^3)\hat{T}}d\frac{t}{\tilde{T}(t)}$$

$$= -\int_0^1 g'(t)\frac{\tilde{T}^2(t)}{\tilde{T}(t)-t\tilde{T}'(t)}\frac{\tilde{W}^3}{\tilde{T}(\tilde{W}^3)\hat{T}}\frac{\tilde{T}(t)-t\tilde{T}'(t)}{\tilde{T}^2(t)}dt$$

$$= -\int_0^1 g'(t)\frac{\tilde{W}^3}{\tilde{T}(\tilde{W}^3)\hat{T}}dt.$$

Exercise 1.8.27. *Prove that*

$$\hat{E}\left(\hat{\int}_0^1 g d\frac{\tilde{W}^3}{\tilde{T}(\tilde{W}^3)\hat{T}}\right) = 0.$$

Exercise 1.8.28. *Prove that*

$$\hat{E}\left(\left(\hat{\int}_0^1 g d\frac{\tilde{W}^3}{\tilde{T}(\tilde{W}^3)\hat{T}}\right)^{\hat{2}}\right) = \int_0^1 g^2(t)dt.$$

Definition 1.8.29. *We suppose that $g \in L^2(0,1)$. Then there exists a sequence $\{g_n\}_{n=1}^{\infty}$ of elements of $C^1(0,1)$, $g_n(0) = g_n(1) = 0$, $n = 1,2,\ldots$, such that*

$$\lim_{n\longrightarrow\infty} g_n = g$$

in $L^2(0,1)$.

Then using the last theorem we have

$$\hat{E}\left(\left(\hat{\int}_0^1 g_m d\frac{\frac{\tilde{W}^3}{\tilde{T}(\tilde{W}^3)}}{\hat{T}} - \hat{\int}_0^1 g_n d\frac{\frac{\tilde{W}^3}{\tilde{T}(\tilde{W}^3)}}{\hat{T}}\right)^{\hat{2}}\right) = \hat{E}\left(\left(\hat{\int}_0^1 (g_m - g_n) d\frac{\frac{\tilde{W}^3}{\tilde{T}(\tilde{W}^3)}}{\hat{T}}\right)^{\hat{2}}\right)$$

$$= \int_0^1 (g_m - g_n)^2 dt$$

$$\longrightarrow_{m,n\longrightarrow\infty} 0.$$

Therefore the sequence

$$\left\{\hat{\int}_0^1 g_m d\frac{\frac{\tilde{W}^3}{\tilde{T}(\tilde{W}^3)}}{\hat{T}}\right\}_{m=1}^{\infty}$$

is a fundamental sequence in $\hat{L}^{\hat{2}}(\Omega)$.

We define

$$\hat{\int}_0^1 g d\frac{\frac{\tilde{W}^3}{\tilde{T}(\tilde{W}^3)}}{\hat{T}} = \lim_{m\longrightarrow\infty} \hat{\int}_0^1 g_m d\frac{\frac{\tilde{W}^3}{\tilde{T}(\tilde{W}^3)}}{\hat{T}}$$

in $\hat{L}^{\hat{2}}(\Omega)$.

Definition 1.8.30. *If*

$$D := \{0 = t_0 < t_2 < \ldots < t_{m-1} < t_m\},$$

then

$$R = R(D,\lambda) := \sum_{k=1}^{m-1} \frac{1}{\hat{T}} \frac{\tilde{W}^3(\tau_k)}{\tilde{T}(\tilde{W}^3(\tau_k))} \left(\frac{\tilde{W}^3(t_{k+1})}{\tilde{T}(\tilde{W}^3(t_{k+1}))} - \frac{\tilde{W}^3(t_k)}{\tilde{T}(\tilde{W}^3(t_k))} \right)$$

will be the corresponding Riemann sum approximation of

$$\hat{\int}_0^1 \frac{\frac{\tilde{W}^3}{\tilde{T}(\tilde{W}^3)}}{\hat{T}} \hat{\times} \hat{d} \frac{\frac{\tilde{W}^3}{\tilde{T}(\tilde{W}^3)}}{\hat{T}}.$$

Exercise 1.8.31. *Let* $[a,b]$ *be an interval in* $[0,\infty)$, *and*

$$D^m := \{0 = t_0 < t_1 < \ldots < t_{m-1} < t_m = T\}$$

be partitions of $[0,T]$ *with* $|D^m| \longrightarrow_{m\longrightarrow\infty} 0$. *Prove that*

$$\sum_{k=0}^{\infty} \frac{\left(\frac{\tilde{W}^3(t_{k+1})}{\tilde{T}(\tilde{W}^3(t_{k+1}))} - \frac{\tilde{W}^3(t_k)}{\tilde{T}(\tilde{W}^3(t_k))} \right)^2}{\hat{T}} \longrightarrow_{m\longrightarrow\infty} (b-a)\frac{\hat{T}(\Omega)}{\hat{T}}$$

in $\hat{L}^{\hat{2}}(\Omega)$.

Exercise 1.8.32. *Let*

$$D^m := \{0 = t_0 < t_1 < \ldots < t_{m-1} < t_m = T\}$$

be partitions of the interval $[0,T]$ *with* $|D^m| \longrightarrow_{m\longrightarrow\infty} 0$, *and*

$$R_m = \sum_{k=0}^{m-1} \frac{\left(\frac{\tilde{W}^3(t_{k+1})}{\tilde{T}(\tilde{W}^3(t_{k+1}))} - \frac{\tilde{W}^3(t_k)}{\tilde{T}(\tilde{W}^3(t_k))} \right) \frac{\tilde{W}^3(\tau_k)}{\tilde{T}(\tilde{W}^3(\tau_k))}}{\hat{T}}.$$

Prove that

$$R_m \longrightarrow_{m\longrightarrow\infty} \frac{\left(\frac{\tilde{W}^3(T)}{\tilde{T}(\tilde{W}^3(T))} \right)^2}{2\hat{T}} + \left(\lambda - \frac{1}{2}\right) \frac{T}{\hat{T}} \hat{T}(\Omega)$$

in $\hat{L}^{\hat{2}}(\Omega)$.

1.9 Definition and Properties of Itô's Iso-Integral

Let $\tilde{W}$ be a 1-dimensional iso-Brownian iso-motion.

Definition 1.9.1. *The σ-algebra*

$$\tilde{W}(t) := U(\tilde{W}(s) : 0 \leq s \leq t)$$

is called the history of the iso-Brownian iso-motion up to and included time t.

Definition 1.9.2. *The σ-algebra*

$$\tilde{W}^{+}(t) := U(\tilde{W}(s) - \tilde{W}(t) : s \geq t)$$

is called the future of the iso-Brownian iso-motion beyond time t.

Definition 1.9.3. *A family $\mathcal{F}(\cdot)$ of σ-algebras $\subset U$ is called non-anticipating with respect to $\tilde{W}(\cdot)$ if*

(i) $\mathcal{F}(s) \subset \mathcal{F}(t)$ *for all* $t \geq s \geq 0$,

(ii) $\tilde{W}(t) \subset \mathcal{F}(t)$ *for all* $t \geq 0$,

(iii) $\mathcal{F}(t)$ *is independent of* $\tilde{W}^{+}(t)$ *for all* $t \geq 0$.

We also refer $\mathcal{F}(\cdot)$ as a filtration.

Definition 1.9.4. *A real-valued iso-stochastic process $\hat{G}(\cdot)$ will be called non-anticipating with respect $\mathcal{F}(\cdot)$ if for each time $t \geq 0$, $\hat{G}(t)$ is $\mathcal{F}(t)$-measurable.*

Definition 1.9.5. *We denote by $\hat{\mathbb{L}}^{\hat{2}}(0,T)$ the space of all real-valued, progressively measurable iso-stochastic process $\hat{G}(\cdot)$ such that*

$$\hat{E}\left(\left|\hat{\int}_0^T \hat{G}^{\hat{2}} \hat{\times} d\hat{t}\right|\right) < \infty.$$

We note that

$$\hat{\int}_0^T \hat{G}^{\hat{2}} \hat{\times} d\hat{t} = \int_0^T \frac{1}{\tilde{T}(t)} \hat{G}^{\hat{2}} \tilde{T}(t) \tilde{T}(t) d\hat{t}$$

$$= \int_0^T \hat{G}^{\hat{2}} \tilde{T}(t) d\frac{t}{\tilde{T}(t)}$$

$$= \int_0^T \hat{G}^{\hat{2}} \tilde{T}(t) \frac{\tilde{T}(t) - \tilde{T}'(t)}{\tilde{T}^2(t)} dt$$

$$= \int_0^T \hat{G}^{\hat{2}} \frac{\tilde{T}(t) - t\tilde{T}'(t)}{\tilde{T}(t)} dt$$

$$= \int_0^t \hat{G}^2 \hat{T} \frac{\tilde{T}(t) - t\tilde{T}'(t)}{\tilde{T}(t)} dt.$$

Therefore $\hat{G} \in \hat{\mathbb{L}}^{\hat{2}}(0,T)$ if and only if

$$\hat{E}\left(\left|\int_0^t \hat{G}^2 \hat{T} \frac{\tilde{T}(t) - t\tilde{T}'(t)}{\tilde{T}(t)} dt\right|\right) < \infty.$$

Definition 1.9.6. *We denote by* $\hat{\mathbb{L}}^1(0,T)$ *the space of all real-valued, progressively measurable iso-stochastic process* $\hat{F}(\cdot)$ *such that*

$$\hat{E}\left(\left|\hat{\int}_0^T \hat{F}\hat{\times}\hat{d}\hat{t}\right|\right) < \infty,$$

which is equivalent to

$$\hat{E}\left(\left|\int_0^T \hat{F}\frac{\tilde{T}(t)-t\tilde{T}'(t)}{\tilde{T}(t)}dt\right|\right) < \infty.$$

Definition 1.9.7. *An iso-stochastic process* $\hat{G} \in \hat{\mathbb{L}}^{\hat{2}}(0,T)$ *is called a step process if there exists a partition*

$$D = \{0 = t_0 < t_1 < t_2 < \ldots < t_{m-1} < t_m = T\}$$

such that

$$\hat{G}(t) \equiv G_k \equiv \text{const} \qquad \text{for} \qquad t_k \le t < t_{k+1}, \qquad k = 0,1,2,\ldots,m-1.$$

Definition 1.9.8. *Let* $\hat{G} \in \hat{\mathbb{L}}^{\hat{2}}(0,T)$ *be a step process, as above. Then*

$$\hat{\int}_0^T \hat{G}\hat{\times}\hat{d}\frac{\tilde{W}^1}{\hat{T}} := \sum_{k=0}^{m-1} G_k\tilde{T}(\tau_k)\frac{\tilde{W}^1(t_{k+1})-\tilde{W}^1(t_k)}{\hat{T}}$$

is called the Itô iso-stochastic iso-integral of the first kind of $\hat{G}$ *on* $[0,T]$.

Theorem 1.9.9. *Let* $\hat{G}$, $\hat{H} \in \hat{\mathbb{L}}^2(0,T)$ *are step processes. Then for all* $a,b \in \mathbb{R}$ *we have*

$$\hat{\int}_0^T \left(a\hat{G}+b\hat{H}\right)\hat{\times}\hat{d}\frac{\tilde{W}^1}{\hat{T}} = a\hat{\int}_0^T \hat{G}\hat{\times}\hat{d}\frac{\tilde{W}^1}{\hat{T}} + b\hat{\int}_0^T \hat{H}\hat{\times}\hat{d}\frac{\tilde{W}^1}{\hat{T}}.$$

Proof. Let

$$D = \{0 = t_0 < t_1 < \ldots < t_{m-1} < t_m = T\}$$

be a partition of $[0,T]$ so that

$$\hat{G} = G_k, \quad \hat{H} = H_k \qquad \text{for} \qquad t_k \le t < t_{k+1}, \quad k = 0,1,2,\ldots,m-1.$$

Then

$$a\hat{G}+b\hat{H} = aG_k + bH_k \qquad \text{for} \qquad t_k \le t < t_{k+1}, \quad k = 0,1,2,\ldots,m-1.$$

Hence,

$$\begin{aligned}
&\hat{\int}_0^T (a\hat{G}+b\hat{H})\hat{\times}\hat{d}\tfrac{\tilde{W}^1}{\hat{T}} = \textstyle\sum_{k=0}^{m-1}(aG_k+bH_k)\tilde{T}(\tau_k)\frac{\tilde{W}^1(t_{k+1})-\tilde{W}^1(t_k)}{\hat{T}}\\
&= \textstyle\sum_{k=0}^{m-1}\left(aG_k\tilde{T}(\tau_k)\frac{\tilde{W}^1(t_{k+1})-\tilde{W}^1(t_k)}{\hat{T}} + bH_k\tilde{T}(\tau_k)\frac{\tilde{W}^1(t_{k+1})-\tilde{W}^1(t_k)}{\hat{T}}\right)\\
&= \textstyle\sum_{k=0}^{m-1} aG_k\tilde{T}(\tau_k)\frac{\tilde{W}^1(t_{k+1})-\tilde{W}^1(t_k)}{\hat{T}} + \sum_{k=0}^{m-1} bH_k\tilde{T}(\tau_k)\frac{\tilde{W}^1(t_{k+1})-\tilde{W}^1(t_k)}{\hat{T}}\\
&= a\textstyle\sum_{k=0}^{m-1} G_k\tilde{T}(\tau_k)\frac{\tilde{W}^1(t_{k+1})-\tilde{W}^1(t_k)}{\hat{T}} + b\sum_{k=0}^{m-1} H_k\tilde{T}(\tau_k)\frac{\tilde{W}^1(t_{k+1})-\tilde{W}^1(t_k)}{\hat{T}}\\
&= a\hat{\int}_0^T \hat{G}\hat{\times}\hat{d}\tfrac{\tilde{W}^1}{\hat{T}} + b\hat{\int}_0^T \hat{H}\hat{\times}\hat{d}\tfrac{\tilde{W}^1}{\hat{T}}.
\end{aligned}$$

□

Theorem 1.9.10. *Let $\hat{G} \in \hat{\mathbb{L}}^{\hat{2}}(0,T)$ be a step process. Then*

$$\hat{E}\left(\hat{\int}_0^T \hat{G}\hat{\times}\hat{d}\frac{\tilde{W}^1}{\hat{T}}\right) = 0.$$

Proof. We have

$$\hat{E}\left(\hat{\int}_0^T \hat{G}\hat{\times}\hat{d}\frac{\tilde{W}^1}{\hat{T}}\right) = \hat{E}\left(\sum_{k=0}^{m-1} G_k\tilde{T}(\tau_k)\frac{\tilde{W}^1(t_{k+1})-\tilde{W}^1(t_k)}{\hat{T}}\right)$$

$$= \sum_{k=0}^{m-1} \hat{E}\left(G_k\tilde{T}(\tau_k)\frac{\tilde{W}^1(t_{k+1})-\tilde{W}^1(t_k)}{\hat{T}}\right)$$

$$= \sum_{k=0}^{m-1} G_k\tilde{T}(\tau_k)\hat{E}\left(\frac{\tilde{W}^1(t_{k+1})-\tilde{W}^1(t_k)}{\hat{T}}\right)$$

$$= 0.$$

□

Theorem 1.9.11. *Let $\hat{G}$, $\hat{H} \in \hat{\mathbb{L}}^{\hat{2}}(0,T)$ be step processes. Then*

$$\hat{E}\left(\hat{\int}_0^T \hat{G}\hat{\times}\hat{d}\frac{\tilde{W}^1}{\hat{T}}\hat{\times}\hat{\int}_0^T \hat{\Pi}\hat{\times}\hat{d}\frac{\tilde{W}^1}{\hat{T}}\right) = \hat{E}\left(\int_0^T \frac{\hat{G}\hat{H}\tilde{T}^2}{\hat{T}}dt\right).$$

Proof. We note that

$$\hat{E}\left(\hat{\int}_0^T \hat{G}\hat{\times}\hat{d}\frac{\tilde{W}^1}{\hat{T}}\hat{\times}\hat{\int}_0^T \hat{H}\hat{\times}\hat{d}\frac{\tilde{W}^1}{\hat{T}}\right)$$

$$= \hat{E}\left(\sum_{k=0}^{m-1} G_k\tilde{T}(\tau_k)\frac{\tilde{W}^1(t_{k+1})-\tilde{W}^1(t_k)}{\hat{T}}\hat{T}\sum_{j=0}^{m-1} H_j\tilde{T}(\tau_j)\frac{\tilde{W}^1(t_{j+1})-\tilde{W}^1(t_j)}{\hat{T}}\right)$$

(26)

$$= \hat{E}\left(\sum_{k=0}^{m-1}\sum_{j=0}^{m-1} G_kH_j\tilde{T}(\tau_k)\tilde{T}(\tau_j)\frac{\left(\tilde{W}^1(t_{k+1})-\tilde{W}^1(t_k)\right)\left(\tilde{W}^1(t_{j+1})-\tilde{W}^1(t_j)\right)}{\hat{T}}\right)$$

$$= \sum_{k=0}^{m-1}\sum_{j=0}^{m-1} \hat{E}\left(G_kH_j\tilde{T}(\tau_k)\tilde{T}(\tau_j)\frac{\left(\tilde{W}^1(t_{k+1})-\tilde{W}^1(t_k)\right)\left(\tilde{W}^1(t_{j+1})-\tilde{W}^1(t_j)\right)}{\hat{T}}\right).$$

For $k \neq j$ we have

$$\hat{E}\left(G_kH_j\tilde{T}(\tau_k)\tilde{T}(\tau_j)\frac{\left(\tilde{W}^1(t_{k+1})-\tilde{W}^1(t_k)\right)\left(\tilde{W}^1(t_{j+1})-\tilde{W}^1(t_j)\right)}{\hat{T}}\right)$$

$$= \hat{E}\left(G_kH_j\tilde{T}(\tau_k)\tilde{T}(\tau_j)\frac{\left(\tilde{W}^1(t_{k+1})-\tilde{W}^1(t_k)\right)}{\hat{T}}\right)\hat{E}\left(\frac{\tilde{W}^1(t_{j+1})-\tilde{W}^1(t_j)}{\hat{T}}\right)$$

$$= 0.$$

Hence and (26) we get

$$\hat{E}\left(\hat{\int}_0^T \hat{G}\hat{\times}\hat{d}\frac{\tilde{W}^1}{\hat{T}}\hat{\times}\hat{\int}_0^T \hat{H}\hat{\times}\hat{d}\frac{\tilde{W}^1}{\hat{T}}\right) = \sum_{k=0}^{m-1}\hat{E}\left(G_k H_k \tilde{T}^2(\tau_k)\frac{\left(\tilde{W}^1(t_{k+1})-\tilde{W}^1(t_k)\right)^2}{\hat{T}}\right)$$

$$= \sum_{k=0}^{m-1}\hat{E}\left(\frac{G_k H_k \tilde{T}^2(\tau_k)}{\hat{T}}\right)\hat{E}\left(\frac{\left(\tilde{W}^1(t_{k+1})-\tilde{W}^1(t_k)\right)}{\hat{T}}\right)$$

$$= \sum_{k=0}^{m-1}\hat{E}\left(\frac{G_k H_k \tilde{T}^2(\tau_k)}{\hat{T}}\right)(t_{k+1}-t_k)$$

$$= \sum_{k=0}^{m-1}\hat{E}\left(\frac{G_k H_k \tilde{T}^2(\tau_k)}{\hat{T}}(t_{k+1}-t_k)\right)$$

$$= \hat{E}\left(\sum_{k=0}^{m-1}\frac{G_k H_k \tilde{T}^2(\tau_k)}{\hat{T}}(t_{k+1}-t_k)\right)$$

$$= \hat{E}\left(\int_0^T \frac{\hat{G}\hat{H}\tilde{T}^2}{\hat{T}}dt\right).$$

□

Corollary 1.9.12. *Let $\hat{G}\in\hat{\mathbb{L}}^{\hat{2}}(0,T)$ ba a step process. Then*

$$\hat{E}\left(\left(\hat{\int}_0^T \hat{G}\hat{\times}\hat{d}\frac{\tilde{W}^1}{\hat{T}}\right)^{\hat{2}}\right) = \hat{E}\left(\int_0^T \frac{\hat{G}^2\tilde{T}^2}{\hat{T}}dt\right).$$

Definition 1.9.13. *Let*

$$M_1 \le \hat{T} \le M_1 \quad \text{on} \quad \Omega, \quad \tilde{T} \le M_3 \quad \text{on} \quad [0,T],$$

where M_1, M_2 and M_3 are some positive constants.

If $\hat{G}\in\hat{\mathbb{L}}^{\hat{2}}(0,T)$, then there exists a sequence $\{\hat{G}_m\}_{m=1}^{\infty}$ of elements of $\hat{\mathbb{L}}^{\hat{2}}(0,T)$ of bounded step processes such that

$$\hat{E}\left(\int_0^T \frac{\left(\hat{G}-\hat{G}_m\right)^2\tilde{T}^2}{\hat{T}}dt\right) \longrightarrow_{m\longrightarrow\infty} 0.$$

We define

$$\hat{\int}_0^T \hat{G}\hat{\times}\hat{d}\frac{\tilde{W}^1}{\hat{T}} := \lim_{m\longrightarrow\infty}\hat{\int}_0^T \hat{G}_m\hat{\times}\hat{d}\frac{\tilde{W}^1}{\hat{T}}.$$

Theorem 1.9.14. *Let $\hat{G}$, $\hat{H}\in\hat{\mathbb{L}}^{\hat{2}}(0,T)$. Then*

(i) *For all real constants we have*

$$\hat{\int}_0^T \left(a\hat{G}+b\hat{H}\right)\hat{\times}\hat{d}\frac{\tilde{W}^1}{\hat{T}} = a\hat{\int}_0^T \hat{G}\hat{\times}\hat{d}\frac{\tilde{W}^1}{\hat{T}} + b\hat{\int}_0^T \hat{H}\hat{\times}\hat{d}\frac{\tilde{W}^1}{\hat{T}}.$$

(ii)

$$\hat{E}\left(\hat{\int}_0^T \hat{G}\hat{\times}\hat{d}\frac{\tilde{W}^1}{\hat{T}}\right) = 0.$$

(iii)

$$\hat{E}\left(\hat{\int}_0^T \hat{G}\hat{\times}\hat{d}\frac{\tilde{W}^1}{\hat{T}}\hat{\times}\hat{\int}_0^T \hat{H}\hat{\times}\hat{d}\frac{\tilde{W}^1}{\hat{T}}\right) = \hat{E}\left(\int_0^T \hat{G}\hat{H}\frac{\tilde{T}^2}{\hat{T}}dt\right).$$

Definition 1.9.15. *For $\hat{G} \in \hat{\mathbb{L}}^{\hat{2}}(0,T)$ we set*

$$\hat{I}^1(t) := \hat{\int}_0^t \hat{G}\hat{\times}\hat{d}\frac{\tilde{W}^1}{\hat{T}}, \qquad 0 \le t \le T,$$

the indefinite Itô iso-integral of the first kind of the iso-stochastic process $\hat{G}$.

We note that

$$\hat{I}^1(0) = 0.$$

Definition 1.9.16. *Suppose that $\hat{X}(\cdot)$ is a real-valued iso-stochastic process satisfying*

$$\hat{X}(r) = \hat{X}(s) + \hat{\int}_s^r \hat{F}\hat{\times}\hat{d}\hat{t} + \hat{\int}_s^r \hat{G}\hat{\times}\hat{d}\frac{\tilde{W}^1}{\hat{T}}$$

for some $\hat{F} \in \hat{\mathbb{L}}^1(0,T)$, $\hat{G} \in \hat{\mathbb{L}}^{\hat{2}}(0,T)$ and all times $0 \le s \le r \le T$. We say that $\hat{X}$ has the iso-stochastic differential of the first kind

$$d\hat{X} = \hat{F}\frac{\tilde{T}(t) - t\tilde{T}'(t)}{\tilde{T}(t)}dt + \hat{G}\tilde{T}d\frac{\tilde{W}^1}{\hat{T}}$$

for $0 \le t \le T$.

Exercise 1.9.17. *Prove that*

$$d\left(a\hat{X} + b\hat{Y}\right) = ad\hat{X} + bd\hat{Y}$$

for all real constants a and b.

Remark 1.9.18. *We will note that for $\hat{G} \in \hat{\mathbb{L}}^{\hat{2}}(0,T)$ we have that*

$$\hat{\int}_0^r \hat{G}\hat{\times}\hat{d}\frac{\tilde{W}^1}{\hat{T}} = \lim_{n \longrightarrow \infty} \sum_{k=0}^{m_n - 1} \hat{G}(t_k^n)\tilde{T}(t_k^n)\frac{\tilde{W}^1(t_{k+1}^n) - \tilde{W}^1(t_k^n)}{\hat{T}},$$

where

$$D^{m_n} := \{0 = t_0^n < t_1^n < \ldots < t_{m_n - 1}^n < t_{m_n}^n = r\}$$

with

$$\lim_{n \longrightarrow \infty} |D^{m_n}| = 0.$$

Remark 1.9.19. *Below we will use the following notation*

$$\hat{\int}_0^1 \hat{\times}\hat{d}(\cdot)$$

instead the notation

$$\hat{\int}_0^r 1\hat{\times}\hat{d}(\cdot).$$

Above we have proved that

$$\hat{\int}_0^T \frac{\tilde{W}^1}{\tilde{T}}\hat{\times}\hat{d}\frac{\tilde{W}^1}{\hat{T}} = \frac{\left(\tilde{W}^1(T)\right)^2}{\hat{T}} + \left(\lambda - \frac{1}{2}\right)\frac{T}{\hat{T}}\hat{T}(\Omega).$$

From here, when $\lambda = 0$ and $T = t$, we get

$$\hat{\int}_0^t \frac{\tilde{W}^1}{\tilde{T}}\hat{\times}\hat{d}\frac{\tilde{W}^1}{\hat{T}} = \frac{\left(\tilde{W}^1(t)\right)^2}{2\hat{T}} - \frac{1}{2}\frac{t}{\hat{T}}\hat{T}(\Omega),$$

whereupon

$$\frac{\left(\tilde{W}^1(t)\right)^2}{\hat{T}} = \frac{t}{\hat{T}}\hat{T}(\Omega) + 2\hat{\int}_0^t \frac{\tilde{W}^1}{\tilde{T}}\hat{\times}\hat{d}\frac{\tilde{W}^1}{\hat{T}}.$$

Using the definition for iso-stochastic differential of the first kind we obtain

$$d\frac{\left(\tilde{W}^1(t)\right)^2}{\hat{T}} = \frac{\hat{T}(\Omega)}{\hat{T}}dt + 2\frac{\tilde{W}^1}{\tilde{T}}\tilde{T}d\frac{\tilde{W}^1}{\hat{T}}$$

$$= \frac{\hat{T}(\Omega)}{\hat{T}}dt + 2\tilde{W}^1 d\frac{\tilde{W}^1}{\hat{T}}$$

Now we will deduct

$$d\left(t\frac{\tilde{W}^1(t)}{\hat{T}}\right).$$

Theorem 1.9.20. *We have*

$$d\left(t\frac{\tilde{W}^1}{\hat{T}}\right) = td\frac{\tilde{W}^1}{\hat{T}} + \frac{\tilde{W}^1}{\hat{T}}dt.$$

Proof. We note that

$$\hat{\int}_0^r \hat{\times}\hat{d}\left(t\frac{\tilde{W}^1}{\hat{T}}\right) = \lim_{n\longrightarrow\infty}\sum_{k=0}^{m_n-1}\left(t_{k+1}^n\frac{\tilde{W}^1(t_{k+1}^n)}{\hat{T}} - t_k^n\frac{\tilde{W}^1(t_k^n)}{\hat{T}}\right)\tilde{T}(t_k^n).$$

Using that $t \longrightarrow \frac{\tilde{W}^1(t)}{\hat{T}}$ is continuous, we get

$$\hat{\int}_0^r \frac{\tilde{W}^1}{\hat{T}}\hat{\times}\hat{d}t = \lim_{n\longrightarrow\infty}\sum_{k=0}^{m_n-1}\tilde{T}(t_k^n)\frac{\tilde{W}^1(t_{k+1}^n)}{\hat{T}}(t_{k+1}^n - t_k^n),$$

$$\hat{\int}_0^r t\hat{\times}\hat{d}\frac{\tilde{W}^1}{\hat{T}} = \lim_{n\longrightarrow\infty}\sum_{k=0}^{m_n-1}t_k^n\tilde{T}(t_k^n)\left(\frac{\tilde{W}^1(t_{k+1}^n)}{\hat{T}} - \frac{\tilde{W}^1(t_k^n)}{\hat{T}}\right).$$

Hence,

$$\hat{\int}_0^r \frac{\tilde{W}^1}{\hat{T}} \hat{\times} \hat{d}t + \hat{\int}_0^r t \hat{\times} \hat{d}\frac{\tilde{W}^1}{\hat{T}} = \lim_{n\longrightarrow\infty} \sum_{k=0}^{m_n-1} \tilde{T}(t_k^n) \frac{\tilde{W}^1(t_{k+1}^n)}{\hat{T}} (t_{k+1}^n - t_k^n)$$

$$+\lim_{n\longrightarrow\infty} \sum_{k=0}^{m_n-1} t_k^n \tilde{T}(t_k^n) \left(\frac{\tilde{W}^1(t_{k+1}^n)}{\hat{T}} - \frac{\tilde{W}^1(t_k^n)}{\hat{T}}\right)$$

$$= \lim_{n\longrightarrow\infty} \sum_{k=0}^{m_n-1} \frac{1}{\hat{T}} \Big(\tilde{T}(t_k^n)\tilde{W}^1(t_{k+1}^n)t_{k+1}^n - \tilde{T}(t_k^n)\tilde{W}^1(t_{k+1}^n)t_k^n$$

$$+t_k^n\tilde{T}(t_k^n)\tilde{W}^1(t_{k+1}^n) - t_k^n\tilde{T}(t_k^n)\tilde{W}^1(t_k^n)\Big)$$

$$= \lim_{n\longrightarrow\infty} \sum_{k=0}^{m_n-1} \frac{1}{\hat{T}} \left(\tilde{T}(t_k^n)\tilde{W}^1(t_{k+1}^n)t_{k+1}^n - t_k^n\tilde{T}(t_k^n)\tilde{W}^1(t_k^n)\right)$$

$$= \lim_{n\longrightarrow\infty} \sum_{k=0}^{m_n-1} \tilde{T}(t_k^n) \frac{\tilde{W}^1(t_{k+1}^n)t_{k+1}^n - \tilde{W}^1(t_k^n)t_k^n}{\hat{T}}$$

$$= \hat{\int}_0^r \hat{\times} \hat{d}\left(t\frac{\tilde{W}^1}{\hat{T}}\right).$$

□

Theorem 1.9.21. *Let $g \in C^1(0,T)$. Then*

$$d\left(g(t)\frac{\tilde{W}^1}{\hat{T}}\right) = g(t)d\frac{\tilde{W}^1}{\hat{T}} + \frac{\tilde{W}^1}{\hat{T}}dg(t).$$

Proof. We note that

$$\hat{\int}_0^r \hat{\times}\hat{d}\left(g(t)\frac{\tilde{W}^1}{\hat{T}}\right) = \lim_{n\longrightarrow\infty} \sum_{k=0}^{m_n-1} \left(g(t_{k+1}^n)\frac{\tilde{W}^1(t_{k+1}^n)}{\hat{T}} - g(t_k^n)\frac{\tilde{W}^1(t_k^n)}{\hat{T}}\right)\tilde{T}(t_k^n).$$

Using that $t \longrightarrow \frac{\tilde{W}^1(t)}{\hat{T}}$ is continuous, we get

$$\hat{\int}_0^r \frac{\tilde{W}^1}{\hat{T}} \hat{\times}\hat{d}g(t) = \lim_{n\longrightarrow\infty} \sum_{k=0}^{m_n-1} \tilde{T}(t_k^n)\frac{\tilde{W}^1(t_{k+1}^n)}{\hat{T}}(g(t_{k+1}^n) - g(t_k^n)),$$

$$\hat{\int}_0^r g(t)\hat{\times}\hat{d}\frac{\tilde{W}^1}{\hat{T}} = \lim_{n\longrightarrow\infty} \sum_{k=0}^{m_n-1} g(t_k^n)\tilde{T}(t_k^n)\left(\frac{\tilde{W}^1(t_{k+1}^n)}{\hat{T}} - \frac{\tilde{W}^1(t_k^n)}{\hat{T}}\right).$$

Hence,

$$\hat{\int}_0^r \frac{\tilde{W}^1}{\hat{T}} \hat{\times}\hat{d}g(t) + \hat{\int}_0^r g(t)\hat{\times}\hat{d}\frac{\tilde{W}^1}{\hat{T}} = \lim_{n\longrightarrow\infty} \sum_{k=0}^{m_n-1} \tilde{T}(t_k^n)\frac{\tilde{W}^1(t_{k+1}^n)}{\hat{T}}(g(t_{k+1}^n) - g(t_k^n))$$

$$+\lim_{n\longrightarrow\infty} \sum_{k=0}^{m_n-1} g(t_k^n)\tilde{T}(t_k^n)\left(\frac{\tilde{W}^1(t_{k+1}^n)}{\hat{T}} - \frac{\tilde{W}^1(t_k^n)}{\hat{T}}\right)$$

$$= \lim_{n\longrightarrow\infty} \sum_{k=0}^{m_n-1} \frac{1}{\hat{T}} \Big(\tilde{T}(t_k^n)\tilde{W}^1(t_{k+1}^n)g(t_{k+1}^n) - \tilde{T}(t_k^n)\tilde{W}^1(t_{k+1}^n)g(t_k^n)$$

$$+g(t_k^n)\tilde{T}(t_k^n)\tilde{W}^1(t_{k+1}^n) - g(t_k^n)\tilde{T}(t_k^n)\tilde{W}^1(t_k^n)\Big)$$

$$= \lim_{n\longrightarrow\infty}\sum_{k=0}^{m_n-1}\frac{1}{\hat{T}}\left(\tilde{T}(t_k^n)\tilde{W}^1(t_{k+1}^n)g(t_{k+1}^n)-g(t_k^n)\tilde{T}(t_k^n)\tilde{W}^1(t_k^n)\right)$$

$$= \lim_{n\longrightarrow\infty}\sum_{k=0}^{m_n-1}\tilde{T}(t_k^n)\frac{\tilde{W}^1(t_{k+1}^n)g(t_{k+1}^n)-\tilde{W}^1(t_k^n)g(t_k^n)}{\hat{T}}$$

$$= \hat{\int}_0^r\hat{\times}\hat{d}\left(g(t)\frac{\tilde{W}^1}{\hat{T}}\right).$$

□

Theorem 1.9.22. *(iso-Itô product rule) Suppose*

$$d\hat{X}_1 = \hat{F}_1 dt + \hat{G}_1 d\frac{\tilde{W}^1}{\hat{T}},$$

$$d\hat{X}_2 = \hat{F}_2 dt + \hat{G}_2 d\frac{\tilde{W}^1}{\hat{T}},$$

for $\hat{F}_i \in \hat{\mathbb{L}}^1(0,T)$, $\hat{G}_i \in \hat{\mathbb{L}}^{\hat{2}}(0,T)$, $i=1,2$. *Then*

$$d(\hat{X}_1\hat{X}_2) = \hat{X}_1 d\hat{X}_2 + \hat{X}_2 d\hat{X}_1 + \hat{G}_1\hat{G}_2\frac{\hat{T}(\Omega)}{\hat{T}}dt.$$

Proof. 1. case. Choose $0 \le r \le T$. First of all, we assume for simplicity that

$$\hat{X}_1(0) = \hat{X}_2(0) = 0,$$

and

$$\hat{F}_i(t) = \hat{F}_i, \qquad \hat{G}_i(t) = \hat{G}_i, \qquad i = 1,2,$$

where $\hat{F}_i$, $\hat{G}_i$ are time-independent $\mathcal{F}(0)$-measurable random iso-variables, $i=1,2$. Thus

$$\hat{X}_1(t) = \hat{F}_1 t + \hat{G}_1\frac{\tilde{W}^1}{\hat{T}},$$

$$\hat{X}_2(t) = \hat{F}_1 t + \hat{G}_2\frac{\tilde{W}^1}{\hat{T}},$$

and

$$\hat{\int}_0^r\frac{1}{\hat{T}}\hat{\times}\hat{d}(\hat{X}_1\hat{X}_2) = \int_0^r\frac{1}{\hat{T}^2}\tilde{T}\tilde{T}d(\hat{X}_1\hat{X}_2)$$

$$= \int_0^r d(\hat{X}_1\hat{X}_2)$$

$$= \hat{X}_1(r)\hat{X}_2(r)$$

$$= \left(\hat{F}_1 r + \hat{G}_1\frac{\tilde{W}^1}{\hat{T}}\right)\left(\hat{F}_2 r + \hat{G}_2\frac{\tilde{W}^1}{\hat{T}}\right)$$

$$= \hat{F}_1\hat{F}_2 r + \hat{F}_1\hat{G}_2 r\frac{\tilde{W}^1}{\hat{T}} + \hat{G}_1\hat{F}_2 r\frac{\tilde{W}^1}{\hat{T}} + \hat{G}_1\hat{G}_2\frac{(\tilde{W}^1)^2}{\hat{T}^2}$$

$$= \hat{F}_1\hat{F}_2 r^2 + (\hat{F}_1\hat{G}_2 + \hat{F}_2\hat{G}_1)r\frac{\tilde{W}^1}{\hat{T}} + \hat{G}_1\hat{G}_2\frac{(\tilde{W}^1)^2}{\hat{T}^2},$$

i.e.,

$$\hat{\int}_0^r\frac{1}{\hat{T}}\hat{\times}\hat{d}(\hat{X}_1\hat{X}_2) = \hat{F}_1\hat{F}_2 r^2 + (\hat{F}_1\hat{G}_2 + \hat{F}_2\hat{G}_1)r\frac{\tilde{W}^1}{\hat{T}} + \hat{G}_1\hat{G}_2\frac{(\tilde{W}^1)^2}{\hat{T}^2}. \tag{27}$$

On the other hand,

$$\hat{\int}_0^r \frac{1}{\hat{T}} \hat{\times} \left(\hat{X}_1 \hat{d}\hat{X}_2 + \hat{X}_2 \hat{d}\hat{X}_1 + \hat{G}_1 \hat{G}_2 \frac{\hat{T}(\Omega)}{\hat{T}} \hat{d}t \right)$$

$$= \int_0^r \frac{1}{\hat{T}} \frac{1}{\hat{T}} \tilde{T} \left(\tilde{T}\hat{X}_1 d\hat{X}_2 + \tilde{T}\hat{X}_2 d\hat{X}_1 + \tilde{T}\hat{G}_1\hat{G}_2 \frac{\hat{T}(\Omega)}{\hat{T}} dt \right)$$

$$= \int_0^r \left(\hat{X}_1 d\hat{X}_2 + \hat{X}_2 d\hat{X}_1 + \hat{G}_1 \hat{G}_2 \frac{\hat{T}(\Omega)}{\hat{T}} dt \right)$$

$$= \int_0^r \left(\left(\hat{F}_1 t + \hat{G}_1 \frac{\tilde{W}^1}{\hat{T}} \right) \left(\hat{F}_2 dt + \hat{G}_2 d\frac{\tilde{W}^1}{\hat{T}} \right) \right.$$

$$\left. + \left(\hat{F}_2 t + \hat{G}_2 \frac{\tilde{W}^1}{\hat{T}} \right) \left(\hat{F}_1 dt + \hat{G}_1 d\frac{\tilde{W}^1}{\hat{T}} \right) + \hat{G}_1 \hat{G}_2 \frac{\hat{T}(\Omega)}{\hat{T}} dt \right)$$

$$= \int_0^r \left(\hat{F}_1 \hat{F}_2 t dt + \hat{F}_1 \hat{G}_2 t d\frac{\tilde{W}^1}{\hat{T}} + \hat{G}_1 \hat{F}_2 \frac{\tilde{W}^1}{\hat{T}} dt \right.$$

$$+ \hat{G}_1 \hat{G}_2 \frac{\tilde{W}^1}{\hat{T}} d\frac{\tilde{W}^1}{\hat{T}} + \hat{F}_1 \hat{F}_2 t dt + \hat{F}_2 \hat{G}_1 t d\frac{\tilde{W}^1}{\hat{T}}$$

$$\left. + \hat{G}_2 \hat{F}_1 \frac{\tilde{W}^1}{\hat{T}} dt + \hat{G}_1 \hat{G}_2 \frac{\tilde{W}^1}{\hat{T}} d\frac{\tilde{W}^1}{\hat{T}} + \hat{G}_1 \hat{G}_2 \frac{\hat{T}(\Omega)}{\hat{T}} dt \right)$$

$$= \int_0^r \left(2\hat{F}_1 \hat{F}_2 t dt + (\hat{F}_1 \hat{G}_2 + \hat{F}_2 \hat{G}_1) t d\frac{\tilde{W}^1}{\hat{T}} + (\hat{G}_1 \hat{F}_2 + \hat{G}_2 \hat{F}_1) \frac{\tilde{W}^1}{\hat{T}} dt \right.$$

$$\left. + 2\hat{G}_1 \hat{G}_2 \frac{\tilde{W}^1}{\hat{T}} d\frac{\tilde{W}^1}{\hat{T}} + \hat{G}_1 \hat{G}_2 \frac{\hat{T}(\Omega)}{\hat{T}} dt \right)$$

$$= 2\hat{F}_1 \hat{F}_2 \int_0^r t dt + (\hat{F}_1 \hat{G}_2 + \hat{F}_2 \hat{G}_1) \int_0^r \left(t d\frac{\tilde{W}^1}{\hat{T}} + \frac{\tilde{W}^1}{\hat{T}} dt \right)$$

$$+ 2\hat{G}_1 \hat{G}_2 \int_0^r \frac{\tilde{W}^1}{\hat{T}} d\frac{\tilde{W}^1}{\hat{T}} + \hat{G}_1 \hat{G}_2 \frac{\hat{T}(\Omega)}{\hat{T}} \int_0^r dt$$

$$= \hat{F}_1 \hat{F}_2 r^2 + (\hat{F}_1 \hat{G}_2 + \hat{F}_2 \hat{G}_1) \int_0^r d\left(t \frac{\tilde{W}^1}{\hat{T}} \right)$$

$$+ \hat{G}_1 \hat{G}_2 \frac{(\tilde{W}^1(r))^2}{\hat{T}^2} - \hat{G}_1 \hat{G}_2 r \frac{\hat{T}(\Omega)}{\hat{T}} + \hat{G}_1 \hat{G}_2 \frac{\hat{T}(\Omega)}{\hat{T}} r$$

$$= \hat{F}_1 \hat{F}_2 r^2 + (\hat{F}_1 \hat{G}_2 + \hat{F}_2 \hat{G}_1) r \frac{\tilde{W}^1}{\hat{T}} + \hat{G}_1 \hat{G}_2 \frac{(\tilde{W}^1(r))^2}{\hat{T}^2}.$$

From the last equality and from (27) it follows that

$$\hat{\int}_0^r \frac{1}{\hat{T}} \hat{\times} \hat{d}(\hat{X}_1 \hat{X}_2) = \hat{\int}_0^r \frac{1}{\hat{T}} \hat{\times} \left(\hat{X}_1 \hat{d}\hat{X}_2 + \hat{X}_2 \hat{d}\hat{X}_1 + \hat{G}_1 \hat{G}_2 \frac{\hat{T}(\Omega)}{\hat{T}} \hat{d}t \right).$$

2. case. If $\hat{F}_i$, $\hat{G}_i$ are step processes, we apply the step 1 in each subinterval $[t_k, t_{k+1})$ on which $\hat{F}_i$ and $\hat{G}_i$ are constant random iso-variables.

3. case. In the general case, we select step processes $\hat{F}_i^m \in \hat{\mathbb{L}}^1(0,T)$, $\hat{G}_i^m \in \hat{\mathbb{L}}^{\hat{2}}(0,T)$ with

$$\hat{E}\left(\left|\int_0^t(\hat{F}_i-\hat{F}_i^m)\frac{\tilde{T}(t)-t\tilde{T}'(t)}{\tilde{T}(t)}dt\right|\right) \longrightarrow_{m\longrightarrow\infty} 0,$$

$$\hat{E}\left(\left|\int_0^t(\hat{G}_i-\hat{G}_i^m)^2\hat{T}\frac{\tilde{T}(t)-t\tilde{T}'(t)}{\tilde{T}(t)}dt\right|\right) \longrightarrow_{m\longrightarrow\infty} 0.$$

We define

$$\hat{X}_i^n := \hat{X}_i(0) + \int_0^t \hat{F}_i^n ds + \int_0^t \hat{G}_i^n d\frac{\tilde{W}^1}{\hat{T}}, \qquad i=1,2.$$

We apply the step 2 to $\hat{X}_i^n$ in (s,r) and pass to limits to obtain the conclusion of the theorem.

□

Corollary 1.9.23. *Let*

$$d\hat{X} = \hat{F}dt + \hat{G}d\frac{\tilde{W}^1}{\hat{T}},$$

where $\hat{F} \in \hat{\mathbb{L}}^1(0,T)$, $\hat{G} \in \hat{\mathbb{L}}^{\hat{2}}(0,T)$. *Then*

$$d\hat{X}^2 = 2\hat{X}d\hat{X} + \hat{G}^2\frac{\hat{T}(\Omega)}{\hat{T}}dt.$$

Proof. Let

$$\hat{X}_1 = \hat{X}, \qquad \hat{X}_2 = \hat{X}.$$

Then

$$\hat{G}_1 = \hat{G}_2 = \hat{G}.$$

Hence,

$$\begin{aligned} d\hat{X}^2 &= d(\hat{X}\hat{X}) \\ &= \hat{X}d\hat{X} + \hat{X}d\hat{X} + \hat{G}^2\tfrac{\hat{T}(\Omega)}{\hat{T}}dt \\ &= 2\hat{X}d\hat{X} + \hat{G}^2\tfrac{\hat{T}(\Omega)}{\hat{T}}dt. \end{aligned}$$

□

Corollary 1.9.24. *Let*

$$d\hat{X} = \hat{F}dt + \hat{G}d\frac{\tilde{W}^1}{\hat{T}},$$

where $\hat{F} \in \hat{\mathbb{L}}^1(0,T)$, $\hat{G} \in \hat{\mathbb{L}}^{\hat{2}}(0,T)$. *Then*

$$d\hat{X}^3 = 3\hat{X}^2d\hat{X} + 3\hat{X}\hat{G}^2\frac{\hat{T}(\Omega)}{\hat{T}}dt.$$

Proof. Let

$$\hat{X}_1 = \hat{X}^2, \qquad \hat{X}_2 = \hat{X}.$$

Then

$$\begin{aligned}
d\hat{X}_1 &= d\hat{X}^2 \\
&= 2\hat{X}d\hat{X} + \hat{G}^2 \tfrac{\hat{T}(\Omega)}{\hat{T}} dt \\
&= 2\hat{X}\left(\hat{F}dt + \hat{G}d\tfrac{\tilde{W}^1}{\hat{T}}\right) + \hat{G}^2 \tfrac{\hat{T}(\Omega)}{\hat{T}} dt \\
&= 2\hat{X}\hat{F}dt + 2\hat{X}\hat{G}d\tfrac{\tilde{W}^1}{\hat{T}} + \hat{G}^2 \tfrac{\hat{T}(\Omega)}{\hat{T}} dt \\
&= \left(2\hat{X}\hat{F} + \hat{G}^2 \tfrac{\hat{T}(\Omega)}{\hat{T}}\right) dt + 2\hat{X}\hat{G}d\tfrac{\tilde{W}^1}{\hat{T}}.
\end{aligned}$$

Therefore

$$\hat{G}_1 = 2\hat{X}\hat{G}, \qquad \hat{G}_2 = \hat{G}.$$

Hence,

$$\begin{aligned}
d\hat{X}^3 &= d(\hat{X}^2\hat{X}) \\
&= d(\hat{X}_1\hat{X}_2) \\
&= \hat{X}_2 d\hat{X}_1 + \hat{X}_1 d\hat{X}_2 + \hat{G}_1\hat{G}_2 \tfrac{\hat{T}(\Omega)}{\hat{T}} dt \\
&= \hat{X}d\hat{X}^2 + \hat{X}^2 d\hat{X} + 2\hat{X}\hat{G}\hat{G}\tfrac{\hat{T}(\Omega)}{\hat{T}} dt \\
&= \hat{X}\left(2\hat{X}d\hat{X} + \hat{G}^2\tfrac{\hat{T}(\Omega)}{\hat{T}} dt\right) + \hat{X}^2 d\hat{X} + 2\hat{X}\hat{G}^2\tfrac{\hat{T}(\Omega)}{\hat{T}} dt \\
&= 2\hat{X}^2 d\hat{X} + \hat{X}\hat{G}^2\tfrac{\hat{T}(\Omega)}{\hat{T}} dt + \hat{X}^2 d\hat{X} + 2\hat{X}\hat{G}^2\tfrac{\hat{T}(\Omega)}{\hat{T}} dt \\
&= 3\hat{X}^2 d\hat{X} + 3\hat{X}\hat{G}^2\tfrac{\hat{T}(\Omega)}{\hat{T}} dt.
\end{aligned}$$

□

Corollary 1.9.25. *Let*

$$d\hat{X} = \hat{F}dt + \hat{G}d\frac{\tilde{W}^1}{\hat{T}},$$

where $\hat{F} \in \hat{\mathbb{L}}^1(0,T)$, $\hat{G} \in \hat{\mathbb{L}}^{\hat{2}}(0,T)$. *Then*

$$d\hat{X}^m = m\hat{X}^{m-1}d\hat{X} + \frac{1}{2}m(m-1)\hat{X}^{m-2}\hat{G}^2\frac{\hat{T}(\Omega)}{\hat{T}}dt, \qquad m \in \mathbb{N}.$$

Proof. We suppose that for some $m \in \mathbb{N}$ we have

$$d\hat{X}^m = m\hat{X}^{m-1}d\hat{X} + \frac{1}{2}m(m-1)\hat{X}^{m-2}\hat{G}^2\frac{\hat{T}(\Omega)}{\hat{T}}dt.$$

We will prove that

$$d\hat{X}^{m+1} = (m+1)\hat{X}^m d\hat{X} + \frac{1}{2}(m+1)m\hat{X}^{m-1}\hat{G}^2\frac{\hat{T}(\Omega)}{\hat{T}}dt.$$

Really, let

$$\hat{X}^m = \hat{X}_1, \qquad \hat{X} = \hat{X}_2.$$

Then

$$\begin{aligned}
d\hat{X}_1^m &= d\hat{X}^m \\
&= m\hat{X}^{m-1}d\hat{X} + \tfrac{1}{2}m(m-1)\hat{X}^{m-2}\hat{G}^2\tfrac{\hat{T}(\Omega)}{\hat{T}}dt \\
&= m\hat{X}^{m-1}\left(\hat{F}dt + \hat{G}d\tfrac{\tilde{W}^1}{\hat{T}}\right) + \tfrac{1}{2}m(m-1)\hat{X}^{m-2}\hat{G}^2\tfrac{\hat{T}(\Omega)}{\hat{T}}dt \\
&= m\hat{X}^{m-1}\hat{F}dt + m\hat{X}^{m-1}\hat{G}d\tfrac{\tilde{W}^1}{\hat{T}} + \tfrac{1}{2}m(m-1)\hat{X}^{m-2}\hat{G}^2\tfrac{\hat{T}(\Omega)}{\hat{T}}dt \\
&= \left(m\hat{X}^{m-1}\hat{F} + \tfrac{1}{2}m(m-1)\hat{X}^{m-2}\hat{G}^2\tfrac{\hat{T}(\Omega)}{\hat{T}}\right)dt + m\hat{X}^{m-1}\hat{G}d\tfrac{\tilde{W}^1}{\hat{T}}.
\end{aligned}$$

Therefore

$$\hat{G}_1 = m\hat{X}^{m-1}\hat{G}, \qquad \hat{G}_2 = \hat{G}.$$

Hence,

$$\begin{aligned}
d\hat{X}^{m+1} &= d(\hat{X}^m\hat{X}) \\
&= d(\hat{X}_1\hat{X}_2) \\
&= \hat{X}_2 d\hat{X}_1 + \hat{X}_1 d\hat{X}_2 + \hat{G}_1\hat{G}_2\tfrac{\hat{T}(\Omega)}{\hat{T}}dt \\
&= \hat{X}d\hat{X}^m + \hat{X}^m d\hat{X} + m\hat{X}^{m-1}\hat{G}\hat{G}\tfrac{\hat{T}(\Omega)}{\hat{T}}dt \\
&= \hat{X}\left(m\hat{X}^{m-1}d\hat{X} + \tfrac{1}{2}m(m-1)\hat{X}^{m-2}\hat{G}^2\tfrac{\hat{T}(\Omega)}{\hat{T}}dt\right) + m\hat{X}^{m-1}\hat{G}^2\tfrac{\hat{T}(\Omega)}{\hat{T}}dt \\
&= m\hat{X}^m d\hat{X} + \tfrac{1}{2}m(m-1)\hat{X}^{m-1}\hat{G}^2\tfrac{\hat{T}(\Omega)}{\hat{T}}dt + m\hat{X}^{m-1}\hat{G}^2\tfrac{\hat{T}(\Omega)}{\hat{T}}dt \\
&= m\hat{X}^m d\hat{X} + m\hat{X}^{m-1}\hat{G}^2\left(\tfrac{m-1}{2}+1\right)\tfrac{\hat{T}(\Omega)}{\hat{T}}dt \\
&= m\hat{X}^m d\hat{X} + \tfrac{1}{2}m(m+1)\hat{G}^2\tfrac{\hat{T}(\Omega)}{\hat{T}}dt.
\end{aligned}$$

□

Corollary 1.9.26. *Let f be a polynomial,*

$$d\hat{X} = \hat{F}dt + \hat{G}d\frac{\tilde{W}^1}{\hat{T}},$$

where $\hat{F} \in \hat{\mathbb{L}}^1(0,T)$, $\hat{G} \in \hat{\mathbb{L}}^{\hat{2}}(0,T)$. *Then*

$$df(\hat{X}) = f'(\hat{X})d\hat{X} + \frac{1}{2}f''(\hat{X})\hat{G}^2\frac{\hat{T}(\Omega)}{\hat{T}}dt.$$

Proof. Let

$$f(\hat{X}) = b_0\hat{X}^m + b_1\hat{X}^{m-1} + \cdots + b_{m-1}\hat{X} + b_m, \qquad b_i \in \mathbb{R}, i = 0,1,2,\ldots,m.$$

Then

$$\begin{aligned}
df(\hat{X}) &= d\left(b_0\hat{X}^m + b_1\hat{X}^{m-1} + \cdots + b_{m-1}\hat{X} + b_m\right)\\
&= b_0 d\hat{X}^m + b_1 d\hat{X}^{m-1} + \cdots + b_{m_1}d\hat{X}\\
&= b_0 m\hat{X}^{m-1}d\hat{X} + b_0\tfrac{1}{2}m(m-1)\hat{X}^{m-2}\hat{G}^2\tfrac{\hat{T}(\Omega)}{\hat{T}}dt\\
&+ b_1(m-1)\hat{X}^{m-2}d\hat{X} + b_1\tfrac{1}{2}(m-1)(m-2)\hat{X}^{m-3}\hat{G}^2\tfrac{\hat{T}(\Omega)}{\hat{T}}dt\\
&+ \cdots\\
&+ b_{m-1}d\hat{X}\\
&= \left(b_0 m\hat{X}^{m-1} + b_1(m-1)\hat{X}^{m-2} + \cdots + b_{m-1}\right)d\hat{X}\\
&+ \tfrac{1}{2}\left(b_0 m(m-1)\hat{X}^{m-2} + b_1(m-1)(m-2)\hat{X}^{m-3} + \cdots + 2b_{m-2}\right)\hat{G}^2\tfrac{\hat{T}(\Omega)}{\hat{T}}dt\\
&= f'(\hat{X})d\hat{X} + \tfrac{1}{2}f''(\hat{X})\hat{G}^2\tfrac{\hat{T}(\Omega)}{\hat{T}}dt.
\end{aligned}$$

□

Theorem 1.9.27. *(iso-Itô's formula for iso-stochastic differential of the first kind) Suppose that* $\hat{X}(\cdot)$ *has an iso-stochastic differential of the first kind*

$$d\hat{X} = \hat{F}dt + \hat{G}d\frac{\tilde{W}^1}{\hat{T}},$$

for $\hat{F} \in \hat{\mathbb{L}}^1(0,T)$, $\hat{G} \in \hat{\mathbb{L}}^2(0,T)$. *Assume* $u : \mathbb{R} \times [0,T] \longrightarrow \mathbb{R}$ *is continuous and*

$$\frac{\partial u}{\partial t}, \qquad \frac{\partial u}{\partial x}, \qquad \frac{\partial^2 u}{\partial x^2}$$

exist and are continuous.

We set

$$\hat{Y}(t) := u(\hat{X}(t),t).$$

Then

$$\begin{aligned}
d\hat{Y} &= \tfrac{\partial u}{\partial t}dt + \tfrac{\partial u}{\partial x}d\hat{X} + \tfrac{1}{2}\tfrac{\partial^2 u}{\partial x^2}\hat{G}^2\tfrac{\hat{T}(\Omega)}{\hat{T}}dt\\
&= \left(\tfrac{\partial u}{\partial t} + \tfrac{\partial u}{\partial x}\hat{F} + \tfrac{1}{2}\tfrac{\partial^2 u}{\partial x^2}\hat{G}^2\tfrac{\hat{T}(\Omega)}{\hat{T}}\right)dt + \tfrac{\partial u}{\partial x}\hat{G}d\tfrac{\tilde{W}^1}{\hat{T}}.
\end{aligned} \tag{28}$$

Definition 1.9.28. *The formula* (28) *will be called iso-Itô's formula for iso-stochastic differential of the first kind.*

Proof. Suppose that

$$u(x,t) = f(x)g(t),$$

where f and g are polynomials. Then

$$\begin{aligned}
d(u(\hat{X},t)) &= d(f(\hat{X})g(t)) \\
&= g(t)df(\hat{X}) + f(\hat{X})dg(t) \\
&= f(\hat{X})g'(t)dt + g(t)\left(f'(\hat{X})d\hat{X} + \tfrac{1}{2}f''(\hat{X})\hat{G}^2\tfrac{\hat{T}(\Omega)}{\hat{T}}dt\right) \\
&= \tfrac{\partial u}{\partial t}dt + \tfrac{\partial u}{\partial x}d\hat{X} + \tfrac{1}{2}\tfrac{\partial^2 u}{\partial x^2}\hat{G}^2\tfrac{\hat{T}(\Omega)}{\hat{T}}dt \\
&= \tfrac{\partial u}{\partial t}dt + \tfrac{\partial u}{\partial x}\left(\hat{F}dt + \hat{G}d\tfrac{\tilde{W}^1}{\hat{T}}\right) + \tfrac{1}{2}\tfrac{\partial^2 u}{\partial x^2}\hat{G}^2\tfrac{\hat{T}(\Omega)}{\hat{T}}dt \\
&= \tfrac{\partial u}{\partial t}dt + \tfrac{\partial u}{\partial x}\hat{F}dt + \tfrac{\partial u}{\partial x}\hat{G}d\tfrac{\tilde{W}^1}{\hat{T}} + \tfrac{1}{2}\tfrac{\partial^2 u}{\partial x^2}\hat{G}^2\tfrac{\hat{T}(\Omega)}{\hat{T}}dt \\
&= \left(\tfrac{\partial u}{\partial t} + \tfrac{\partial u}{\partial x}\hat{F} + \tfrac{1}{2}\tfrac{\partial^2 u}{\partial x^2}\hat{G}^2\tfrac{\hat{T}(\Omega)}{\hat{T}}\right)dt + \tfrac{\partial u}{\partial x}\hat{G}d\tfrac{\tilde{W}^1}{\hat{T}}.
\end{aligned}$$

Thus the iso-Itô's formula for iso-stochastic differential of the first kind is true as well for any function u having the form

$$u(x,t) = \sum_{i=1}^{m} f^i(x)g^i(t),$$

where f^i and g^i are polynomials.

Given u as in the iso-Itô's formula for iso-stochastic differential of the first kind, there exists a sequence of polynomials u^n such that

$$u^n \longrightarrow u, \quad \frac{\partial u^n}{\partial t} \longrightarrow \frac{\partial u}{\partial t}, \quad \frac{\partial u^n}{\partial x} \longrightarrow \frac{\partial u}{\partial x}, \quad \frac{\partial^2 u^n}{\partial x^2} \longrightarrow \frac{\partial^2 u}{\partial x^2}$$

uniformly on compact subsets of $\mathbb{R}\times[0,T]$. As in above, we have that for all $0 \le r \le T$

$$\begin{aligned}
u^n(\hat{X}(r),r) &= u^n(\hat{X}(0),0) \\
&+ \int_0^r \left(\tfrac{\partial u^n}{\partial t} + \tfrac{\partial u^n}{\partial x}\hat{F} + \tfrac{1}{2}\tfrac{\partial^2 u^n}{\partial x^2}\hat{G}^2\tfrac{\hat{T}(\Omega)}{\hat{T}}\right)dt + \int_0^r \tfrac{\partial u^n}{\partial x}\hat{G}d\tfrac{\tilde{W}^1}{\hat{T}}
\end{aligned}$$

almost surely.

We may pass to limits as $n \longrightarrow \infty$ in this expression, thereby proving the iso-Itô's formula for iso-stochastic differential of the first kind. □

Example 1.9.29. *Let*

$$\hat{X}(\cdot) = \frac{\tilde{W}^1(\cdot)}{\hat{T}}.$$

Then

$$d\hat{X}(\cdot) = d\frac{\tilde{W}^1(\cdot)}{\hat{T}}.$$

Therefore

$$\hat{G} = 1, \qquad \hat{F} = 0.$$

Hence,

$$d\left(\left(\frac{\tilde{W}^1}{\hat{T}}\right)^m\right) = m\left(\frac{\tilde{W}^1}{\hat{T}}\right)^{m-1} d\hat{X} + \frac{1}{2}m(m-1)\left(\frac{\tilde{W}^1}{\hat{T}}\right)^{m-2}\frac{\hat{T}(\Omega)}{\hat{T}}dt$$

$$= m\left(\frac{\tilde{W}^1}{\hat{T}}\right)^{m-1} d\frac{\tilde{W}^1}{\hat{T}} + \frac{1}{2}m(m-1)\left(\frac{\tilde{W}^1}{\hat{T}}\right)^{m-2}\frac{\hat{T}(\Omega)}{\hat{T}}dt.$$

Example 1.9.30. *Let $\hat{X}(\cdot)$ be an iso-stochastic process such that*

$$d\hat{X} = \sqrt{\frac{\hat{T}}{\hat{T}(\Omega)}}d\frac{\tilde{W}^1}{\hat{T}}.$$

Let also,

$$u(x,t) = e^{\lambda x - \frac{\lambda^2}{2}t}, \qquad \hat{Y} = u(\hat{X},t).$$

Then

$$\hat{F} = 0, \qquad \hat{G} = \sqrt{\frac{\hat{T}}{\hat{T}(\Omega)}},$$

$$\frac{\partial u}{\partial x} = \lambda e^{\lambda x - \frac{\lambda^2}{2}t}, \qquad \frac{\partial^2 u}{\partial x^2} = \lambda^2 e^{\lambda x - \frac{\lambda^2}{2}t},$$

$$\frac{\partial u}{\partial t} = -\frac{\lambda^2}{2}e^{\lambda x - \frac{\lambda^2}{2}t}.$$

Hence,

$$d\hat{Y} = d\left(e^{\lambda\hat{X} - \frac{\lambda^2}{2}t}\right)$$

$$= -\frac{\lambda^2}{2}\hat{Y}dt + \lambda\hat{Y}d\hat{X} + \frac{1}{2}\lambda^2\hat{Y}\frac{\hat{T}}{\hat{T}(\Omega)}\frac{\hat{T}(\Omega)}{\hat{T}}dt$$

$$= \lambda\hat{Y}d\hat{X},$$

i.e.,

$$\begin{cases} d\hat{Y} = \lambda\hat{Y}d\hat{X} \\ \hat{Y}(0) = e^{\lambda\hat{X}(0)}, \end{cases}$$

which is an iso-stochastic differential equation.

Definition 1.9.31. *Let $\hat{G} \in \hat{\mathbb{L}}^{\hat{2}}(0,T)$ be a step process,*

$$\hat{G}(t) = G_k, \qquad t_k \le t < t_{k+1}, \qquad k = 0,1,2,\ldots,m-1,$$

$$G_k = \text{const}, \qquad 0 = t_0 < t_1 < t_2 < \ldots < t_{m_1} < t_m = T.$$

Then

$$\hat{\int}_0^T \hat{G}\hat{\times}\hat{d}\frac{\tilde{W}^2\tilde{T}(\tilde{W}^2)}{\hat{T}} := \sum_{k=0}^{m-1} G_k\tilde{T}(\tau_k)\frac{\tilde{W}^2(t_{k+1})\tilde{T}(\tilde{W}^2(t_{k+1})) - \tilde{W}^2(t_k)\tilde{T}(\tilde{W}^2(t_k))}{\hat{T}}$$

is called the Itô iso-stochastic iso-integral of the second kind. Here $\tau_k = (1-\lambda)t_k + \lambda t_{k+1}$, $\lambda \in [0,1]$, $k = 0,1,2,\ldots,m-1$.

Theorem 1.9.32. *Let $\hat{G}$, $\hat{H} \in \hat{\mathbb{L}}^{\hat{2}}(0,T)$ are step processes. Then for all a, $b \in \mathbb{R}$ we have*

$$\hat{\int}_0^T (a\hat{G}+b\hat{H})\hat{\times}\hat{d}\frac{\tilde{W}^2\tilde{T}(\tilde{W}^2)}{\hat{T}} = a\hat{\int}_0^T \hat{G}\hat{\times}\hat{d}\frac{\tilde{W}^2\tilde{T}(\tilde{W}^2)}{\hat{T}} + b\hat{\int}_0^T \hat{H}\hat{\times}\hat{d}\frac{\tilde{W}^2\tilde{T}(\tilde{W}^2)}{\hat{T}}.$$

Proof. Let

$$D = \{0 = t_0 < t_1 < t_2 < \ldots < t_{m_1} < t_m\}$$

be a partition of the interval $[0,T]$ so that

$$\hat{G}(t) = G_k, \qquad \hat{H}(t) = H_k, \qquad t_k \le t < t_{k+1}, \qquad k = 0,1,2,\ldots,m-1.$$

Then

$$\hat{\int}_0^T \hat{G}\hat{\times}\hat{d}\frac{\tilde{W}^2\tilde{T}(\tilde{W}^2)}{\hat{T}} = \sum_{k=0}^{m-1} G_k\tilde{T}(\tau_k)\frac{\tilde{W}^2(t_{k+1})\tilde{T}(\tilde{W}^2(t_{k+1})) - \tilde{W}^2(t_k)\tilde{T}(\tilde{W}^2(t_k))}{\hat{T}},$$

$$\hat{\int}_0^T \hat{H}\hat{\times}\hat{d}\frac{\tilde{W}^2\tilde{T}(\tilde{W}^2)}{\hat{T}} = \sum_{k=0}^{m-1} H_k\tilde{T}(\tau_k)\frac{\tilde{W}^2(t_{k+1})\tilde{T}(\tilde{W}^2(t_{k+1})) - \tilde{W}^2(t_k)\tilde{T}(\tilde{W}^2(t_k))}{\hat{T}},$$

$$a\hat{G}(t) + b\hat{H}(t) = aG_k + bH_k, \qquad t_k \le t < t_{k+1}, \qquad k = 0,1,2,\ldots,m-1.$$

Hence,

$$\hat{\int}_0^T (a\hat{G}+b\hat{H})\hat{\times}\hat{d}\frac{\tilde{W}^2\tilde{T}(\tilde{W}^2)}{\hat{T}} = \sum_{k=0}^{m-1} (aG_k+bH_k)\tilde{T}(\tau_k)\frac{\tilde{W}^2(t_{k+1})\tilde{T}(\tilde{W}^2(t_{k+1})) - \tilde{W}^2(t_k)\tilde{T}(\tilde{W}^2(t_k))}{\hat{T}}$$

$$= \sum_{k=0}^{m-1} aG_k\tilde{T}(\tau_k)\frac{\tilde{W}^2(t_{k+1})\tilde{T}(\tilde{W}^2(t_{k+1})) - \tilde{W}^2(t_k)\tilde{T}(\tilde{W}^2(t_k))}{\hat{T}}$$

$$+ \sum_{k=0}^{m-1} bH_k\tilde{T}(\tau_k)\frac{\tilde{W}^2(t_{k+1})\tilde{T}(\tilde{W}^2(t_{k+1})) - \tilde{W}^2(t_k)\tilde{T}(\tilde{W}^2(t_k))}{\hat{T}}$$

$$= a\sum_{k=0}^{m-1} G_k\tilde{T}(\tau_k)\frac{\tilde{W}^2(t_{k+1})\tilde{T}(\tilde{W}^2(t_{k+1})) - \tilde{W}^2(t_k)\tilde{T}(\tilde{W}^2(t_k))}{\hat{T}}$$

$$+ b\sum_{k=0}^{m-1} H_k\tilde{T}(\tau_k)\frac{\tilde{W}^2(t_{k+1})\tilde{T}(\tilde{W}^2(t_{k+1})) - \tilde{W}^2(t_k)\tilde{T}(\tilde{W}^2(t_k))}{\hat{T}}$$

$$= a\hat{\int}_0^T \hat{G}\hat{\times}\hat{d}\frac{\tilde{W}^2\tilde{T}(\tilde{W}^2)}{\hat{T}} + b\hat{\int}_0^T \hat{H}\hat{\times}\hat{d}\frac{\tilde{W}^2\tilde{T}(\tilde{W}^2)}{\hat{T}}.$$

□

Theorem 1.9.33. *Let $\hat{G} \in \hat{\mathbb{L}}^{\hat{2}}(0,T)$ be a step process. Then*

$$\hat{E}\left(\hat{\int}_0^T \hat{G}\hat{\times}\hat{d}\frac{\tilde{W}^2\tilde{T}(\tilde{W}^2)}{\hat{T}}\right) = 0.$$

Proof. First of all, we will note that

$$\hat{E}\left(\frac{\tilde{W}^2(t_{k+1})\tilde{T}(\tilde{W}^2(t_{k+1}))-\tilde{W}^2(t_k)\tilde{T}(\tilde{W}^2(t_k))}{\hat{T}}\right) = 0,$$

$$\hat{\int}_0^T \hat{G}\hat{\times}\hat{d}\frac{\tilde{W}^2\tilde{T}(\tilde{W}^2)}{\hat{T}} = \sum_{k=0}^{m-1} G_k\tilde{T}(\tau_k)\frac{\tilde{W}^2(t_{k+1})\tilde{T}(\tilde{W}^2(t_{k+1}))-\tilde{W}^2(t_k)\tilde{T}(\tilde{W}^2(t_k))}{\hat{T}}.$$

Therefore

$$\hat{E}\left(\hat{\int}_0^T \hat{G}\hat{\times}\hat{d}\frac{\tilde{W}^2\tilde{T}(\tilde{W}^2)}{\hat{T}}\right) = \hat{E}\left(\sum_{k=0}^{m-1} G_k\tilde{T}(\tau_k)\frac{\tilde{W}^2(t_{k+1})\tilde{T}(\tilde{W}^2(t_{k+1}))-\tilde{W}^2(t_k)\tilde{T}(\tilde{W}^2(t_k))}{\hat{T}}\right)$$

$$= \sum_{k=0}^{m-1}\hat{E}\left(G_k\tilde{T}(\tau_k)\frac{\tilde{W}^2(t_{k+1})\tilde{T}(\tilde{W}^2(t_{k+1}))-\tilde{W}^2(t_k)\tilde{T}(\tilde{W}^2(t_k))}{\hat{T}}\right) = 0.$$

□

Theorem 1.9.34. *Let $\hat{G}$, $\hat{H} \in \hat{\mathbb{L}}^{\hat{2}}(0,T)$ are step processes. Then*

$$\hat{E}\left(\hat{\int}_0^T \hat{G}\hat{\times}\hat{d}\frac{\tilde{W}^2\tilde{T}(\tilde{W}^2)}{\hat{T}}\hat{\times}\hat{\int}_0^T \hat{H}\hat{\times}\hat{d}\frac{\tilde{W}^2\tilde{T}(\tilde{W}^2)}{\hat{T}}\right) = \hat{E}\left(\int_0^T \frac{\hat{G}\hat{H}\tilde{T}^2}{\hat{T}}dt\right).$$

Proof. Let

$$D = \{0 = t_0 < t_1 < t_2 \ldots < t_{m-1} < t_m = T\}$$

be a partition of the interval $[0,T]$ so that

$$\hat{G}(t) = G_k, \qquad \hat{H}(t) = H_k, \qquad t_k \le t < t_{k+1}, \qquad k = 0,1,2,\ldots,m-1.$$

Then

$$\hat{E}\left(\hat{\int}_0^T \hat{G}\hat{\times}\hat{d}\frac{\tilde{W}^2\tilde{T}(\tilde{W}^2)}{\hat{T}}\hat{\times}\hat{\int}_0^T \hat{H}\hat{\times}\hat{d}\frac{\tilde{W}^2\tilde{T}(\tilde{W}^2)}{\hat{T}}\right)$$

$$= \hat{E}\left(\sum_{k=0}^{m-1} G_k\tilde{T}(\tau_k)\frac{\tilde{W}^2(t_{k+1})\tilde{T}(\tilde{W}^2(t_{k+1}))-\tilde{W}^2(t_k)\tilde{T}(\tilde{W}^2(t_k))}{\hat{T}}\hat{T}\right.$$

$$\left.\times\sum_{j=0}^{m-1}\tilde{T}(\tau_j)\frac{\tilde{W}^2(t_{j+1})\tilde{T}(\tilde{W}^2(t_{j+1}))-\tilde{W}^2(t_j)\tilde{T}(\tilde{W}^2(t_j))}{\hat{T}}\right) \qquad (29)$$

$$= \hat{E}\left(\sum_{k=0}^{m-1}\sum_{j=0}^{m-1} G_kH_j\tilde{T}(\tau_k)\tilde{T}(\tau_j)\right.$$

$$\left.\times\frac{\left(\tilde{W}^2(t_{k+1})\tilde{T}(\tilde{W}^2(t_{k+1}))-\tilde{W}^2(t_k)\tilde{T}(\tilde{W}^2(t_k))\right)\left(\tilde{W}^2(t_{j+1})\tilde{T}(\tilde{W}^2(t_{j+1}))-\tilde{W}^2(t_j)\tilde{T}(\tilde{W}^2(t_j))\right)}{\hat{T}}\right)$$

$$= \sum_{k=0}^{m-1}\sum_{j=0}^{m-1}\hat{E}\left(G_kH_j\tilde{T}(\tau_k)\tilde{T}(\tau_j)\right.$$

$$\left.\times\frac{\left(\tilde{W}^2(t_{k+1})\tilde{T}(\tilde{W}^2(t_{k+1}))-\tilde{W}^2(t_k)\tilde{T}(\tilde{W}^2(t_k))\right)\left(\tilde{W}^2(t_{j+1})\tilde{T}(\tilde{W}^2(t_{j+1}))-\tilde{W}^2(t_j)\tilde{T}(\tilde{W}^2(t_j))\right)}{\hat{T}}\right).$$

For $k \neq j$ we have that

$$\hat{E}\left(G_k H_j \tilde{T}(\tau_k)\tilde{T}(\tau_j)\frac{\left(\tilde{W}^2(t_{k+1})\tilde{T}(\tilde{W}^2(t_{k+1}))-\tilde{W}^2(t_k)\tilde{T}(\tilde{W}^2(t_k))\right)\left(\tilde{W}^2(t_{j+1})\tilde{T}(\tilde{W}^2(t_{j+1}))-\tilde{W}^2(t_j)\tilde{T}(\tilde{W}^2(t_j))\right)}{\hat{T}}\right)$$

$$= \hat{E}\left(G_k H_j \tilde{T}(\tau_k)\tilde{T}(\tau_j)\frac{\tilde{W}^2(t_{k+1})\tilde{T}(\tilde{W}^2(t_{k+1}))-\tilde{W}^2(t_k)\tilde{T}(\tilde{W}^2(t_k))}{\hat{T}}\right)$$

$$\times \hat{E}\left(\frac{\tilde{W}^2(t_{j+1})\tilde{T}(\tilde{W}^2(t_{j+1}))-\tilde{W}^2(t_j)\tilde{T}(\tilde{W}^2(t_j))}{\hat{T}}\right)$$

$$= 0.$$

Hence and (29) we get

$$\hat{E}\left(\hat{\int}_0^T \hat{G}\hat{\times}\hat{d}\frac{\tilde{W}^2\tilde{T}(\tilde{W}^2)}{\hat{T}}\hat{\times}\hat{\int}_0^T \hat{H}\hat{\times}\hat{d}\frac{\tilde{W}^2\tilde{T}(\tilde{W}^2)}{\hat{T}}\right)$$

$$= \sum_{k=0}^{m-1}\hat{E}\left(G_k H_k \tilde{T}^2(\tau_k)\frac{\left(\tilde{W}^2(t_{k+1})\tilde{T}(\tilde{W}^2(t_{k+1}))-\tilde{W}^2(t_k)\tilde{T}(\tilde{W}^2(t_k))\right)^2}{\hat{T}}\right)$$

$$= \sum_{k=0}^{m-1}\hat{E}\left(\frac{G_k H_k \tilde{T}^2(\tau_k)}{\hat{T}}\right)\hat{E}\left(\frac{\left(\tilde{W}^2(t_{k+1})\tilde{T}(\tilde{W}^2(t_{k+1}))-\tilde{W}^2(t_k)\tilde{T}(\tilde{W}^2(t_k))\right)^2}{\hat{T}}\right)$$

$$= \sum_{k=0}^{m-1}\hat{E}\left(\frac{G_k H_k \tilde{T}^2(\tau_k)}{\hat{T}}\right)(t_{k+1}-t_k)$$

$$= \sum_{k=0}^{m-1}\hat{E}\left(\frac{G_k H_k \tilde{T}^2(\tau_k)}{\hat{T}}(t_{k+1}-t_k)\right)$$

$$= \hat{E}\left(\sum_{k=0}^{m-1}\frac{G_k H_k \tilde{T}^2(\tau_k)}{\hat{T}}(t_{k+1}-t_k)\right)$$

$$= \hat{E}\left(\int_0^T \frac{\hat{G}\hat{H}\tilde{T}^2}{\hat{T}}dt\right).$$

□

Corollary 1.9.35. *Let $\hat{G} \in \hat{\mathbb{L}}^{\hat{2}}(0,T)$ be a step process. Then*

$$\hat{E}\left(\left(\hat{\int}_0^T \hat{G}\hat{\times}\hat{d}\frac{\tilde{W}^2\tilde{T}(\tilde{W}^2)}{\hat{T}}\right)^{\hat{2}}\right) = \hat{E}\left(\int_0^T \frac{\hat{G}^2\tilde{T}^2}{\hat{T}}dt\right).$$

Definition 1.9.36. *Let*

$$M_1 \leq \hat{T} \leq M_2 \quad \text{on} \quad \Omega,$$

$$\tilde{T} \leq M_3 \quad \text{for} \quad t \in [0,T],$$

where M_1, M_2 and M_3 are some positive constants. Let also, $\hat{G} \in \hat{\mathbb{L}}^{\hat{2}}(0,T)$. Then there exists a sequence of bounded step processes $\hat{G}_m \in \hat{\mathbb{L}}^{\hat{2}}(0,T)$ such that

$$\lim_{m\longrightarrow\infty}\hat{E}\left(\int_0^T \frac{(\hat{G}-\hat{G}_m)^2\tilde{T}^2}{\hat{T}}dt\right) = 0.$$

We define

$$\hat{\int}_0^T \hat{G}\hat{\times}\hat{d}\frac{\tilde{W}^2\tilde{T}(\tilde{W}^2)}{\hat{T}} := \lim_{m\longrightarrow\infty}\hat{\int}_0^T \hat{G}_m\hat{\times}\hat{d}\frac{\tilde{W}^2\tilde{T}(\tilde{W}^2)}{\hat{T}}.$$

Theorem 1.9.37. *Let $\hat{G}$, $\hat{H}\in\hat{\mathbb{L}}^{\hat{2}}(0,T)$. Then*

(i)

$$\hat{\int}_0^T (a\hat{G}+b\hat{H})\hat{\times}\hat{d}\frac{\tilde{W}^2\tilde{T}(\tilde{W}^2)}{\hat{T}} = a\hat{\int}_0^T \hat{G}\hat{\times}\hat{d}\frac{\tilde{W}^2\tilde{T}(\tilde{W}^2)}{\hat{T}} + b\hat{\int}_0^T \hat{H}\hat{\times}\hat{d}\frac{\tilde{W}^2\tilde{T}(\tilde{W}^2)}{\hat{T}}$$

for all a, $b\in\mathbb{R}$.

(ii)

$$\hat{E}\left(\hat{\int}_0^T \hat{G}\hat{\times}\hat{d}\frac{\tilde{W}^2\tilde{T}(\tilde{W}^2)}{\hat{T}}\right) = 0.$$

(iii)

$$\hat{E}\left(\hat{\int}_0^T \hat{G}\hat{\times}\hat{d}\frac{\tilde{W}^2\tilde{T}(\tilde{W}^2)}{\hat{T}}\hat{\times}\hat{\int}_0^T \hat{H}\hat{\times}\hat{d}\frac{\tilde{W}^2\tilde{T}(\tilde{W}^2)}{\hat{T}}\right) = \hat{E}\left(\int_0^T \frac{\hat{G}\hat{H}\tilde{T}^2}{\hat{T}}dt\right).$$

Definition 1.9.38. *For $\hat{G}\in\hat{\mathbb{L}}^{\hat{2}}(0,T)$ we set*

$$\hat{I}^2(t) := \hat{\int}_0^t \hat{G}\hat{\times}\hat{d}\frac{\tilde{W}^2\tilde{T}(\tilde{W}^2)}{\hat{T}}, \qquad 0\le t\le T,$$

the indefinite Itö iso-integral of the second kind of $\hat{G}(\cdot)$.

We note that

$$\hat{I}^2(0) = 0.$$

Definition 1.9.39. *Let $\hat{X}(\cdot)$ be a real-valued iso-stochastic process such that*

$$\hat{X}(r) = \hat{X}(s) + \hat{\int}_s^r \hat{F}\hat{\times}\hat{d}\hat{t} + \hat{\int}_s^r \hat{G}\hat{\times}\hat{d}\frac{\tilde{W}^2\tilde{T}(\tilde{W}^2)}{\hat{T}}$$

for some $\hat{F}\in\hat{\mathbb{L}}^1(0,T)$ and $\hat{G}\in\hat{\mathbb{L}}^{\hat{2}}(0,T)$, and all times $0\le s\le r\le t$. We say that $\hat{X}$ has the iso-stochastic differential of the second kind

$$d\hat{X} = \hat{F}\frac{\tilde{T}(t)-t\tilde{T}'(t)}{\tilde{T}(t)}dt + \hat{G}\tilde{T}d\frac{\tilde{W}^2\tilde{T}(\tilde{W}^2)}{\hat{T}}$$

for $0\le t\le T$.

Remark 1.9.40. *In fact, if*

$$\hat{X}(r) = \hat{X}(s) + \int_s^r \hat{F}_1 dt + \int_s^r \hat{G}_1 d\frac{\tilde{W}^2\tilde{T}(\tilde{W}^2)}{\hat{T}}$$

for some $\hat{F}_1\in\hat{\mathbb{L}}^1(0,T)$ and $\hat{G}_1\in\hat{\mathbb{L}}^{\hat{2}}(0,T)$, then

$$d\hat{X} = \hat{F}_1 dt + \hat{G}_1 d\frac{\tilde{W}^2\tilde{T}(\tilde{W}^2)}{\hat{T}}$$

is the iso-stochastic differential of the second kind of $\hat{X}$.

Exercise 1.9.41. *Let $\hat{X}$ and $\hat{Y}$ be real-valued iso-stochastic process. Prove that*

$$d(a\hat{X}+b\hat{Y}) = ad\hat{X}+bd\hat{Y}$$

for all a, $b \in \mathbb{R}$.

We have proved that

$$\hat{\int_0^t} \frac{\tilde{W}^2\tilde{T}(\tilde{W}^2)}{\tilde{T}} \hat{\times} \hat{d}\frac{\tilde{W}^2\tilde{T}(\tilde{W}^2)}{\hat{T}} = \frac{\left(\tilde{W}^2(t)\tilde{T}(\tilde{W}^2(t))\right)^2}{2\hat{T}} + \left(\lambda-\frac{1}{2}\right)t\frac{\hat{T}(\Omega)}{\hat{T}}$$

for all $\lambda \in [0,1]$.

When $\lambda = 0$ we have that

$$\hat{\int_0^t} \frac{\tilde{W}^2\tilde{T}(\tilde{W}^2)}{\tilde{T}} \hat{\times} \hat{d}\frac{\tilde{W}^2\tilde{T}(\tilde{W}^2)}{\hat{T}} = \frac{\left(\tilde{W}^2(t)\tilde{T}(\tilde{W}^2(t))\right)^2}{2\hat{T}} - \frac{1}{2}t\frac{\hat{T}(\Omega)}{\hat{T}},$$

whereupon

$$\frac{\left(\tilde{W}^2(t)\tilde{T}(\tilde{W}^2(t))\right)^2}{\hat{T}} = t\frac{\hat{T}(\Omega)}{\hat{T}} + \int_0^t \frac{1}{\tilde{T}}\frac{\tilde{W}^2\tilde{T}(\tilde{W}^2)}{\tilde{T}}\tilde{T}\tilde{T}d\frac{\tilde{W}^2\tilde{T}(\tilde{W}^2)}{\hat{T}},$$

or

$$\frac{\left(\tilde{W}^2(t)\tilde{T}(\tilde{W}^2(t))\right)^2}{\hat{T}} t\frac{\hat{T}(\Omega)}{\hat{T}} + \int_0^t \tilde{W}^2\tilde{T}(\tilde{W}^2)d\frac{\tilde{W}^2\tilde{T}(\tilde{W}^2)}{\hat{T}},$$

or

$$\frac{\left(\tilde{W}^2(t)\tilde{T}(\tilde{W}^2(t))\right)^2}{\hat{T}} = \frac{\hat{T}(\Omega)}{\hat{T}}\int_0^t ds + \int_0^t \tilde{W}^2\tilde{T}(\tilde{W}^2)d\frac{\tilde{W}^2\tilde{T}(\tilde{W}^2)}{\hat{T}}.$$

Therefore

$$d\frac{\left(\tilde{W}^2(t)\tilde{T}(\tilde{W}^2(t))\right)^2}{\hat{T}} = \frac{\hat{T}(\Omega)}{\hat{T}}dt + \tilde{W}^2\tilde{T}(\tilde{W}^2)d\frac{\tilde{W}^2\tilde{T}(\tilde{W}^2)}{\hat{T}}.$$

Remark 1.9.42. *We note that*

$$\hat{\int_0^t} \hat{G}\hat{\times}\hat{d}\frac{\tilde{W}^2\tilde{T}(\tilde{W}^2)}{\hat{T}} := \lim_{n\longrightarrow\infty}\sum_{k=0}^{m_n-1}\hat{G}(t_k^n)\tilde{T}(\tau_k^n)\frac{\tilde{W}^2(t_{k+1}^n)\tilde{T}(\tilde{W}^2(t_{k+1}^n)) - \tilde{W}^2(t_k^n)\tilde{T}(\tilde{W}^2(t_k^n))}{\hat{T}},$$

where

$$D^{m_n} := \{0 = t_0^n < t_1^n < \ldots t_{m_n-1}^n < t_{m_n}^n = t\}$$

with

$$|D^{m_n}| \longrightarrow_{n\longrightarrow\infty} 0.$$

Theorem 1.9.43. *We have*

$$d\left(t\frac{\tilde{W}^2\tilde{T}(\tilde{W}^2)}{\hat{T}}\right) = td\frac{\tilde{W}^2\tilde{T}(\tilde{W}^2)}{\hat{T}} + \frac{\tilde{W}^2\tilde{T}(\tilde{W}^2)}{\hat{T}}dt.$$

Proof. We note that

$$\hat{\int}_0^r 1\hat{\times}\hat{d}\left(t\frac{\tilde{W}^2\tilde{T}(\tilde{W}^2)}{\hat{T}}\right)$$

(30)

$$= \lim_{n\longrightarrow\infty}\sum_{k=0}^{m_n-1}\tilde{T}(\tau_k^n)\frac{t_{k+1}^n\tilde{W}^2(t_{k+1}^n)\tilde{T}(\tilde{W}^2(t_{k+1}^n))-t_k^n\tilde{W}^2(t_k^n)\tilde{T}(\tilde{W}^2(t_k^n))}{\hat{T}}.$$

Also,

$$\hat{\int}_0^r\frac{\tilde{W}^2\tilde{T}(\tilde{W}^2)}{\hat{T}}\hat{\times}\hat{d}t = \lim_{n\longrightarrow\infty}\sum_{k=0}^{m_n-1}\tilde{T}(\tau_k^n)\frac{\tilde{W}^2(t_{k+1}^n)\tilde{T}(\tilde{W}^2(t_{k+1}^n))}{\hat{T}}(t_{k+1}^n-t_k^n),$$

$$\hat{\int}_0^r t\hat{\times}\hat{d}\frac{\tilde{W}^2\tilde{T}(\tilde{W}^2)}{\hat{T}} = \lim_{n\longrightarrow\infty}\sum_{k=0}^{m_n-1}t_k^n\tilde{T}(\tau_k^n)\frac{\tilde{W}^2(t_{k+1}^n)\tilde{T}(\tilde{W}^2(t_{k+1}^n))-\tilde{W}^2(t_k^n)\tilde{T}(\tilde{W}^2(t_k^n))}{\hat{T}}.$$

Hence, using (30),

$$\hat{\int}_0^r\frac{\tilde{W}^2\tilde{T}(\tilde{W}^2)}{\hat{T}}\hat{\times}\hat{d}t+\hat{\int}_0^r t\hat{\times}\hat{d}\frac{\tilde{W}^2\tilde{T}(\tilde{W}^2)}{\hat{T}}$$

$$= \lim_{n\longrightarrow\infty}\sum_{k=0}^{m_n-1}\tilde{T}(\tau_k^n)\frac{\tilde{W}^2(t_{k+1}^n)\tilde{T}(\tilde{W}^2(t_{k+1}^n))}{\hat{T}}(t_{k+1}^n-t_k^n)$$

$$+\lim_{n\longrightarrow\infty}\sum_{k=0}^{m_n-1}t_k^n\tilde{T}(\tau_k^n)\frac{\tilde{W}^2(t_{k+1}^n)\tilde{T}(\tilde{W}^2(t_{k+1}^n))-\tilde{W}^2(t_k^n)\tilde{T}(\tilde{W}^2(t_k^n))}{\hat{T}}$$

$$= \lim_{n\longrightarrow\infty}\sum_{k=0}^{m_n-1}\tilde{T}(\tau_k^n)\frac{1}{\hat{T}}\Big(t_{k+1}^n\tilde{W}^2(t_{k+1}^n)\tilde{T}(\tilde{W}^2(t_{k+1}^n))$$

$$-t_k^n\tilde{W}^2(t_{k+1}^n)\tilde{T}(\tilde{W}^2(t_{k+1}^n))+t_k^n\tilde{W}^2(t_{k+1}^n)\tilde{T}(\tilde{W}^2(t_{k+1}^n))-t_k^n\tilde{W}^2(t_k^n)\tilde{T}(\tilde{W}^2(t_k^n))\Big)$$

$$= \lim_{n\longrightarrow\infty}\sum_{k=0}^{m_n-1}\tilde{T}(\tau_k^n)\frac{1}{\hat{T}}\left(t_{k+1}^n\tilde{W}^2(t_{k+1}^n)\tilde{T}(\tilde{W}^2(t_{k+1}^n))-t_k^n\tilde{W}^2(t_k^n)\tilde{T}(\tilde{W}^2(t_k^n))\right)$$

$$= \hat{\int}_0^r 1\hat{\times}\hat{d}\left(t\frac{\tilde{W}^2\tilde{T}(\tilde{W}^2)}{\hat{T}}\right),$$

i.e.,

$$\hat{\int}_0^r 1\hat{\times}\hat{d}\left(t\frac{\tilde{W}^2\tilde{T}(\tilde{W}^2)}{\hat{T}}\right) = \hat{\int}_0^r\frac{\tilde{W}^2\tilde{T}(\tilde{W}^2)}{\hat{T}}\hat{\times}\hat{d}t+\hat{\int}_0^r t\hat{\times}\hat{d}\frac{\tilde{W}^2\tilde{T}(\tilde{W}^2)}{\hat{T}},$$

or

$$\int_0^r\frac{1}{\tilde{T}}\tilde{T}\tilde{T}d\left(t\frac{\tilde{W}^2\tilde{T}(\tilde{W}^2)}{\hat{T}}\right) = \int_0^r\frac{1}{\tilde{T}}\frac{\tilde{W}^2\tilde{T}(\tilde{W}^2)}{\hat{T}}\tilde{T}\tilde{T}dt+\int_0^r\frac{1}{\tilde{T}}t\tilde{T}\tilde{T}d\frac{\tilde{W}^2\tilde{T}(\tilde{W}^2)}{\hat{T}},$$

or

$$\int_0^r\tilde{T}d\left(t\frac{\tilde{W}^2\tilde{T}(\tilde{W}^2)}{\hat{T}}\right) = \int_0^r\tilde{T}\frac{\tilde{W}^2\tilde{T}(\tilde{W}^2)}{\hat{T}}dt+\int_0^r t\tilde{T}d\frac{\tilde{W}^2\tilde{T}(\tilde{W}^2)}{\hat{T}}.$$

Therefore

$$\tilde{T}d\left(t\frac{\tilde{W}^2\tilde{T}(\tilde{W}^2)}{\hat{T}}\right) = \tilde{T}\frac{\tilde{W}^2\tilde{T}(\tilde{W}^2)}{\hat{T}}dt+t\tilde{T}d\frac{\tilde{W}^2\tilde{T}(\tilde{W}^2)}{\hat{T}}$$

or

$$d\left(t\frac{\tilde{W}^2\tilde{T}(\tilde{W}^2)}{\hat{T}}\right) = \frac{\tilde{W}^2\tilde{T}(\tilde{W}^2)}{\hat{T}}dt+td\frac{\tilde{W}^2\tilde{T}(\tilde{W}^2)}{\hat{T}},$$

which is the main conclusion of the theorem. □

Theorem 1.9.44. *Let $g \in \mathcal{C}^1([0,T])$. Then*

$$d\left(g(t)\frac{\tilde{W}^2\tilde{T}(\tilde{W}^2)}{\hat{T}}\right) = g(t)d\frac{\tilde{W}^2\tilde{T}(\tilde{W}^2)}{\hat{T}} + \frac{\tilde{W}^2\tilde{T}(\tilde{W}^2)}{\hat{T}}dg(t).$$

Remark 1.9.45. *We can rewrite the above result as follows*

$$d\left(g(t)\frac{\tilde{W}^2\tilde{T}(\tilde{W}^2)}{\hat{T}}\right) = g(t)d\frac{\tilde{W}^2\tilde{T}(\tilde{W}^2)}{\hat{T}} + \frac{\tilde{W}^2\tilde{T}(\tilde{W}^2)}{\hat{T}}g'(t)dt.$$

Proof. We have that

$$\hat{\int}_0^r 1 \hat{\times} \hat{d}\left(g(t)\frac{\tilde{W}^2\tilde{T}(\tilde{W}^2)}{\hat{T}}\right)$$

$$= \lim_{n\longrightarrow\infty}\sum_{k=0}^{m_n-1}\tilde{T}(\tau_k^n)\frac{g(t_{k+1}^n)\tilde{W}^2(t_{k+1}^n)\tilde{T}(\tilde{W}^2(t_{k+1}^n)) - g(t_k^n)\tilde{W}^2(t_k^n)\tilde{T}(\tilde{W}^2(t_k^n))}{\hat{T}}.$$

Also,

$$\hat{\int}_0^r \frac{\tilde{W}^2\tilde{T}(\tilde{W}^2)}{\hat{T}} \hat{\times} \hat{d}g(t) = \lim_{n\longrightarrow\infty}\sum_{k=0}^{m_n-1}\tilde{T}(\tau_k^n)\frac{\tilde{W}^2(t_{k+1}^n)\tilde{T}(\tilde{W}^2(t_{k+1}^n))}{\hat{T}}(g(t_{k+1}^n) - g(t_k^n)),$$

$$\hat{\int}_0^r g(t) \hat{\times} \hat{d}\frac{\tilde{W}^2\tilde{T}(\tilde{W}^2)}{\hat{T}} = \lim_{n\longrightarrow\infty}\sum_{k=0}^{m_n-1}g(t_k^n)\tilde{T}(\tau_k^n)\frac{\tilde{W}^2(t_{k+1}^n)\tilde{T}(\tilde{W}^2(t_{k+1}^n)) - \tilde{W}^2(t_k^n)\tilde{T}(\tilde{W}^2(t_k^n))}{\hat{T}}.$$

From here,

$$\hat{\int}_0^r \frac{\tilde{W}^2\tilde{T}(\tilde{W}^2)}{\hat{T}} \hat{\times} \hat{d}g(t) + \hat{\int}_0^r g(t) \hat{\times} \hat{d}\frac{\tilde{W}^2\tilde{T}(\tilde{W}^2)}{\hat{T}}$$

$$= \lim_{n\longrightarrow\infty}\sum_{k=0}^{m_n-1}\tilde{T}(\tau_k^n)\frac{\tilde{W}^2(t_{k+1}^n)\tilde{T}(\tilde{W}^2(t_{k+1}^n))}{\hat{T}}(g(t_{k+1}^n) - g(t_k^n))$$

$$+ \lim_{n\longrightarrow\infty}\sum_{k=0}^{m_n-1}g(t_k^n)\tilde{T}(\tau_k^n)\frac{\tilde{W}^2(t_{k+1}^n)\tilde{T}(\tilde{W}^2(t_{k+1}^n)) - \tilde{W}^2(t_k^n)\tilde{T}(\tilde{W}^2(t_k^n))}{\hat{T}}$$

$$= \lim_{n\longrightarrow\infty}\sum_{k=0}^{m_n-1}\tilde{T}(\tau_k^n)\frac{1}{\hat{T}}\Big(g(t_{k+1}^n)\tilde{W}^2(t_{k+1}^n)\tilde{T}(\tilde{W}^2(t_{k+1}^n))$$

$$-g(t_k^n)\tilde{W}^2(t_{k+1}^n)\tilde{T}(\tilde{W}^2(t_{k+1}^n)) + g(t_k^n)\tilde{W}^2(t_{k+1}^n)\tilde{T}(\tilde{W}^2(t_{k+1}^n)) - g(t_k^n)\tilde{W}^2(t_k^n)\tilde{T}(\tilde{W}^2(t_k^n))\Big)$$

$$= \lim_{n\longrightarrow\infty}\sum_{k=0}^{m_n-1}\tilde{T}(\tau_k^n)\frac{1}{\hat{T}}\left(g(t_{k+1}^n)\tilde{W}^2(t_{k+1}^n)\tilde{T}(\tilde{W}^2(t_{k+1}^n)) - g(t_k^n)\tilde{W}^2(t_k^n)\tilde{T}(\tilde{W}^2(t_k^n))\right)$$

$$= \hat{\int}_0^r 1 \hat{\times} \hat{d}\left(g(t)\frac{\tilde{W}^2\tilde{T}(\tilde{W}^2)}{\hat{T}}\right),$$

i.e.,

$$\hat{\int}_0^r 1 \hat{\times} \hat{d}\left(g(t)\frac{\tilde{W}^2\tilde{T}(\tilde{W}^2)}{\hat{T}}\right) = \hat{\int}_0^r \frac{\tilde{W}^2\tilde{T}(\tilde{W}^2)}{\hat{T}} \hat{\times} \hat{d}g(t) + \hat{\int}_0^r g(t) \hat{\times} \hat{d}\frac{\tilde{W}^2\tilde{T}(\tilde{W}^2)}{\hat{T}},$$

or

$$\int_0^r \frac{1}{\tilde{T}}\tilde{T}\tilde{T}d\left(g(t)\frac{\tilde{W}^2\tilde{T}(\tilde{W}^2)}{\hat{T}}\right) = \int_0^r \frac{1}{\tilde{T}}\frac{\tilde{W}^2\tilde{T}(\tilde{W}^2)}{\hat{T}}\tilde{T}\tilde{T}dg(t) + \int_0^r \frac{1}{\tilde{T}}g(t)\tilde{T}\tilde{T}d\frac{\tilde{W}^2\tilde{T}(\tilde{W}^2)}{\hat{T}},$$

or

$$\int_0^r \tilde{T}d\left(g(t)\frac{\tilde{W}^2\tilde{T}(\tilde{W}^2)}{\hat{T}}\right) = \int_0^r \tilde{T}\frac{\tilde{W}^2\tilde{T}(\tilde{W}^2)}{\hat{T}}dg(t) + \int_0^r g(t)\tilde{T}d\frac{\tilde{W}^2\tilde{T}(\tilde{W}^2)}{\hat{T}}.$$

Consequently

$$\tilde{T}d\left(g(t)\frac{\tilde{W}^2\tilde{T}(\tilde{W}^2)}{\hat{T}}\right) = \tilde{T}\frac{\tilde{W}^2\tilde{T}(\tilde{W}^2)}{\hat{T}}dg(t) + g(t)\tilde{T}d\frac{\tilde{W}^2\tilde{T}(\tilde{W}^2)}{\hat{T}}$$

or

$$d\left(g(t)\frac{\tilde{W}^2\tilde{T}(\tilde{W}^2)}{\hat{T}}\right) = \frac{\tilde{W}^2\tilde{T}(\tilde{W}^2)}{\hat{T}}dg(t) + g(t)d\frac{\tilde{W}^2\tilde{T}(\tilde{W}^2)}{\hat{T}}.$$

□

Theorem 1.9.46. *(iso-Itô product rule) Suppose*

$$d\hat{X}_1 = \hat{F}_1dt + \hat{G}_1d\frac{\tilde{W}^2\tilde{T}(\tilde{W}^2)}{\hat{T}},$$

$$d\hat{X}_2 = \hat{F}_2dt + \hat{G}_2d\frac{\tilde{W}^2\tilde{T}(\tilde{W}^2)}{\hat{T}},$$

for $\hat{F}_i \in \hat{\mathbb{L}}^1(0,T)$, $\hat{G}_i \in \hat{\mathbb{L}}^{\hat{2}}(0,T)$, $i = 1,2$. *Then*

$$d(\hat{X}_1\hat{X}_2) = \hat{X}_1d\hat{X}_2 + \hat{X}_2d\hat{X}_1 + \hat{G}_1\hat{G}_2\frac{\hat{T}(\Omega)}{\hat{T}}dt.$$

Proof. 1. case. Choose $0 \leq r \leq T$. First of all, we assume for simplicity that

$$\hat{X}_1(0) = \hat{X}_2(0) = 0,$$

and

$$\hat{F}_i(t) = \hat{F}_i, \qquad \hat{G}_i(t) = \hat{G}_i, \qquad i = 1,2,$$

where $\hat{F}_i$, $\hat{G}_i$ are time-independent $\mathcal{F}(0)$-measurable random iso-variables, $i = 1,2$. Thus

$$\hat{X}_1(t) = \hat{F}_1t + \hat{G}_1\frac{\tilde{W}^2\tilde{T}(\tilde{W}^2)}{\hat{T}},$$

$$\hat{X}_2(t) = \hat{F}_1t + \hat{G}_2\frac{\tilde{W}^2\tilde{T}(\tilde{W}^2)}{\hat{T}},$$

and

$$\hat{\int}_0^r \frac{1}{\hat{T}}\hat{\times}\hat{d}(\hat{X}_1\hat{X}_2) = \int_0^r \frac{1}{\hat{T}^2}\hat{T}\tilde{T}d(\hat{X}_1\hat{X}_2)$$

$$= \int_0^r d(\hat{X}_1\hat{X}_2)$$

$$= \hat{X}_1(r)\hat{X}_2(r)$$

$$= \left(\hat{F}_1r + \hat{G}_1\frac{\tilde{W}^2\tilde{T}(\tilde{W}^2)}{\hat{T}}\right)\left(\hat{F}_2r + \hat{G}_2\frac{\tilde{W}^2\tilde{T}(\tilde{W}^2)}{\hat{T}}\right)$$

$$= \hat{F}_1\hat{F}_2r + \hat{F}_1\hat{G}_2r\frac{\tilde{W}^2\tilde{T}(\tilde{W}^2)}{\hat{T}} + \hat{G}_1\hat{F}_2r\frac{\tilde{W}^2\tilde{T}(\tilde{W}^2)}{\hat{T}} + \hat{G}_1\hat{G}_2\frac{(\tilde{W}^2\tilde{T}(\tilde{W}^2))^2}{\hat{T}^2}$$

$$= \hat{F}_1\hat{F}_2r^2 + (\hat{F}_1\hat{G}_2 + \hat{F}_2\hat{G}_1)r\frac{\tilde{W}^2\tilde{T}(\tilde{W}^2)}{\hat{T}} + \hat{G}_1\hat{G}_2\frac{(\tilde{W}^2\tilde{T}(\tilde{W}^2))^2}{\hat{T}^2},$$

i.e.,

$$\hat{\int}_0^r \frac{1}{\tilde{T}} \hat{\times} \hat{d}(\hat{X}_1\hat{X}_2) = \hat{F}_1\hat{F}_2 r^2 + (\hat{F}_1\hat{G}_2 + \hat{F}_2\hat{G}_1) r \frac{\tilde{W}^2\tilde{T}(\tilde{W}^2)}{\hat{T}} + \hat{G}_1\hat{G}_2 \frac{(\tilde{W}^2\tilde{T}(\tilde{W}^2))^2}{\hat{T}^2}. \tag{31}$$

On the other hand,

$$\hat{\int}_0^r \frac{1}{\tilde{T}} \hat{\times} \left(\hat{X}_1\hat{d}\hat{X}_2 + \hat{X}_2\hat{d}\hat{X}_1 + \hat{G}_1\hat{G}_2 \frac{\hat{T}(\Omega)}{\hat{T}} \hat{d}t\right)$$

$$= \int_0^r \frac{1}{\tilde{T}} \frac{1}{\tilde{T}} \tilde{T} \left(\tilde{T}\hat{X}_1 d\hat{X}_2 + \tilde{T}\hat{X}_2 d\hat{X}_1 + \tilde{T}\hat{G}_1\hat{G}_2 \frac{\hat{T}(\Omega)}{\hat{T}} dt\right)$$

$$= \int_0^r \left(\hat{X}_1 d\hat{X}_2 + \hat{X}_2 d\hat{X}_1 + \hat{G}_1\hat{G}_2 \frac{\hat{T}(\Omega)}{\hat{T}} dt\right)$$

$$= \int_0^r \left(\left(\hat{F}_1 t + \hat{G}_1 \frac{\tilde{W}^2\tilde{T}(\tilde{W}^2)}{\hat{T}}\right)\left(\hat{F}_2 dt + \hat{G}_2 d\frac{\tilde{W}^2\tilde{T}(\tilde{W}^2)}{\hat{T}}\right)\right.$$

$$\left. + \left(\hat{F}_2 t + \hat{G}_2 \frac{\tilde{W}^2\tilde{T}(\tilde{W}^2)}{\hat{T}}\right)\left(\hat{F}_1 dt + \hat{G}_1 d\frac{\tilde{W}^2\tilde{T}(\tilde{W}^2)}{\hat{T}}\right) + \hat{G}_1\hat{G}_2 \frac{\hat{T}(\Omega)}{\hat{T}} dt\right)$$

$$= \int_0^r \left(\hat{F}_1\hat{F}_2 t dt + \hat{F}_1\hat{G}_2 t d\frac{\tilde{W}^2\tilde{T}(\tilde{W}^2)}{\hat{T}} + \hat{G}_1\hat{F}_2 \frac{\tilde{W}^2\tilde{T}(\tilde{W}^2)}{\hat{T}} dt\right.$$

$$+ \hat{G}_1\hat{G}_2 \frac{\tilde{W}^2\tilde{T}(\tilde{W}^2)}{\hat{T}} d\frac{\tilde{W}^2\tilde{T}(\tilde{W}^2)}{\hat{T}} + \hat{F}_1\hat{F}_2 t dt + \hat{F}_2\hat{G}_1 t d\frac{\tilde{W}^2\tilde{T}(\tilde{W}^2)}{\hat{T}}$$

$$\left. + \hat{G}_2\hat{F}_1 \frac{\tilde{W}^2\tilde{T}(\tilde{W}^2)}{\hat{T}} dt + \hat{G}_1\hat{G}_2 \frac{\tilde{W}^2\tilde{T}(\tilde{W}^2)}{\hat{T}} d\frac{\tilde{W}^2\tilde{T}(\tilde{W}^2)}{\hat{T}} + \hat{G}_1\hat{G}_2 \frac{\hat{T}(\Omega)}{\hat{T}} dt\right)$$

$$= \int_0^r \left(2\hat{F}_1\hat{F}_2 t dt + (\hat{F}_1\hat{G}_2 + \hat{F}_2\hat{G}_1) t d\frac{\tilde{W}^2\tilde{T}(\tilde{W}^2)}{\hat{T}} + (\hat{G}_1\hat{F}_2 + \hat{G}_2\hat{F}_1) \frac{\tilde{W}^2\tilde{T}(\tilde{W}^2)}{\hat{T}} dt\right.$$

$$\left. + 2\hat{G}_1\hat{G}_2 \frac{\tilde{W}^2\tilde{T}(\tilde{W}^2)}{\hat{T}} d\frac{\tilde{W}^2\tilde{T}(\tilde{W}^2)}{\hat{T}} + \hat{G}_1\hat{G}_2 \frac{\hat{T}(\Omega)}{\hat{T}} dt\right)$$

$$= 2\hat{F}_1\hat{F}_2 \int_0^r t dt + (\hat{F}_1\hat{G}_2 + \hat{F}_2\hat{G}_1) \int_0^r \left(t d\frac{\tilde{W}^2\tilde{T}(\tilde{W}^2)}{\hat{T}} + \frac{\tilde{W}^2\tilde{T}(\tilde{W}^2)}{\hat{T}} dt\right)$$

$$+ 2\hat{G}_1\hat{G}_2 \int_0^r \frac{\tilde{W}^2\tilde{T}(\tilde{W}^2)}{\hat{T}} d\frac{\tilde{W}^2\tilde{T}(\tilde{W}^2)}{\hat{T}} + \hat{G}_1\hat{G}_2 \frac{\hat{T}(\Omega)}{\hat{T}} \int_0^r dt$$

$$= \hat{F}_1\hat{F}_2 r^2 + (\hat{F}_1\hat{G}_2 + \hat{F}_2\hat{G}_1) \int_0^r d\left(t \frac{\tilde{W}^2\tilde{T}(\tilde{W}^2)}{\hat{T}}\right)$$

$$+ \hat{G}_1\hat{G}_2 \frac{(\tilde{W}^2\tilde{T}(\tilde{W}^2)(r))^2}{\hat{T}^2} - \hat{G}_1\hat{G}_2 r \frac{\hat{T}(\Omega)}{\hat{T}} + \hat{G}_1\hat{G}_2 \frac{\hat{T}(\Omega)}{\hat{T}} r$$

$$= \hat{F}_1\hat{F}_2 r^2 + (\hat{F}_1\hat{G}_2 + \hat{F}_2\hat{G}_1) r \frac{\tilde{W}^2\tilde{T}(\tilde{W}^2)}{\hat{T}} + \hat{G}_1\hat{G}_2 \frac{(\tilde{W}^2\tilde{T}(\tilde{W}^2)(r))^2}{\hat{T}^2}.$$

From the last equality and from (31) it follows that

$$\hat{\int}_0^r \frac{1}{\tilde{T}} \hat{\times} \hat{d}(\hat{X}_1\hat{X}_2) = \hat{\int}_0^r \frac{1}{\tilde{T}} \hat{\times} \left(\hat{X}_1\hat{d}\hat{X}_2 + \hat{X}_2\hat{d}\hat{X}_1 + \hat{G}_1\hat{G}_2 \frac{\hat{T}(\Omega)}{\hat{T}} \hat{d}t\right).$$

2. case. If $\hat{F}_i$, $\hat{G}_i$ are step processes, we apply the step 1 in each subinterval $[t_k, t_{k+1})$ on which $\hat{F}_i$ and $\hat{G}_i$ are constant random iso-variables.

3. case. In the general case, we select step processes $\hat{F}_i^m \in \hat{\mathbb{L}}^1(0,T)$, $\hat{G}_i^m \in \hat{\mathbb{L}}^{\hat{2}}(0,T)$ with

$$\hat{E}\left(\left|\int_0^t (\hat{F}_i - \hat{F}_i^m)\frac{\tilde{T}(t)-t\tilde{T}'(t)}{\tilde{T}(t)}dt\right|\right) \longrightarrow_{m\longrightarrow\infty} 0,$$

$$\hat{E}\left(\left|\int_0^t (\hat{G}_i - \hat{G}_i^m)^2 \hat{T}\frac{\tilde{T}(t)-t\tilde{T}'(t)}{\tilde{T}(t)}dt\right|\right) \longrightarrow_{m\longrightarrow\infty} 0.$$

We define

$$\hat{X}_i^n := \hat{X}_i(0) + \int_0^t \hat{F}_i^n ds + \int_0^t \hat{G}_i^n d\frac{\tilde{W}^2\tilde{T}(\tilde{W}^2)}{\hat{T}}, \qquad i = 1,2.$$

We apply the step 2 to $\hat{X}_i^n$ in (s,r) and pass to limits to obtain the conclusion of the theorem.

□

Corollary 1.9.47. *Let*

$$d\hat{X} = \hat{F}dt + \hat{G}d\frac{\tilde{W}^2\tilde{T}(\tilde{W}^2)}{\hat{T}},$$

where $\hat{F} \in \hat{\mathbb{L}}^1(0,T)$, $\hat{G} \in \hat{\mathbb{L}}^{\hat{2}}(0,T)$. *Then*

$$d\hat{X}^2 = 2\hat{X}d\hat{X} + \hat{G}^2\frac{\hat{T}(\Omega)}{\hat{T}}dt.$$

Proof. Let

$$\hat{X}_1 = \hat{X}, \qquad \hat{X}_2 = \hat{X}.$$

Then

$$\hat{G}_1 = \hat{G}_2 = \hat{G}.$$

Hence,

$$d\hat{X}^2 = d(\hat{X}\hat{X})$$

$$= \hat{X}d\hat{X} + \hat{X}d\hat{X} + \hat{G}^2\tfrac{\hat{T}(\Omega)}{\hat{T}}dt$$

$$= 2\hat{X}d\hat{X} + \hat{G}^2\tfrac{\hat{T}(\Omega)}{\hat{T}}dt.$$

□

Corollary 1.9.48. *Let*

$$d\hat{X} = \hat{F}dt + \hat{G}d\frac{\tilde{W}^2\tilde{T}(\tilde{W}^2)}{\hat{T}},$$

where $\hat{F} \in \hat{\mathbb{L}}^1(0,T)$, $\hat{G} \in \hat{\mathbb{L}}^{\hat{2}}(0,T)$. *Then*

$$d\hat{X}^3 = 3\hat{X}^2d\hat{X} + 3\hat{X}\hat{G}^2\frac{\hat{T}(\Omega)}{\hat{T}}dt.$$

Proof. Let

$$\hat{X}_1 = \hat{X}^2, \qquad \hat{X}_2 = \hat{X}.$$

Then

$$\begin{aligned} d\hat{X}_1 &= d\hat{X}^2 \\ &= 2\hat{X}d\hat{X} + \hat{G}^2\tfrac{\hat{T}(\Omega)}{\hat{T}}dt \\ &= 2\hat{X}\left(\hat{F}dt + \hat{G}d\tfrac{\tilde{W}^2\tilde{T}(\tilde{W}^2)}{\hat{T}}\right) + \hat{G}^2\tfrac{\hat{T}(\Omega)}{\hat{T}}dt \\ &= 2\hat{X}\hat{F}dt + 2\hat{X}\hat{G}d\tfrac{\tilde{W}^2\tilde{T}(\tilde{W}^2)}{\hat{T}} + \hat{G}^2\tfrac{\hat{T}(\Omega)}{\hat{T}}dt \\ &= \left(2\hat{X}\hat{F} + \hat{G}^2\tfrac{\hat{T}(\Omega)}{\hat{T}}\right)dt + 2\hat{X}\hat{G}d\tfrac{\tilde{W}^2\tilde{T}(\tilde{W}^2)}{\hat{T}}. \end{aligned}$$

Therefore

$$\hat{G}_1 = 2\hat{X}\hat{G}, \qquad \hat{G}_2 = \hat{G}.$$

Hence,

$$\begin{aligned} d\hat{X}^3 &= d(\hat{X}^2\hat{X}) \\ &= d(\hat{X}_1\hat{X}_2) \\ &= \hat{X}_2 d\hat{X}_1 + \hat{X}_1 d\hat{X}_2 + \hat{G}_1\hat{G}_2\tfrac{\hat{T}(\Omega)}{\hat{T}}dt \\ &= \hat{X}d\hat{X}^2 + \hat{X}^2 d\hat{X} + 2\hat{X}\hat{G}\hat{G}\tfrac{\hat{T}(\Omega)}{\hat{T}}dt \\ &= \hat{X}\left(2\hat{X}d\hat{X} + \hat{G}^2\tfrac{\hat{T}(\Omega)}{\hat{T}}dt\right) + \hat{X}^2 d\hat{X} + 2\hat{X}\hat{G}^2\tfrac{\hat{T}(\Omega)}{\hat{T}}dt \\ &= 2\hat{X}^2 d\hat{X} + \hat{X}\hat{G}^2\tfrac{\hat{T}(\Omega)}{\hat{T}}dt + \hat{X}^2 d\hat{X} + 2\hat{X}\hat{G}^2\tfrac{\hat{T}(\Omega)}{\hat{T}}dt \\ &= 3\hat{X}^2 d\hat{X} + 3\hat{X}\hat{G}^2\tfrac{\hat{T}(\Omega)}{\hat{T}}dt. \end{aligned}$$

□

Corollary 1.9.49. *Let*

$$d\hat{X} = \hat{F}dt + \hat{G}d\frac{\tilde{W}^2\tilde{T}(\tilde{W}^2)}{\hat{T}},$$

where $\hat{F} \in \hat{\mathbb{L}}^1(0,T)$, $\hat{G} \in \hat{\mathbb{L}}^2(0,T)$. *Then*

$$d\hat{X}^m = m\hat{X}^{m-1}d\hat{X} + \frac{1}{2}m(m-1)\hat{X}^{m-2}\hat{G}^2\frac{\hat{T}(\Omega)}{\hat{T}}dt, \qquad m \in \mathbb{N}.$$

Proof. We suppose that for some $m \in \mathbb{N}$ we have

$$d\hat{X}^m = m\hat{X}^{m-1}d\hat{X} + \frac{1}{2}m(m-1)\hat{X}^{m-2}\hat{G}^2\frac{\hat{T}(\Omega)}{\hat{T}}dt.$$

We will prove that

$$d\hat{X}^{m+1} = (m+1)\hat{X}^m d\hat{X} + \frac{1}{2}(m+1)m\hat{X}^{m-1}\hat{G}^2\frac{\hat{T}(\Omega)}{\hat{T}}dt.$$

Really, let

$$\hat{X}^m = \hat{X}_1, \qquad \hat{X} = \hat{X}_2.$$

Then

$$\begin{aligned}
d\hat{X}_1^m &= d\hat{X}^m \\
&= m\hat{X}^{m-1}d\hat{X} + \tfrac{1}{2}m(m-1)\hat{X}^{m-2}\hat{G}^2\tfrac{\hat{T}(\Omega)}{\hat{T}}dt \\
&= m\hat{X}^{m-1}\left(\hat{F}dt + \hat{G}d\tfrac{\tilde{W}^2\tilde{T}(\tilde{W}^2)}{\hat{T}}\right) + \tfrac{1}{2}m(m-1)\hat{X}^{m-2}\hat{G}^2\tfrac{\hat{T}(\Omega)}{\hat{T}}dt \\
&= m\hat{X}^{m-1}\hat{F}dt + m\hat{X}^{m-1}\hat{G}d\tfrac{\tilde{W}^2\tilde{T}(\tilde{W}^2)}{\hat{T}} + \tfrac{1}{2}m(m-1)\hat{X}^{m-2}\hat{G}^2\tfrac{\hat{T}(\Omega)}{\hat{T}}dt \\
&= \left(m\hat{X}^{m-1}\hat{F} + \tfrac{1}{2}m(m-1)\hat{X}^{m-2}\hat{G}^2\tfrac{\hat{T}(\Omega)}{\hat{T}}\right)dt + m\hat{X}^{m-1}\hat{G}d\tfrac{\tilde{W}^2\tilde{T}(\tilde{W}^2)}{\hat{T}}.
\end{aligned}$$

Therefore

$$\hat{G}_1 = m\hat{X}^{m-1}\hat{G}, \qquad \hat{G}_2 = \hat{G}.$$

Hence,

$$\begin{aligned}
d\hat{X}^{m+1} &= d(\hat{X}^m\hat{X}) \\
&= d(\hat{X}_1\hat{X}_2) \\
&= \hat{X}_2 d\hat{X}_1 + \hat{X}_1 d\hat{X}_2 + \hat{G}_1\hat{G}_2\tfrac{\hat{T}(\Omega)}{\hat{T}}dt \\
&= \hat{X}d\hat{X}^m + \hat{X}^m d\hat{X} + m\hat{X}^{m-1}\hat{G}\hat{G}\tfrac{\hat{T}(\Omega)}{\hat{T}}dt \\
&= \hat{X}\left(m\hat{X}^{m-1}d\hat{X} + \tfrac{1}{2}m(m-1)\hat{X}^{m-2}\hat{G}^2\tfrac{\hat{T}(\Omega)}{\hat{T}}dt\right) + m\hat{X}^{m-1}\hat{G}^2\tfrac{\hat{T}(\Omega)}{\hat{T}}dt \\
&= m\hat{X}^m d\hat{X} + \tfrac{1}{2}m(m-1)\hat{X}^{m-1}\hat{G}^2\tfrac{\hat{T}(\Omega)}{\hat{T}}dt + m\hat{X}^{m-1}\hat{G}^2\tfrac{\hat{T}(\Omega)}{\hat{T}}dt \\
&= m\hat{X}^m d\hat{X} + m\hat{X}^{m-1}\hat{G}^2\left(\tfrac{m-1}{2}+1\right)\tfrac{\hat{T}(\Omega)}{\hat{T}}dt \\
&= m\hat{X}^m d\hat{X} + \tfrac{1}{2}m(m+1)\hat{G}^2\tfrac{\hat{T}(\Omega)}{\hat{T}}dt.
\end{aligned}$$

□

Corollary 1.9.50. *Let f be a polynomial,*

$$d\hat{X} = \hat{F}dt + \hat{G}d\frac{\tilde{W}^2\tilde{T}(\tilde{W}^2)}{\hat{T}},$$

where $\hat{F} \in \hat{\mathbb{L}}^1(0,T)$, $\hat{G} \in \hat{\mathbb{L}}^{\hat{2}}(0,T)$. *Then*

$$df(\hat{X}) = f'(\hat{X})d\hat{X} + \frac{1}{2}f''(\hat{X})\hat{G}^2\frac{\hat{T}(\Omega)}{\hat{T}}dt.$$

Proof. Let

$$f(\hat{X}) = b_0\hat{X}^m + b_1\hat{X}^{m-1} + \cdots + b_{m-1}\hat{X} + b_m, \qquad b_i \in \mathbb{R}, i = 0,1,2,\ldots,m.$$

Then

$$df(\hat{X}) = d\left(b_0\hat{X}^m + b_1\hat{X}^{m-1} + \cdots + b_{m-1}\hat{X} + b_m\right)$$

$$= b_0 d\hat{X}^m + b_1 d\hat{X}^{m-1} + \cdots + b_{m_1}d\hat{X}$$

$$= b_0 m\hat{X}^{m-1}d\hat{X} + b_0\tfrac{1}{2}m(m-1)\hat{X}^{m-2}\hat{G}^2\tfrac{\hat{T}(\Omega)}{\hat{T}}dt$$

$$+ b_1(m-1)\hat{X}^{m-2}d\hat{X} + b_1\tfrac{1}{2}(m-1)(m-2)\hat{X}^{m-3}\hat{G}^2\tfrac{\hat{T}(\Omega)}{\hat{T}}dt$$

$$+\cdots$$

$$+ b_{m-1}d\hat{X}$$

$$= \left(b_0 m\hat{X}^{m-1} + b_1(m-1)\hat{X}^{m-2} + \cdots + b_{m-1}\right)d\hat{X}$$

$$+ \tfrac{1}{2}\left(b_0 m(m-1)\hat{X}^{m-2} + b_1(m-1)(m-2)\hat{X}^{m-3} + \cdots + 2b_{m-2}\right)\hat{G}^2\tfrac{\hat{T}(\Omega)}{\hat{T}}dt$$

$$= f'(\hat{X})d\hat{X} + \tfrac{1}{2}f''(\hat{X})\hat{G}^2\tfrac{\hat{T}(\Omega)}{\hat{T}}dt.$$

□

Theorem 1.9.51. *(iso-Itô's formula for iso-stochastic differential of the second kind) Suppose that* $\hat{X}(\cdot)$ *has an iso-stochastic differential of the second kind*

$$d\hat{X} = \hat{F}dt + \hat{G}d\frac{\tilde{W}^2\tilde{T}(\tilde{W}^2)}{\hat{T}},$$

for $\hat{F} \in \hat{\mathbb{L}}^1(0,T)$, $\hat{G} \in \hat{\mathbb{L}}^2(0,T)$. *Assume* $u: \mathbb{R} \times [0,T] \longrightarrow \mathbb{R}$ *is continuous and*

$$\frac{\partial u}{\partial t}, \qquad \frac{\partial u}{\partial x}, \qquad \frac{\partial^2 u}{\partial x^2}$$

exist and are continuous.

We set

$$\hat{Y}(t) := u(\hat{X}(t),t).$$

Then

$$(32) \qquad \begin{aligned} d\hat{Y} &= \tfrac{\partial u}{\partial t}dt + \tfrac{\partial u}{\partial x}d\hat{X} + \tfrac{1}{2}\tfrac{\partial^2 u}{\partial x^2}\hat{G}^2\tfrac{\hat{T}(\Omega)}{\hat{T}}dt \\ &= \left(\tfrac{\partial u}{\partial t} + \tfrac{\partial u}{\partial x}\hat{F} + \tfrac{1}{2}\tfrac{\partial^2 u}{\partial x^2}\hat{G}^2\tfrac{\hat{T}(\Omega)}{\hat{T}}\right)dt + \tfrac{\partial u}{\partial x}\hat{G}d\tfrac{\tilde{W}^2\tilde{T}(\tilde{W}^2)}{\hat{T}}. \end{aligned}$$

Definition 1.9.52. *The formula* (32) *will be called iso-Itô's formula for iso-stochastic differential of the second kind.*

Proof. Suppose that

$$u(x,t) = f(x)g(t),$$

where f and g are polynomials. Then

$$d(u(\hat{X},t)) = d(f(\hat{X})g(t))$$

$$= g(t)df(\hat{X}) + f(\hat{X})dg(t)$$

$$= f(\hat{X})g'(t)dt + g(t)\left(f'(\hat{X})d\hat{X} + \tfrac{1}{2}f''(\hat{X})\hat{G}^2\tfrac{\hat{T}(\Omega)}{\hat{T}}dt\right)$$

$$= \tfrac{\partial u}{\partial t}dt + \tfrac{\partial u}{\partial x}d\hat{X} + \tfrac{1}{2}\tfrac{\partial^2 u}{\partial x^2}\hat{G}^2\tfrac{\hat{T}(\Omega)}{\hat{T}}dt$$

$$= \tfrac{\partial u}{\partial t}dt + \tfrac{\partial u}{\partial x}\left(\hat{F}dt + \hat{G}d\tfrac{\tilde{W}^2\tilde{T}(\tilde{W}^2)}{\hat{T}}\right) + \tfrac{1}{2}\tfrac{\partial^2 u}{\partial x^2}\hat{G}^2\tfrac{\hat{T}(\Omega)}{\hat{T}}dt$$

$$= \tfrac{\partial u}{\partial t}dt + \tfrac{\partial u}{\partial x}\hat{F}dt + \tfrac{\partial u}{\partial x}\hat{G}d\tfrac{\tilde{W}^2\tilde{T}(\tilde{W}^2)}{\hat{T}} + \tfrac{1}{2}\tfrac{\partial^2 u}{\partial x^2}\hat{G}^2\tfrac{\hat{T}(\Omega)}{\hat{T}}dt$$

$$= \left(\tfrac{\partial u}{\partial t} + \tfrac{\partial u}{\partial x}\hat{F} + \tfrac{1}{2}\tfrac{\partial^2 u}{\partial x^2}\hat{G}^2\tfrac{\hat{T}(\Omega)}{\hat{T}}\right)dt + \tfrac{\partial u}{\partial x}\hat{G}d\tfrac{\tilde{W}^2\tilde{T}(\tilde{W}^2)}{\hat{T}}.$$

Thus the iso-Itô's formula for iso-stochastic differential of the second kind is true as well for any function u having the form

$$u(x,t) = \sum_{i=1}^{m} f^i(x)g^i(t),$$

where f^i and g^i are polynomials.

Given u as in the iso-Itô's formula for iso-stochastic differential of the second kind, there exists a sequence of polynomials u^n such that

$$u^n \longrightarrow u, \quad \frac{\partial u^n}{\partial t} \longrightarrow \frac{\partial u}{\partial t}, \quad \frac{\partial u^n}{\partial x} \longrightarrow \frac{\partial u}{\partial x}, \quad \frac{\partial^2 u^n}{\partial x^2} \longrightarrow \frac{\partial^2 u}{\partial x^2}$$

uniformly on compact subsets of $\mathbb{R} \times [0,T]$. As in above, we have that for all $0 \le r \le T$

$$u^n(\hat{X}(r),r) = u^n(\hat{X}(0),0)$$

$$+ \int_0^r \left(\tfrac{\partial u^n}{\partial t} + \tfrac{\partial u^n}{\partial x}\hat{F} + \tfrac{1}{2}\tfrac{\partial^2 u^n}{\partial x^2}\hat{G}^2\tfrac{\hat{T}(\Omega)}{\hat{T}}\right)dt + \int_0^r \tfrac{\partial u^n}{\partial x}\hat{G}d\tfrac{\tilde{W}^2\tilde{T}(\tilde{W}^2)}{\hat{T}}$$

almost surely.

We may pass to limits as $n \longrightarrow \infty$ in this expression, thereby proving the iso-Itô's formula for iso-stochastic differential of the second kind. □

Example 1.9.53. *Let*

$$\hat{X}=\sqrt{\tfrac{1}{\hat{T}(\Omega)\hat{T}}}\tilde{W}^2\tilde{T}(\tilde{W}^2),$$

$$u(x,t)=e^{x-\frac{t}{2}},\qquad \hat{Y}=u(\hat{X},t).$$

Then

$$d\hat{X}=\sqrt{\frac{\hat{T}}{\hat{T}(\Omega)}}d\frac{\tilde{W}^2\tilde{T}(\tilde{W}^2)}{\hat{T}},$$

whereupon

$$\hat{F}=0,\qquad \hat{G}=\sqrt{\frac{\hat{T}}{\hat{T}(\Omega)}}.$$

We have

$$\frac{\partial u}{\partial x}=e^{x-\frac{t}{2}},\qquad \frac{\partial^2 u}{\partial x^2}=e^{x-\frac{t}{2}},\qquad \frac{\partial u}{\partial t}=-\frac{1}{2}e^{x-\frac{t}{2}}.$$

Therefore

$$\begin{aligned}d\hat{Y}&=-\tfrac{1}{2}\hat{Y}dt+\hat{Y}d\hat{X}+\tfrac{1}{2}\hat{Y}\tfrac{\hat{T}}{\hat{T}(\Omega)}\tfrac{\hat{T}(\Omega)}{\hat{T}}dt\\&=\hat{Y}d\hat{X}\\&=\hat{Y}\sqrt{\tfrac{\hat{T}}{\hat{T}(\Omega)}}d\tfrac{\tilde{W}^2\tilde{T}(\tilde{W}^2)}{\hat{T}},\end{aligned}$$

i.e.,

$$d\hat{Y}=\hat{Y}\sqrt{\frac{\hat{T}}{\hat{T}(\Omega)}}d\frac{\tilde{W}^2\tilde{T}(\tilde{W}^2)}{\hat{T}},$$

which is an iso-stochastic differential equation.

Definition 1.9.54. *Let $\hat{G}\in\hat{\mathbb{L}}^{\hat{2}}(0,T)$ be a step process,*

$$\hat{G}(t)=G_k,\qquad t_k\le t<t_{k+1},\qquad k=0,1,2,\ldots,m-1,$$

$$G_k=\text{const},\qquad 0=t_0<t_1<t_2<\ldots<t_{m_1}<t_m=T.$$

Then

$$\hat{\int}_0^T\hat{G}\hat{\times}\hat{d}\frac{\frac{\tilde{W}^3}{\tilde{T}(\tilde{W}^3)}}{\hat{T}}:=\sum_{k=0}^{m-1}G_k\tilde{T}(\tau_k)\frac{\frac{\tilde{W}^3(t_{k+1})}{\tilde{T}(\tilde{W}^3(t_{k+1}))}-\frac{\tilde{W}^3(t_k)}{\tilde{T}(\tilde{W}^3(t_k))}}{\hat{T}}$$

is called the Itô iso-stochastic iso-integral of the third kind. Here $\tau_k=(1-\lambda)t_k+\lambda t_{k+1}$, $\lambda\in[0,1]$, $k=0,1,2,\ldots,m-1$.

Theorem 1.9.55. *Let $\hat{G}$, $\hat{H}\in\hat{\mathbb{L}}^{\hat{2}}(0,T)$ are step processes. Then for all a, $b\in\mathbb{R}$ we have*

$$\hat{\int}_0^T(a\hat{G}+b\hat{H})\hat{\times}\hat{d}\frac{\frac{\tilde{W}^3}{\tilde{T}(\tilde{W}^3)}}{\hat{T}}=a\hat{\int}_0^T\hat{G}\hat{\times}\hat{d}\frac{\frac{\tilde{W}^3}{\tilde{T}(\tilde{W}^3)}}{\hat{T}}+b\hat{\int}_0^T\hat{H}\hat{\times}\hat{d}\frac{\frac{\tilde{W}^3}{\tilde{T}(\tilde{W}^3)}}{\hat{T}}.$$

Proof. Let

$$D = \{0 = t_0 < t_1 < t_2 < \ldots < t_{m_1} < t_m\}$$

be a partition of the interval $[0,T]$ so that

$$\hat{G}(t) = G_k, \qquad \hat{H}(t) = H_k, \qquad t_k \le t < t_{k+1}, \qquad k = 0,1,2,\ldots,m-1.$$

Then

$$\hat{\int}_0^T \hat{G}\hat{\times}d\frac{\frac{\tilde{W}^3}{\tilde{T}(\tilde{W}^3)}}{\hat{T}} = \sum_{k=0}^{m-1} G_k\tilde{T}(\tau_k)\frac{\frac{\tilde{W}^3(t_{k+1})}{\tilde{T}(\tilde{W}^3(t_{k+1}))}-\frac{\tilde{W}^3(t_k)}{\tilde{T}(\tilde{W}^3(t_k))}}{\hat{T}},$$

$$\hat{\int}_0^T \hat{H}\hat{\times}d\frac{\frac{\tilde{W}^3}{\tilde{T}(\tilde{W}^3)}}{\hat{T}} = \sum_{k=0}^{m-1} H_k\tilde{T}(\tau_k)\frac{\frac{\tilde{W}^3(t_{k+1})}{\tilde{T}(\tilde{W}^3(t_{k+1}))}-\frac{\tilde{W}^3(t_k)}{\tilde{T}(\tilde{W}^3(t_k))}}{\hat{T}},$$

$$a\hat{G}(t) + b\hat{H}(t) = aG_k + bH_k, \qquad t_k \le t < t_{k+1}, \qquad k = 0,1,2,\ldots,m-1.$$

Hence,

$$\hat{\int}_0^T (a\hat{G}+b\hat{H})\hat{\times}\hat{d}\frac{\frac{\tilde{W}^3}{\tilde{T}(\tilde{W}^3)}}{\hat{T}} = \sum_{k=0}^{m-1}(aG_k+bH_k)\tilde{T}(\tau_k)\frac{\frac{\tilde{W}^3(t_{k+1})}{\tilde{T}(\tilde{W}^3(t_{k+1}))}-\frac{\tilde{W}^3(t_k)}{\tilde{T}(\tilde{W}^3(t_k))}}{\hat{T}}$$

$$= \sum_{k=0}^{m-1} aG_k\tilde{T}(\tau_k)\frac{\frac{\tilde{W}^3(t_{k+1})}{\tilde{T}(\tilde{W}^3(t_{k+1}))}-\frac{\tilde{W}^3(t_k)}{\tilde{T}(\tilde{W}^3(t_k))}}{\hat{T}}$$

$$+ \sum_{k=0}^{m-1} bH_k\tilde{T}(\tau_k)\frac{\frac{\tilde{W}^3(t_{k+1})}{\tilde{T}(\tilde{W}^3(t_{k+1}))}-\frac{\tilde{W}^3(t_k)}{\tilde{T}(\tilde{W}^3(t_k))}}{\hat{T}}$$

$$= a\sum_{k=0}^{m-1} G_k\tilde{T}(\tau_k)\frac{\frac{\tilde{W}^3(t_{k+1})}{\tilde{T}(\tilde{W}^3(t_{k+1}))}-\frac{\tilde{W}^3(t_k)}{\tilde{T}(\tilde{W}^3(t_k))}}{\hat{T}}$$

$$+ b\sum_{k=0}^{m-1} H_k\tilde{T}(\tau_k)\frac{\frac{\tilde{W}^3(t_{k+1})}{\tilde{T}(\tilde{W}^3(t_{k+1}))}-\frac{\tilde{W}^3(t_k)}{\tilde{T}(\tilde{W}^3(t_k))}}{\hat{T}}$$

$$= a\hat{\int}_0^T \hat{G}\hat{\times}\hat{d}\frac{\frac{\tilde{W}^3}{\tilde{T}(\tilde{W}^3)}}{\hat{T}} + b\hat{\int}_0^T \hat{H}\hat{\times}\hat{d}\frac{\frac{\tilde{W}^3}{\tilde{T}(\tilde{W}^3)}}{\hat{T}}.$$

□

Theorem 1.9.56. *Let $\hat{G} \in \hat{\mathbb{L}}^{\hat{2}}(0,T)$ be a step process. Then*

$$\hat{E}\left(\hat{\int}_0^T \hat{G}\hat{\times}\hat{d}\frac{\frac{\tilde{W}^3}{\tilde{T}(\tilde{W}^3)}}{\hat{T}}\right) = 0.$$

Proof. First of all, we will note that

$$\hat{E}\left(\frac{\frac{\tilde{W}^3(t_{k+1})}{\tilde{T}(\tilde{W}^3(t_{k+1}))}-\frac{\tilde{W}^3(t_k)}{\tilde{T}(\tilde{W}^3(t_k))}}{\hat{T}}\right) = 0,$$

$$\hat{\int}_0^T \hat{G}\hat{\times}\hat{d}\frac{\frac{\tilde{W}^3}{\tilde{T}(\tilde{W}^3)}}{\hat{T}} = \sum_{k=0}^{m-1} G_k\tilde{T}(\tau_k)\frac{\frac{\tilde{W}^3(t_{k+1})}{\tilde{T}(\tilde{W}^3(t_{k+1}))}-\frac{\tilde{W}^3(t_k)}{\tilde{T}(\tilde{W}^3(t_k))}}{\hat{T}}.$$

Therefore

$$\hat{E}\left(\hat{\int}_0^T \hat{G}\hat{\times}\hat{d}\frac{\frac{\tilde{W}^3}{\tilde{T}(\tilde{W}^3)}}{\hat{T}}\right) = \hat{E}\left(\sum_{k=0}^{m-1} G_k\tilde{T}(\tau_k)\frac{\frac{\tilde{W}^3(t_{k+1})}{\tilde{T}(\tilde{W}^3(t_{k+1}))}-\frac{\tilde{W}^3(t_k)}{\tilde{T}(\tilde{W}^3(t_k))}}{\hat{T}}\right)$$

$$= \sum_{k=0}^{m-1}\hat{E}\left(G_k\tilde{T}(\tau_k)\frac{\frac{\tilde{W}^3(t_{k+1})}{\tilde{T}(\tilde{W}^3(t_{k+1}))}-\frac{\tilde{W}^3(t_k)}{\tilde{T}(\tilde{W}^3(t_k))}}{\hat{T}}\right) = 0.$$

□

Theorem 1.9.57. *Let $\hat{G}$, $\hat{H} \in \hat{\mathbb{L}}^2(0,T)$ are step processes. Then*

$$\hat{E}\left(\hat{\int}_0^T \hat{G}\hat{\times}\hat{d}\frac{\frac{\tilde{W}^3}{\tilde{T}(\tilde{W}^3)}}{\hat{T}}\hat{\times}\hat{\int}_0^T \hat{H}\hat{\times}\hat{d}\frac{\frac{\tilde{W}^3}{\tilde{T}(\tilde{W}^3)}}{\hat{T}}\right) = \hat{E}\left(\int_0^T \frac{\hat{G}\hat{H}\tilde{T}^2}{\hat{T}}dt\right).$$

Proof. Let

$$D = \{0 = t_0 < t_1 < t_2 \ldots < t_{m-1} < t_m = T\}$$

be a partition of the interval $[0,T]$ so that

$$\hat{G}(t) = G_k, \qquad \hat{H}(t) = H_k, \qquad t_k \le t < t_{k+1}, \qquad k = 0,1,2,\ldots,m-1.$$

Then

$$\hat{E}\left(\hat{\int}_0^T \hat{G}\hat{\times}\hat{d}\frac{\frac{\tilde{W}^3}{\tilde{T}(\tilde{W}^3)}}{\hat{T}}\hat{\times}\hat{\int}_0^T \hat{H}\hat{\times}\hat{d}\frac{\frac{\tilde{W}^3}{\tilde{T}(\tilde{W}^3)}}{\hat{T}}\right)$$

$$= \hat{E}\left(\sum_{k=0}^{m-1} G_k\tilde{T}(\tau_k)\frac{\frac{\tilde{W}^3(t_{k+1})}{\tilde{T}(\tilde{W}^3(t_{k+1}))}-\frac{\tilde{W}^3(t_k)}{\tilde{T}(\tilde{W}^3(t_k))}}{\hat{T}}\hat{T}\right.$$

(33)
$$\left.\times\sum_{j=0}^{m-1}\tilde{T}(\tau_j)\frac{\frac{\tilde{W}^3(t_{j+1})}{\tilde{T}(\tilde{W}^3(t_{j+1}))}-\frac{\tilde{W}^3(t_j)}{\tilde{T}(\tilde{W}^3(t_j))}}{\hat{T}}\right)$$

$$= \hat{E}\left(\sum_{k=0}^{m-1}\sum_{j=0}^{m-1} G_kH_j\tilde{T}(\tau_k)\tilde{T}(\tau_j)\right.$$

$$\left.\times\frac{\left(\frac{\tilde{W}^3(t_{k+1})}{\tilde{T}(\tilde{W}^3(t_{k+1}))}-\frac{\tilde{W}^3(t_k)}{\tilde{T}(\tilde{W}^3(t_k))}\right)\left(\frac{\tilde{W}^3(t_{j+1})}{\tilde{T}(\tilde{W}^3(t_{j+1}))}-\frac{\tilde{W}^3(t_j)}{\tilde{T}(\tilde{W}^3(t_j))}\right)}{\hat{T}}\right)$$

$$= \sum_{k=0}^{m-1}\sum_{j=0}^{m-1}\hat{E}\left(G_kH_j\tilde{T}(\tau_k)\tilde{T}(\tau_j)\right.$$

$$\left.\times\frac{\left(\frac{\tilde{W}^3(t_{k+1})}{\tilde{T}(\tilde{W}^3(t_{k+1}))}-\frac{\tilde{W}^3(t_k)}{\tilde{T}(\tilde{W}^3(t_k))}\right)\left(\frac{\tilde{W}^3(t_{j+1})}{\tilde{T}(\tilde{W}^3(t_{j+1}))}-\frac{\tilde{W}^3(t_j)}{\tilde{T}(\tilde{W}^3(t_j))}\right)}{\hat{T}}\right).$$

For $k \neq j$ we have that

$$\hat{E}\left(G_k H_j \tilde{T}(\tau_k)\tilde{T}(\tau_j)\frac{\left(\frac{\tilde{W}^3(t_{k+1})}{\tilde{T}(\tilde{W}^3(t_{k+1}))}-\frac{\tilde{W}^3(t_k)}{\tilde{T}(\tilde{W}^3(t_k))}\right)\left(\frac{\tilde{W}^3(t_{j+1})}{\tilde{T}(\tilde{W}^3(t_{j+1}))}-\frac{\tilde{W}^3(t_j)}{\tilde{T}(\tilde{W}^3(t_j))}\right)}{\hat{T}}\right)$$

$$= \hat{E}\left(G_k H_j \tilde{T}(\tau_k)\tilde{T}(\tau_j)\frac{\frac{\tilde{W}^3(t_{k+1})}{\tilde{T}(\tilde{W}^3(t_{k+1}))}-\frac{\tilde{W}^3(t_k)}{\tilde{T}(\tilde{W}^3(t_k))}}{\hat{T}}\right)$$

$$\times \hat{E}\left(\frac{\frac{\tilde{W}^3(t_{j+1})}{\tilde{T}(\tilde{W}^3(t_{j+1}))}-\frac{\tilde{W}^3(t_j)}{\tilde{T}(\tilde{W}^3(t_j))}}{\hat{T}}\right)$$

$$= 0.$$

Hence and (33) we get

$$\hat{E}\left(\hat{\int}_0^T \hat{G}\hat{\times}\hat{d}\frac{\frac{\tilde{W}^3}{\tilde{T}(\tilde{W}^3)}}{\hat{T}}\hat{\times}\hat{\int}_0^T \hat{H}\hat{\times}\hat{d}\frac{\frac{\tilde{W}^3}{\tilde{T}(\tilde{W}^3)}}{\hat{T}}\right)$$

$$= \sum_{k=0}^{m-1}\hat{E}\left(G_k H_k \tilde{T}^2(\tau_k)\frac{\left(\frac{\tilde{W}^3(t_{k+1})}{\tilde{T}(\tilde{W}^3(t_{k+1}))}-\frac{\tilde{W}^3(t_k)}{\tilde{T}(\tilde{W}^3(t_k))}\right)^2}{\hat{T}}\right)$$

$$= \sum_{k=0}^{m-1}\hat{E}\left(\frac{G_k H_k \tilde{T}^2(\tau_k)}{\hat{T}}\right)\hat{E}\left(\frac{\left(\frac{\tilde{W}^3(t_{k+1})}{\tilde{T}(\tilde{W}^3(t_{k+1}))}-\frac{\tilde{W}^3(t_k)}{\tilde{T}(\tilde{W}^3(t_k))}\right)^2}{\hat{T}}\right)$$

$$= \sum_{k=0}^{m-1}\hat{E}\left(\frac{G_k H_k \tilde{T}^2(\tau_k)}{\hat{T}}\right)(t_{k+1}-t_k)$$

$$= \sum_{k=0}^{m-1}\hat{E}\left(\frac{G_k H_k \tilde{T}^2(\tau_k)}{\hat{T}}(t_{k+1}-t_k)\right)$$

$$= \hat{E}\left(\sum_{k=0}^{m-1}\frac{G_k H_k \tilde{T}^2(\tau_k)}{\hat{T}}(t_{k+1}-t_k)\right)$$

$$= \hat{E}\left(\int_0^T \frac{\hat{G}\hat{H}\tilde{T}^2}{\hat{T}}dt\right).$$

□

Corollary 1.9.58. *Let $\hat{G} \in \hat{\mathbb{L}}^{\hat{2}}(0,T)$ be a step process. Then*

$$\hat{E}\left(\left(\hat{\int}_0^T \hat{G}\hat{\times}\hat{d}\frac{\frac{\tilde{W}^3}{\tilde{T}(\tilde{W}^3)}}{\hat{T}}\right)^{\hat{2}}\right) = \hat{E}\left(\int_0^T \frac{\hat{G}^2\tilde{T}^2}{\hat{T}}dt\right).$$

Definition 1.9.59. *Let*

$$M_1 \leq \hat{T} \leq M_2 \qquad \text{on} \qquad \Omega,$$

$$\tilde{T} \leq M_3 \qquad \text{for} \qquad t \in [0,T],$$

where M_1, M_2 and M_3 are some positive constants. Let also, $\hat{G} \in \hat{\mathbb{L}}^{\hat{2}}(0,T)$. Then there exists a sequence of bounded step processes $\hat{G}_m \in \hat{\mathbb{L}}^{\hat{2}}(0,T)$ such that

$$\lim_{m\longrightarrow\infty} \hat{E}\left(\int_0^T \frac{(\hat{G}-\hat{G}_m)^2\tilde{T}^2}{\hat{T}}dt\right) = 0.$$

We define

$$\hat{\int}_0^T \hat{G}\hat{\times}\hat{d}\frac{\frac{\tilde{W}^3}{\tilde{T}(\tilde{W}^3)}}{\hat{T}} := \lim_{m\longrightarrow\infty}\hat{\int}_0^T \hat{G}_m\hat{\times}\hat{d}\frac{\frac{\tilde{W}^3}{\tilde{T}(\tilde{W}^3)}}{\hat{T}}.$$

Theorem 1.9.60. *Let $\hat{G}$, $\hat{H} \in \hat{\mathbb{L}}^{\hat{2}}(0,T)$. Then*

(i)

$$\hat{\int}_0^T (a\hat{G}+b\hat{H})\hat{\times}\hat{d}\frac{\frac{\tilde{W}^3}{\tilde{T}(\tilde{W}^3)}}{\hat{T}} = a\hat{\int}_0^T \hat{G}\hat{\times}\hat{d}\frac{\frac{\tilde{W}^3}{\tilde{T}(\tilde{W}^3)}}{\hat{T}} + b\hat{\int}_0^T \hat{H}\hat{\times}\hat{d}\frac{\frac{\tilde{W}^3}{\tilde{T}(\tilde{W}^3)}}{\hat{T}}$$

for all a, $b \in \mathbb{R}$.

(ii)

$$\hat{E}\left(\hat{\int}_0^T \hat{G}\hat{\times}\hat{d}\frac{\frac{\tilde{W}^3}{\tilde{T}(\tilde{W}^3)}}{\hat{T}}\right) = 0.$$

(iii)

$$\hat{E}\left(\hat{\int}_0^T \hat{G}\hat{\times}\hat{d}\frac{\frac{\tilde{W}^3}{\tilde{T}(\tilde{W}^3)}}{\hat{T}}\hat{\times}\hat{\int}_0^T \hat{H}\hat{\times}\hat{d}\frac{\frac{\tilde{W}^3}{\tilde{T}(\tilde{W}^3)}}{\hat{T}}\right) = \hat{E}\left(\int_0^T \frac{\hat{G}\hat{H}\tilde{T}^2}{\hat{T}}dt\right).$$

Definition 1.9.61. *For $\hat{G} \in \hat{\mathbb{L}}^{\hat{2}}(0,T)$ we set*

$$\hat{I}^3(t) := \hat{\int}_0^t \hat{G}\hat{\times}\hat{d}\frac{\frac{\tilde{W}^3}{\tilde{T}(\tilde{W}^3)}}{\hat{T}}, \qquad 0 \leq t \leq T,$$

the indefinite Itô iso-integral of the third kind of $\hat{G}(\cdot)$.

We note that

$$\hat{I}^2(0) = 0.$$

Definition 1.9.62. *Let $\hat{X}(\cdot)$ be a real-valued iso-stochastic process such that*

$$\hat{X}(r) = \hat{X}(s) + \hat{\int}_s^r \hat{F}\hat{\times}\hat{d}\hat{t} + \hat{\int}_s^r \hat{G}\hat{\times}\hat{d}\frac{\frac{\tilde{W}^3}{\tilde{T}(\tilde{W}^3)}}{\hat{T}}$$

for some $\hat{F} \in \hat{\mathbb{L}}^1(0,T)$ *and* $\hat{G} \in \hat{\mathbb{L}}^{\hat{2}}(0,T)$, *and all times* $0 \leq s \leq r \leq t$. *We say that* $\hat{X}$ *has the iso-stochastic differential of the third kind*

$$d\hat{X} = \hat{F}\frac{\tilde{T}(t) - t\tilde{T}'(t)}{\tilde{T}(t)}dt + \hat{G}\tilde{T}d\frac{\frac{\tilde{W}^3}{\tilde{T}(\tilde{W}^3)}}{\hat{T}}$$

for $0 \leq t \leq T$.

Remark 1.9.63. *In fact, if*

$$\hat{X}(r) = \hat{X}(s) + \int_s^r \hat{F}_1 dt + \int_s^r \hat{G}_1 d\frac{\frac{\tilde{W}^3}{\tilde{T}(\tilde{W}^3)}}{\hat{T}}$$

for some $\hat{F}_1 \in \hat{\mathbb{L}}^1(0,T)$ *and* $\hat{G}_1 \in \hat{\mathbb{L}}^{\hat{2}}(0,T)$, *then*

$$d\hat{X} = \hat{F}_1 dt + \hat{G}_1 d\frac{\frac{\tilde{W}^3}{\tilde{T}(\tilde{W}^3)}}{\hat{T}}$$

is the iso-stochastic differential of the third kind of $\hat{X}$.

Exercise 1.9.64. *Let* $\hat{X}$ *and* $\hat{Y}$ *be real-valued iso-stochastic process. Prove that*

$$d(a\hat{X} + b\hat{Y}) = ad\hat{X} + bd\hat{Y}$$

for all a, $b \in \mathbb{R}$.

We have proved that

$$\hat{\int_0^t} \frac{\frac{\tilde{W}^3}{\tilde{T}(\tilde{W}^3)}}{\tilde{T}}\hat{\times}\hat{d}\frac{\frac{\tilde{W}^3}{\tilde{T}(\tilde{W}^3)}}{\hat{T}} = \frac{\left(\frac{\tilde{W}^3(t)}{\tilde{T}(\tilde{W}^3(t))}\right)^{\hat{2}}}{2\hat{T}} + \left(\lambda - \frac{1}{2}\right)t\frac{\hat{T}(\Omega)}{\hat{T}}$$

for all $\lambda \in [0,1]$.

When $\lambda - 0$ we have that

$$\hat{\int_0^t} \frac{\frac{\tilde{W}^3}{\tilde{T}(\tilde{W}^3)}}{\tilde{T}}\hat{\times}\hat{d}\frac{\frac{\tilde{W}^3}{\tilde{T}(\tilde{W}^3)}}{\hat{T}} = \frac{\left(\frac{\tilde{W}^3(t)}{\tilde{T}(\tilde{W}^3(t))}\right)^2}{2\hat{T}} - \frac{1}{2}t\frac{\hat{T}(\Omega)}{\hat{T}},$$

whereupon

$$\frac{\left(\frac{\tilde{W}^3(t)}{\tilde{T}(\tilde{W}^3(t))}\right)^2}{\hat{T}} = t\frac{\hat{T}(\Omega)}{\hat{T}} + \int_0^t \frac{1}{\tilde{T}}\frac{\frac{\tilde{W}^3}{\tilde{T}(\tilde{W}^3)}}{\tilde{T}}\tilde{T}\tilde{T}d\frac{\frac{\tilde{W}^3}{\tilde{T}(\tilde{W}^3)}}{\hat{T}},$$

or

$$\frac{\left(\frac{\tilde{W}^3(t)}{\tilde{T}(\tilde{W}^3(t))}\right)^2}{\hat{T}}t\frac{\hat{T}(\Omega)}{\hat{T}} + \int_0^t \frac{\tilde{W}^3}{\tilde{T}(\tilde{W}^3)}d\frac{\frac{\tilde{W}^3}{\tilde{T}(\tilde{W}^3)}}{\hat{T}},$$

or

$$\frac{\left(\frac{\tilde{W}^3(t)}{\tilde{T}(\tilde{W}^3(t))}\right)^2}{\hat{T}} = \frac{\hat{T}(\Omega)}{\hat{T}}\int_0^t ds + \int_0^t \frac{\tilde{W}^3}{\tilde{T}(\tilde{W}^3)}d\frac{\frac{\tilde{W}^3}{\tilde{T}(\tilde{W}^3)}}{\hat{T}}.$$

Therefore

$$d\frac{\left(\frac{\tilde{W}^3(t)}{\tilde{T}(\tilde{W}^3(t))}\right)^2}{\hat{T}} = \frac{\hat{T}(\Omega)}{\hat{T}}dt + \frac{\tilde{W}^3}{\tilde{T}(\tilde{W}^3)}d\frac{\frac{\tilde{W}^3}{\tilde{T}(\tilde{W}^3)}}{\hat{T}}.$$

Remark 1.9.65. *We note that*

$$\hat{\int}_0^t \hat{G}\hat{\times}\hat{d}\frac{\frac{\tilde{W}^3}{\tilde{T}(\tilde{W}^3)}}{\hat{T}} := \lim_{n\longrightarrow\infty}\sum_{k=0}^{m_n-1}\hat{G}(t_k^n)\tilde{T}(\tau_k^n)\frac{\frac{\tilde{W}^3(t_{k+1}^n)}{\tilde{T}(\tilde{W}^3(t_{k+1}^n))}-\frac{\tilde{W}^3(t_k^n)}{\tilde{T}(\tilde{W}^3(t_k^n))}}{\hat{T}},$$

where

$$D^{m_n} := \{0 = t_0^n < t_1^n < \ldots t_{m_n-1}^n < t_{m_n}^n = t\}$$

with

$$|D^{m_n}| \longrightarrow_{n\longrightarrow\infty} 0.$$

Theorem 1.9.66. *We have*

$$d\left(t\frac{\frac{\tilde{W}^3}{\tilde{T}(\tilde{W}^3)}}{\hat{T}}\right) = td\frac{\frac{\tilde{W}^3}{\tilde{T}(\tilde{W}^3)}}{\hat{T}} + \frac{\frac{\tilde{W}^3}{\tilde{T}(\tilde{W}^3)}}{\hat{T}}dt.$$

Proof. We note that

$$\hat{\int}_0^r 1\hat{\times}\hat{d}\left(t\frac{\frac{\tilde{W}^3}{\tilde{T}(\tilde{W}^3)}}{\hat{T}}\right) = \lim_{n\longrightarrow\infty}\sum_{k=0}^{m_n-1}\tilde{T}(\tau_k^n)\frac{t_{k+1}^n\frac{\tilde{W}^3(t_{k+1}^n)}{\tilde{T}(\tilde{W}^3(t_{k+1}^n))}-t_k^n\frac{\tilde{W}^3(t_k^n)}{\tilde{T}(\tilde{W}^3(t_k^n))}}{\hat{T}}. \tag{34}$$

Also,

$$\hat{\int}_0^r \frac{\frac{\tilde{W}^3}{\tilde{T}(\tilde{W}^3)}}{\hat{T}}\hat{\times}\hat{d}t = \lim_{n\longrightarrow\infty}\sum_{k=0}^{m_n-1}\tilde{T}(\tau_k^n)\frac{\frac{\tilde{W}^3(t_{k+1}^n)}{\tilde{T}(\tilde{W}^3(t_{k+1}^n))}}{\hat{T}}(t_{k+1}^n - t_k^n),$$

$$\hat{\int}_0^r t\hat{\times}\hat{d}\frac{\frac{\tilde{W}^3}{\tilde{T}(\tilde{W}^3)}}{\hat{T}} = \lim_{n\longrightarrow\infty}\sum_{k=0}^{m_n-1}t_k^n\tilde{T}(\tau_k^n)\frac{\frac{\tilde{W}^3(t_{k+1}^n)}{\tilde{T}(\tilde{W}^3(t_{k+1}^n))}-\frac{\tilde{W}^3(t_k^n)}{\tilde{T}(\tilde{W}^3(t_k^n))}}{\hat{T}}.$$

Hence, using (34),

$$\hat{\int}_0^r \frac{\frac{\tilde{W}^3}{\tilde{T}(\tilde{W}^3)}}{\hat{T}}\hat{\times}\hat{d}t + \hat{\int}_0^r t\hat{\times}\hat{d}\frac{\frac{\tilde{W}^3}{\tilde{T}(\tilde{W}^3)}}{\hat{T}} = \lim_{n\longrightarrow\infty}\sum_{k=0}^{m_n-1}\tilde{T}(\tau_k^n)\frac{\frac{\tilde{W}^3(t_{k+1}^n)}{\tilde{T}(\tilde{W}^3(t_{k+1}^n))}}{\hat{T}}(t_{k+1}^n - t_k^n)$$

$$+\lim_{n\longrightarrow\infty}\sum_{k=0}^{m_n-1}t_k^n\tilde{T}(\tau_k^n)\frac{\frac{\tilde{W}^3(t_{k+1}^n)}{\tilde{T}(\tilde{W}^3(t_{k+1}^n))}-\frac{\tilde{W}^3(t_k^n)}{\tilde{T}(\tilde{W}^3(t_k^n))}}{\hat{T}}$$

$$= \lim_{n\longrightarrow\infty}\sum_{k=0}^{m_n-1}\tilde{T}(\tau_k^n)\frac{1}{\hat{T}}\left(t_{k+1}^n\frac{\tilde{W}^3(t_{k+1}^n)}{\tilde{T}(\tilde{W}^3(t_{k+1}^n))}\right.$$

$$\left.-t_k^n\frac{\tilde{W}^3(t_{k+1}^n)}{\tilde{T}(\tilde{W}^3(t_{k+1}^n))}+t_k^n\frac{\tilde{W}^3(t_{k+1}^n)}{\tilde{T}(\tilde{W}^3(t_{k+1}^n))}-t_k^n\frac{\tilde{W}^3(t_k^n)}{\tilde{T}(\tilde{W}^3(t_k^n))}\right)$$

$$= \lim_{n\longrightarrow\infty}\sum_{k=0}^{m_n-1}\tilde{T}(\tau_k^n)\frac{1}{\hat{T}}\left(t_{k+1}^n\frac{\tilde{W}^3(t_{k+1}^n)}{\tilde{T}(\tilde{W}^3(t_{k+1}^n))}-t_k^n\frac{\tilde{W}^3(t_k^n)}{\tilde{T}(\tilde{W}^3(t_k^n))}\right)$$

$$= \hat{\int}_0^r 1\hat{\times}\hat{d}\left(t\frac{\frac{\tilde{W}^3}{\tilde{T}(\tilde{W}^3)}}{\hat{T}}\right),$$

i.e.,

$$\hat{\int}_0^r 1\hat{\times}\hat{d}\left(t\frac{\tilde{W}^3}{\frac{\tilde{T}(\tilde{W}^3)}{\hat{T}}}\right) = \hat{\int}_0^r \frac{\tilde{W}^3}{\frac{\tilde{T}(\tilde{W}^3)}{\hat{T}}}\hat{\times}\hat{d}t + \hat{\int}_0^r t\hat{\times}\hat{d}\frac{\tilde{W}^3}{\frac{\tilde{T}(\tilde{W}^3)}{\hat{T}}},$$

or

$$\int_0^r \frac{1}{\tilde{\tilde{T}}}\tilde{T}\tilde{T}d\left(t\frac{\tilde{W}^3}{\frac{\tilde{T}(\tilde{W}^3)}{\hat{T}}}\right) = \int_0^r \frac{1}{\tilde{T}}\frac{\tilde{W}^3}{\frac{\tilde{T}(\tilde{W}^3)}{\hat{T}}}\tilde{T}\tilde{T}dt + \int_0^r \frac{1}{\tilde{\tilde{T}}}t\tilde{T}\tilde{T}d\frac{\tilde{W}^3}{\frac{\tilde{T}(\tilde{W}^3)}{\hat{T}}},$$

or

$$\int_0^r \tilde{T}d\left(t\frac{\tilde{W}^3}{\frac{\tilde{T}(\tilde{W}^3)}{\hat{T}}}\right) = \int_0^r \tilde{T}\frac{\tilde{W}^3}{\frac{\tilde{T}(\tilde{W}^3)}{\hat{T}}}dt + \int_0^r t\tilde{T}d\frac{\tilde{W}^3}{\frac{\tilde{T}(\tilde{W}^3)}{\hat{T}}}.$$

Therefore

$$\tilde{T}d\left(t\frac{\tilde{W}^3}{\frac{\tilde{T}(\tilde{W}^3)}{\hat{T}}}\right) = \tilde{T}\frac{\tilde{W}^3}{\frac{\tilde{T}(\tilde{W}^3)}{\hat{T}}}dt + t\tilde{T}d\frac{\tilde{W}^3}{\frac{\tilde{T}(\tilde{W}^3)}{\hat{T}}}$$

or

$$d\left(t\frac{\tilde{W}^3}{\frac{\tilde{T}(\tilde{W}^3)}{\hat{T}}}\right) = \frac{\tilde{W}^3}{\frac{\tilde{T}(\tilde{W}^3)}{\hat{T}}}dt + td\frac{\tilde{W}^3}{\frac{\tilde{T}(\tilde{W}^3)}{\hat{T}}},$$

which is the main conclusion of the theorem. □

Theorem 1.9.67. *Let $g \in \mathcal{C}^1([0,T])$. Then*

$$d\left(g(t)\frac{\tilde{W}^3}{\frac{\tilde{T}(\tilde{W}^3)}{\hat{T}}}\right) = g(t)d\frac{\tilde{W}^3}{\frac{\tilde{T}(\tilde{W}^3)}{\hat{T}}} + \frac{\tilde{W}^3}{\frac{\tilde{T}(\tilde{W}^3)}{\hat{T}}}dg(t).$$

Remark 1.9.68. *We can rewrite the above result as follows*

$$d\left(g(t)\frac{\tilde{W}^3}{\frac{\tilde{T}(\tilde{W}^3)}{\hat{T}}}\right) = g(t)d\frac{\tilde{W}^3}{\frac{\tilde{T}(\tilde{W}^3)}{\hat{T}}} + \frac{\tilde{W}^3}{\frac{\tilde{T}(\tilde{W}^3)}{\hat{T}}}g'(t)dt.$$

Proof. We have that

$$\hat{\int}_0^r 1\hat{\times}\hat{d}\left(g(t)\frac{\tilde{W}^3}{\frac{\tilde{T}(\tilde{W}^3)}{\hat{T}}}\right)$$

$$= \lim_{n\longrightarrow\infty}\sum_{k=0}^{m_n-1}\tilde{T}(\tau_k^n)\frac{g(t_{k+1}^n)\frac{\tilde{W}^3(t_{k+1}^n)}{\tilde{T}(\tilde{W}^3(t_{k+1}^n))} - g(t_k^n)\frac{\tilde{W}^3(t_k^n)}{\tilde{T}(\tilde{W}^3(t_k^n))}}{\hat{T}}.$$

Also,

$$\hat{\int}_0^r \frac{\tilde{W}^3}{\frac{\tilde{T}(\tilde{W}^3)}{\hat{T}}}\hat{\times}\hat{d}g(t) = \lim_{n\longrightarrow\infty}\sum_{k=0}^{m_n-1}\tilde{T}(\tau_k^n)\frac{\frac{\tilde{W}^3(t_{k+1}^n)}{\tilde{T}(\tilde{W}^3(t_{k+1}^n))}}{\hat{T}}(g(t_{k+1}^n) - g(t_k^n)),$$

$$\hat{\int}_0^r g(t)\hat{\times}\hat{d}\frac{\tilde{W}^3}{\frac{\tilde{T}(\tilde{W}^3)}{\hat{T}}} = \lim_{n\longrightarrow\infty}\sum_{k=0}^{m_n-1}g(t_k^n)\tilde{T}(\tau_k^n)\frac{\frac{\tilde{W}^3(t_{k+1}^n)}{\tilde{T}(\tilde{W}^3(t_{k+1}^n))} - \frac{\tilde{W}^3(t_k^n)}{\tilde{T}(\tilde{W}^3(t_k^n))}}{\hat{T}}.$$

From here,

$$\hat{\int}_0^r \frac{\frac{\tilde{W}^3}{\tilde{T}(\tilde{W}^3)}}{\hat{T}} \hat{\times} \hat{d} g(t) + \hat{\int}_0^r g(t) \hat{\times} \hat{d} \frac{\frac{\tilde{W}^3}{\tilde{T}(\tilde{W}^3)}}{\hat{T}}$$

$$= \lim_{n \longrightarrow \infty} \sum_{k=0}^{m_n - 1} \tilde{T}(\tau_k^n) \frac{\frac{\tilde{W}^3(t_{k+1}^n)}{\tilde{T}(\tilde{W}^3(t_{k+1}^n))}}{\hat{T}} (g(t_{k+1}^n) - g(t_k^n))$$

$$+ \lim_{n \longrightarrow \infty} \sum_{k=0}^{m_n - 1} g(t_k^n) \tilde{T}(\tau_k^n) \frac{\frac{\tilde{W}^3(t_{k+1}^n)}{\tilde{T}(\tilde{W}^3(t_{k+1}^n))} - \frac{\tilde{W}^3(t_k^n)}{\tilde{T}(\tilde{W}^3(t_k^n))}}{\hat{T}}$$

$$= \lim_{n \longrightarrow \infty} \sum_{k=0}^{m_n - 1} \tilde{T}(\tau_k^n) \frac{1}{\hat{T}} \Bigg(g(t_{k+1}^n) \frac{\tilde{W}^3(t_{k+1}^n)}{\tilde{T}(\tilde{W}^3(t_{k+1}^n))}$$

$$- g(t_k^n) \frac{\tilde{W}^3(t_{k+1}^n)}{\tilde{T}(\tilde{W}^3(t_{k+1}^n))} + g(t_k^n) \frac{\tilde{W}^3(t_{k+1}^n)}{\tilde{T}(\tilde{W}^3(t_{k+1}^n))} - g(t_k^n) \frac{\tilde{W}^3(t_k^n)}{\tilde{T}(\tilde{W}^3(t_k^n))} \Bigg)$$

$$= \lim_{n \longrightarrow \infty} \sum_{k=0}^{m_n - 1} \tilde{T}(\tau_k^n) \frac{1}{\hat{T}} \left(g(t_{k+1}^n) \frac{\tilde{W}^3(t_{k+1}^n)}{\tilde{T}(\tilde{W}^3(t_{k+1}^n))} - g(t_k^n) \frac{\tilde{W}^3(t_k^n)}{\tilde{T}(\tilde{W}^3(t_k^n))} \right)$$

$$= \hat{\int}_0^r 1 \hat{\times} \hat{d} \left(g(t) \frac{\frac{\tilde{W}^3}{\tilde{T}(\tilde{W}^3)}}{\hat{T}} \right),$$

i.e.,

$$\hat{\int}_0^r 1 \hat{\times} \hat{d} \left(g(t) \frac{\frac{\tilde{W}^3}{\tilde{T}(\tilde{W}^3)}}{\hat{T}} \right) = \hat{\int}_0^r \frac{\frac{\tilde{W}^3}{\tilde{T}(\tilde{W}^3)}}{\hat{T}} \hat{\times} \hat{d} g(t) + \hat{\int}_0^r g(t) \hat{\times} \hat{d} \frac{\frac{\tilde{W}^3}{\tilde{T}(\tilde{W}^3)}}{\hat{T}},$$

or

$$\int_0^r \frac{1}{\tilde{T}} \tilde{T} \tilde{T} d \left(g(t) \frac{\frac{\tilde{W}^3}{\tilde{T}(\tilde{W}^3)}}{\hat{T}} \right) = \int_0^r \frac{1}{\tilde{T}} \frac{\frac{\tilde{W}^3}{\tilde{T}(\tilde{W}^3)}}{\hat{T}} \tilde{T} \tilde{T} d g(t) + \int_0^r \frac{1}{\tilde{T}} g(t) \tilde{T} \tilde{T} d \frac{\frac{\tilde{W}^3}{\tilde{T}(\tilde{W}^3)}}{\hat{T}},$$

or

$$\int_0^r \tilde{T} d \left(g(t) \frac{\frac{\tilde{W}^3}{\tilde{T}(\tilde{W}^3)}}{\hat{T}} \right) = \int_0^r \tilde{T} \frac{\frac{\tilde{W}^3}{\tilde{T}(\tilde{W}^3)}}{\hat{T}} d g(t) + \int_0^r g(t) \tilde{T} d \frac{\frac{\tilde{W}^3}{\tilde{T}(\tilde{W}^3)}}{\hat{T}}.$$

Consequently

$$\tilde{T} d \left(g(t) \frac{\frac{\tilde{W}^3}{\tilde{T}(\tilde{W}^3)}}{\hat{T}} \right) = \tilde{T} \frac{\frac{\tilde{W}^3}{\tilde{T}(\tilde{W}^3)}}{\hat{T}} d g(t) + g(t) \tilde{T} d \frac{\frac{\tilde{W}^3}{\tilde{T}(\tilde{W}^3)}}{\hat{T}}$$

or

$$d \left(g(t) \frac{\frac{\tilde{W}^3}{\tilde{T}(\tilde{W}^3)}}{\hat{T}} \right) = \frac{\frac{\tilde{W}^3}{\tilde{T}(\tilde{W}^3)}}{\hat{T}} d g(t) + g(t) d \frac{\frac{\tilde{W}^3}{\tilde{T}(\tilde{W}^3)}}{\hat{T}}.$$

□

Theorem 1.9.69. *(iso-Itô product rule) Suppose*

$$d\hat{X}_1 = \hat{F}_1 dt + \hat{G}_1 d\frac{\frac{\tilde{W}^3}{\tilde{T}(\tilde{W}^3)}}{\hat{T}},$$

$$d\hat{X}_2 = \hat{F}_2 dt + \hat{G}_2 d\frac{\frac{\tilde{W}^3}{\tilde{T}(\tilde{W}^3)}}{\hat{T}},$$

for $\hat{F}_i \in \hat{\mathbb{L}}^1(0,T)$, $\hat{G}_i \in \hat{\mathbb{L}}^{\hat{2}}(0,T)$, $i = 1,2$. *Then*

$$d(\hat{X}_1\hat{X}_2) = \hat{X}_1 d\hat{X}_2 + \hat{X}_2 d\hat{X}_1 + \hat{G}_1\hat{G}_2\frac{\hat{T}(\Omega)}{\hat{T}}dt.$$

Proof. 1. case. Choose $0 \leq r \leq T$. First of all, we assume for simplicity that

$$\hat{X}_1(0) = \hat{X}_2(0) = 0,$$

and

$$\hat{F}_i(t) = \hat{F}_i, \qquad \hat{G}_i(t) = \hat{G}_i, \qquad i = 1,2,$$

where $\hat{F}_i$, $\hat{G}_i$ are time-independent $\mathcal{F}(0)$-measurable random iso-variables, $i = 1,2$. Thus

$$\hat{X}_1(t) = \hat{F}_1 t + \hat{G}_1\frac{\frac{\tilde{W}^3}{\tilde{T}(\tilde{W}^3)}}{\hat{T}},$$

$$\hat{X}_2(t) = \hat{F}_1 t + \hat{G}_2\frac{\frac{\tilde{W}^3}{\tilde{T}(\tilde{W}^3)}}{\hat{T}},$$

and

$$\hat{\int}_0^r \frac{1}{\hat{T}}\hat{\times}\hat{d}(\hat{X}_1\hat{X}_2) = \int_0^r \frac{1}{\tilde{T}^2}\tilde{T}\tilde{T}d(\hat{X}_1\hat{X}_2)$$

$$= \int_0^r d(\hat{X}_1\hat{X}_2)$$

$$= \hat{X}_1(r)\hat{X}_2(r)$$

$$= \left(\hat{F}_1 r + \hat{G}_1\frac{\frac{\tilde{W}^3}{\tilde{T}(\tilde{W}^3)}}{\hat{T}}\right)\left(\hat{F}_2 r + \hat{G}_2\frac{\frac{\tilde{W}^3}{\tilde{T}(\tilde{W}^3)}}{\hat{T}}\right)$$

$$= \hat{F}_1\hat{F}_2 r + \hat{F}_1\hat{G}_2 r\frac{\frac{\tilde{W}^3}{\tilde{T}(\tilde{W}^3)}}{\hat{T}} + \hat{G}_1\hat{F}_2 r\frac{\frac{\tilde{W}^3}{\tilde{T}(\tilde{W}^3)}}{\hat{T}} + \hat{G}_1\hat{G}_2\frac{\left(\frac{\tilde{W}^3}{\tilde{T}(\tilde{W}^3)}\right)^2}{\hat{T}^2}$$

$$= \hat{F}_1\hat{F}_2 r^2 + (\hat{F}_1\hat{G}_2 + \hat{F}_2\hat{G}_1)r\frac{\frac{\tilde{W}^3}{\tilde{T}(\tilde{W}^3)}}{\hat{T}} + \hat{G}_1\hat{G}_2\frac{\left(\frac{\tilde{W}^3}{\tilde{T}(\tilde{W}^3)}\right)^2}{\hat{T}^2},$$

i.e.,

$$\hat{\int}_0^r \frac{1}{\hat{T}}\hat{\times}\hat{d}(\hat{X}_1\hat{X}_2) = \hat{F}_1\hat{F}_2 r^2 + (\hat{F}_1\hat{G}_2 + \hat{F}_2\hat{G}_1)r\frac{\frac{\tilde{W}^3}{\tilde{T}(\tilde{W}^3)}}{\hat{T}} + \hat{G}_1\hat{G}_2\frac{\left(\frac{\tilde{W}^3}{\tilde{T}(\tilde{W}^3)}\right)^2}{\hat{T}^2}. \tag{35}$$

On the other hand,

$$\hat{\int}_0^r \frac{1}{\tilde{T}} \hat{\times} \left(\hat{X}_1 \hat{d}\hat{X}_2 + \hat{X}_2 \hat{d}\hat{X}_1 + \hat{G}_1 \hat{G}_2 \frac{\hat{T}(\Omega)}{\hat{T}} \hat{d}t \right)$$

$$= \int_0^r \frac{1}{\tilde{T}} \frac{1}{\tilde{T}} \tilde{T} \left(\tilde{T}\hat{X}_1 d\hat{X}_2 + \tilde{T}\hat{X}_2 d\hat{X}_1 + \tilde{T}\hat{G}_1 \hat{G}_2 \frac{\hat{T}(\Omega)}{\hat{T}} dt \right)$$

$$= \int_0^r \left(\hat{X}_1 d\hat{X}_2 + \hat{X}_2 d\hat{X}_1 + \hat{G}_1 \hat{G}_2 \frac{\hat{T}(\Omega)}{\hat{T}} dt \right)$$

$$= \int_0^r \left(\left(\hat{F}_1 t + \hat{G}_1 \frac{\frac{\tilde{W}^3}{\tilde{T}(\tilde{W}^3)}}{\hat{T}} \right) \left(\hat{F}_2 dt + \hat{G}_2 d \frac{\frac{\tilde{W}^3}{\tilde{T}(\tilde{W}^3)}}{\hat{T}} \right) \right.$$

$$\left. + \left(\hat{F}_2 t + \hat{G}_2 \frac{\frac{\tilde{W}^3}{\tilde{T}(\tilde{W}^3)}}{\hat{T}} \right) \left(\hat{F}_1 dt + \hat{G}_1 d \frac{\frac{\tilde{W}^3}{\tilde{T}(\tilde{W}^3)}}{\hat{T}} \right) + \hat{G}_1 \hat{G}_2 \frac{\hat{T}(\Omega)}{\hat{T}} dt \right)$$

$$= \int_0^r \left(\hat{F}_1 \hat{F}_2 t dt + \hat{F}_1 \hat{G}_2 t d \frac{\frac{\tilde{W}^3}{\tilde{T}(\tilde{W}^3)}}{\hat{T}} + \hat{G}_1 \hat{F}_2 \frac{\frac{\tilde{W}^3}{\tilde{T}(\tilde{W}^3)}}{\hat{T}} dt \right.$$

$$+ \hat{G}_1 \hat{G}_2 \frac{\frac{\tilde{W}^3}{\tilde{T}(\tilde{W}^3)}}{\hat{T}} d \frac{\frac{\tilde{W}^3}{\tilde{T}(\tilde{W}^3)}}{\hat{T}} + \hat{F}_1 \hat{F}_2 t dt + \hat{F}_2 \hat{G}_1 t d \frac{\frac{\tilde{W}^3}{\tilde{T}(\tilde{W}^3)}}{\hat{T}}$$

$$\left. + \hat{G}_2 \hat{F}_1 \frac{\frac{\tilde{W}^3}{\tilde{T}(\tilde{W}^3)}}{\hat{T}} dt + \hat{G}_1 \hat{G}_2 \frac{\frac{\tilde{W}^3}{\tilde{T}(\tilde{W}^3)}}{\hat{T}} d \frac{\frac{\tilde{W}^3}{\tilde{T}(\tilde{W}^3)}}{\hat{T}} + \hat{G}_1 \hat{G}_2 \frac{\hat{T}(\Omega)}{\hat{T}} dt \right)$$

$$= \int_0^r \left(2\hat{F}_1 \hat{F}_2 t dt + (\hat{F}_1 \hat{G}_2 + \hat{F}_2 \hat{G}_1) t d \frac{\frac{\tilde{W}^3}{\tilde{T}(\tilde{W}^3)}}{\hat{T}} + (\hat{G}_1 \hat{F}_2 + \hat{G}_2 \hat{F}_1) \frac{\frac{\tilde{W}^3}{\tilde{T}(\tilde{W}^3)}}{\hat{T}} dt \right.$$

$$\left. + 2\hat{G}_1 \hat{G}_2 \frac{\frac{\tilde{W}^3}{\tilde{T}(\tilde{W}^3)}}{\hat{T}} d \frac{\frac{\tilde{W}^3}{\tilde{T}(\tilde{W}^3)}}{\hat{T}} + \hat{G}_1 \hat{G}_2 \frac{\hat{T}(\Omega)}{\hat{T}} dt \right)$$

$$= 2\hat{F}_1 \hat{F}_2 \int_0^r t dt + (\hat{F}_1 \hat{G}_2 + \hat{F}_2 \hat{G}_1) \int_0^r \left(t d \frac{\frac{\tilde{W}^3}{\tilde{T}(\tilde{W}^3)}}{\hat{T}} + \frac{\frac{\tilde{W}^3}{\tilde{T}(\tilde{W}^3)}}{\hat{T}} dt \right)$$

$$+ 2\hat{G}_1 \hat{G}_2 \int_0^r \frac{\frac{\tilde{W}^3}{\tilde{T}(\tilde{W}^3)}}{\hat{T}} d \frac{\frac{\tilde{W}^3}{\tilde{T}(\tilde{W}^3)}}{\hat{T}} + \hat{G}_1 \hat{G}_2 \frac{\hat{T}(\Omega)}{\hat{T}} \int_0^r dt$$

$$= \hat{F}_1 \hat{F}_2 r^2 + (\hat{F}_1 \hat{G}_2 + \hat{F}_2 \hat{G}_1) \int_0^r d \left(t \frac{\frac{\tilde{W}^3}{\tilde{T}(\tilde{W}^3)}}{\hat{T}} \right)$$

$$+ \hat{G}_1 \hat{G}_2 \frac{(\frac{\tilde{W}^3}{\tilde{T}(\tilde{W}^3)}(r))^2}{\hat{T}^2} - \hat{G}_1 \hat{G}_2 r \frac{\hat{T}(\Omega)}{\hat{T}} + \hat{G}_1 \hat{G}_2 \frac{\hat{T}(\Omega)}{\hat{T}} r$$

$$= \hat{F}_1 \hat{F}_2 r^2 + (\hat{F}_1 \hat{G}_2 + \hat{F}_2 \hat{G}_1) r \frac{\frac{\tilde{W}^3}{\tilde{T}(\tilde{W}^3)}}{\hat{T}} + \hat{G}_1 \hat{G}_2 \frac{(\frac{\tilde{W}^3}{\tilde{T}(\tilde{W}^3)}(r))^2}{\hat{T}^2}.$$

From the last equality and from (35) it follows that

$$\hat{\int}_0^r \frac{1}{\tilde{T}} \hat{\times} \hat{d}(\hat{X}_1\hat{X}_2) = \hat{\int}_0^r \frac{1}{\tilde{T}} \hat{\times} \left(\hat{X}_1 \hat{d}\hat{X}_2 + \hat{X}_2 \hat{d}\hat{X}_1 + \hat{G}_1\hat{G}_2 \frac{\hat{T}(\Omega)}{\hat{T}} \hat{d}t\right).$$

2. case. If $\hat{F}_i$, $\hat{G}_i$ are step processes, we apply the step 1 in each subinterval $[t_k, t_{k+1})$ on which $\hat{F}_i$ and $\hat{G}_i$ are constant random iso-variables.

3. case. In the general case, we select step processes $\hat{F}_i^m \in \hat{\mathbb{L}}^1(0,T)$, $\hat{G}_i^m \in \hat{\mathbb{L}}^{\hat{2}}(0,T)$ with

$$\hat{E}\left(\left|\int_0^t (\hat{F}_i - \hat{F}_i^m) \frac{\tilde{T}(t) - t\tilde{T}'(t)}{\tilde{T}(t)} dt\right|\right) \longrightarrow_{m \longrightarrow \infty} 0,$$

$$\hat{E}\left(\left|\int_0^t (\hat{G}_i - \hat{G}_i^m)^2 \hat{T} \frac{\tilde{T}(t) - t\tilde{T}'(t)}{\tilde{T}(t)} dt\right|\right) \longrightarrow_{m \longrightarrow \infty} 0.$$

We define

$$\hat{X}_i^n := \hat{X}_i(0) + \int_0^t \hat{F}_i^n ds + \int_0^t \hat{G}_i^n d \frac{\frac{\tilde{W}^3}{\tilde{T}(\tilde{W}^3)}}{\hat{T}}, \qquad i = 1, 2.$$

We apply the step 2 to $\hat{X}_i^n$ in (s, r) and pass to limits to obtain the conclusion of the theorem.

□

Corollary 1.9.70. *Let*

$$d\hat{X} = \hat{F}dt + \hat{G}d \frac{\frac{\tilde{W}^3}{\tilde{T}(\tilde{W}^3)}}{\hat{T}},$$

where $\hat{F} \in \hat{\mathbb{L}}^1(0,T)$, $\hat{G} \in \hat{\mathbb{L}}^{\hat{2}}(0,T)$. *Then*

$$d\hat{X}^2 = 2\hat{X}d\hat{X} + \hat{G}^2 \frac{\hat{T}(\Omega)}{\hat{T}} dt.$$

Proof. Let

$$\hat{X}_1 = \hat{X}, \qquad \hat{X}_2 = \hat{X}.$$

Then

$$\hat{G}_1 = \hat{G}_2 = \hat{G}.$$

Hence,

$$d\hat{X}^2 = d(\hat{X}\hat{X})$$

$$= \hat{X}d\hat{X} + \hat{X}d\hat{X} + \hat{G}^2 \frac{\hat{T}(\Omega)}{\hat{T}} dt$$

$$= 2\hat{X}d\hat{X} + \hat{G}^2 \frac{\hat{T}(\Omega)}{\hat{T}} dt.$$

□

Corollary 1.9.71. *Let*

$$d\hat{X} = \hat{F}dt + \hat{G}d\frac{\frac{\tilde{W}^3}{\tilde{T}(\tilde{W}^3)}}{\hat{T}},$$

where $\hat{F} \in \hat{\mathbb{L}}^1(0,T)$, $\hat{G} \in \hat{\mathbb{L}}^{\hat{2}}(0,T)$. *Then*

$$d\hat{X}^3 = 3\hat{X}^2 d\hat{X} + 3\hat{X}\hat{G}^2\frac{\hat{T}(\Omega)}{\hat{T}}dt.$$

Proof. Let

$$\hat{X}_1 = \hat{X}^2, \qquad \hat{X}_2 = \hat{X}.$$

Then

$$\begin{aligned}
d\hat{X}_1 &= d\hat{X}^2 \\
&= 2\hat{X}d\hat{X} + \hat{G}^2\frac{\hat{T}(\Omega)}{\hat{T}}dt \\
&= 2\hat{X}\left(\hat{F}dt + \hat{G}d\frac{\frac{\tilde{W}^3}{\tilde{T}(\tilde{W}^3)}}{\hat{T}}\right) + \hat{G}^2\frac{\hat{T}(\Omega)}{\hat{T}}dt \\
&= 2\hat{X}\hat{F}dt + 2\hat{X}\hat{G}d\frac{\frac{\tilde{W}^3}{\tilde{T}(\tilde{W}^3)}}{\hat{T}} + \hat{G}^2\frac{\hat{T}(\Omega)}{\hat{T}}dt \\
&= \left(2\hat{X}\hat{F} + \hat{G}^2\frac{\hat{T}(\Omega)}{\hat{T}}\right)dt + 2\hat{X}\hat{G}d\frac{\frac{\tilde{W}^3}{\tilde{T}(\tilde{W}^3)}}{\hat{T}}.
\end{aligned}$$

Therefore

$$\hat{G}_1 = 2\hat{X}\hat{G}, \qquad \hat{G}_2 = \hat{G}.$$

Hence,

$$\begin{aligned}
d\hat{X}^3 &= d(\hat{X}^2\hat{X}) \\
&= d(\hat{X}_1\hat{X}_2) \\
&= \hat{X}_2 d\hat{X}_1 + \hat{X}_1 d\hat{X}_2 + \hat{G}_1\hat{G}_2\frac{\hat{T}(\Omega)}{\hat{T}}dt \\
&= \hat{X}d\hat{X}^2 + \hat{X}^2 d\hat{X} + 2\hat{X}\hat{G}\hat{G}\frac{\hat{T}(\Omega)}{\hat{T}}dt \\
&= \hat{X}\left(2\hat{X}d\hat{X} + \hat{G}^2\frac{\hat{T}(\Omega)}{\hat{T}}dt\right) + \hat{X}^2 d\hat{X} + 2\hat{X}\hat{G}^2\frac{\hat{T}(\Omega)}{\hat{T}}dt \\
&= 2\hat{X}^2 d\hat{X} + \hat{X}\hat{G}^2\frac{\hat{T}(\Omega)}{\hat{T}}dt + \hat{X}^2 d\hat{X} + 2\hat{X}\hat{G}^2\frac{\hat{T}(\Omega)}{\hat{T}}dt \\
&= 3\hat{X}^2 d\hat{X} + 3\hat{X}\hat{G}^2\frac{\hat{T}(\Omega)}{\hat{T}}dt.
\end{aligned}$$

□

Corollary 1.9.72. *Let*

$$d\hat{X} = \hat{F}dt + \hat{G}d\frac{\frac{\tilde{W}^3}{\hat{T}(\tilde{W}^3)}}{\hat{T}},$$

where $\hat{F} \in \hat{\mathbb{L}}^1(0,T)$, $\hat{G} \in \hat{\mathbb{L}}^{\hat{2}}(0,T)$. *Then*

$$d\hat{X}^m = m\hat{X}^{m-1}d\hat{X} + \frac{1}{2}m(m-1)\hat{X}^{m-2}\hat{G}^2\frac{\hat{T}(\Omega)}{\hat{T}}dt, \qquad m \in \mathbb{N}.$$

Proof. We suppose that for some $m \in \mathbb{N}$ we have

$$d\hat{X}^m = m\hat{X}^{m-1}d\hat{X} + \frac{1}{2}m(m-1)\hat{X}^{m-2}\hat{G}^2\frac{\hat{T}(\Omega)}{\hat{T}}dt.$$

We will prove that

$$d\hat{X}^{m+1} = (m+1)\hat{X}^m d\hat{X} + \frac{1}{2}(m+1)m\hat{X}^{m-1}\hat{G}^2\frac{\hat{T}(\Omega)}{\hat{T}}dt.$$

Really, let

$$\hat{X}^m = \hat{X}_1, \qquad \hat{X} = \hat{X}_2.$$

Then

$$\begin{aligned}
d\hat{X}_1^m &= d\hat{X}^m \\
&= m\hat{X}^{m-1}d\hat{X} + \tfrac{1}{2}m(m-1)\hat{X}^{m-2}\hat{G}^2\tfrac{\hat{T}(\Omega)}{\hat{T}}dt \\
&= m\hat{X}^{m-1}\left(\hat{F}dt + \hat{G}d\tfrac{\frac{\tilde{W}^3}{\hat{T}(\tilde{W}^3)}}{\hat{T}}\right) + \tfrac{1}{2}m(m-1)\hat{X}^{m-2}\hat{G}^2\tfrac{\hat{T}(\Omega)}{\hat{T}}dt \\
&= m\hat{X}^{m-1}\hat{F}dt + m\hat{X}^{m-1}\hat{G}d\tfrac{\frac{\tilde{W}^3}{\hat{T}(\tilde{W}^3)}}{\hat{T}} + \tfrac{1}{2}m(m-1)\hat{X}^{m-2}\hat{G}^2\tfrac{\hat{T}(\Omega)}{\hat{T}}dt \\
&= \left(m\hat{X}^{m-1}\hat{F} + \tfrac{1}{2}m(m-1)\hat{X}^{m-2}\hat{G}^2\tfrac{\hat{T}(\Omega)}{\hat{T}}\right)dt + m\hat{X}^{m-1}\hat{G}d\tfrac{\frac{\tilde{W}^3}{\hat{T}(\tilde{W}^3)}}{\hat{T}}.
\end{aligned}$$

Therefore

$$\hat{G}_1 = m\hat{X}^{m-1}\hat{G}, \qquad \hat{G}_2 = \hat{G}.$$

Hence,

$$
\begin{aligned}
d\hat{X}^{m+1} &= d(\hat{X}^m\hat{X}) \\
&= d(\hat{X}_1\hat{X}_2) \\
&= \hat{X}_2 d\hat{X}_1 + \hat{X}_1 d\hat{X}_2 + \hat{G}_1\hat{G}_2\tfrac{\hat{T}(\Omega)}{\hat{T}}dt \\
&= \hat{X}d\hat{X}^m + \hat{X}^m d\hat{X} + m\hat{X}^{m-1}\hat{G}\hat{G}\tfrac{\hat{T}(\Omega)}{\hat{T}}dt \\
&= \hat{X}\left(m\hat{X}^{m-1}d\hat{X} + \tfrac{1}{2}m(m-1)\hat{X}^{m-2}\hat{G}^2\tfrac{\hat{T}(\Omega)}{\hat{T}}dt\right) + m\hat{X}^{m-1}\hat{G}^2\tfrac{\hat{T}(\Omega)}{\hat{T}}dt \\
&= m\hat{X}^m d\hat{X} + \tfrac{1}{2}m(m-1)\hat{X}^{m-1}\hat{G}^2\tfrac{\hat{T}(\Omega)}{\hat{T}}dt + m\hat{X}^{m-1}\hat{G}^2\tfrac{\hat{T}(\Omega)}{\hat{T}}dt \\
&= m\hat{X}^m d\hat{X} + m\hat{X}^{m-1}\hat{G}^2\left(\tfrac{m-1}{2}+1\right)\tfrac{\hat{T}(\Omega)}{\hat{T}}dt \\
&= m\hat{X}^m d\hat{X} + \tfrac{1}{2}m(m+1)\hat{G}^2\tfrac{\hat{T}(\Omega)}{\hat{T}}dt.
\end{aligned}
$$

□

Corollary 1.9.73. *Let f be a polynomial,*

$$d\hat{X} = \hat{F}dt + \hat{G}d\frac{\frac{\tilde{W}^3}{\hat{T}(\tilde{W}^3)}}{\hat{T}},$$

where $\hat{F} \in \hat{\mathbb{L}}^1(0,T)$, $\hat{G} \in \hat{\mathbb{L}}^2(0,T)$. Then

$$df(\hat{X}) = f'(\hat{X})d\hat{X} + \frac{1}{2}f''(\hat{X})\hat{G}^2\frac{\hat{T}(\Omega)}{\hat{T}}dt.$$

Proof. Let

$$f(\hat{X}) = b_0\hat{X}^m + b_1\hat{X}^{m-1} + \cdots + b_{m-1}\hat{X} + b_m, \qquad b_i \in \mathbb{R}, i = 0,1,2,\ldots,m.$$

Then

$$df(\hat{X}) = d\left(b_0\hat{X}^m + b_1\hat{X}^{m-1} + \cdots + b_{m-1}\hat{X} + b_m\right)$$

$$= b_0 d\hat{X}^m + b_1 d\hat{X}^{m-1} + \cdots + b_{m_1} d\hat{X}$$

$$= b_0 m\hat{X}^{m-1}d\hat{X} + b_0\tfrac{1}{2}m(m-1)\hat{X}^{m-2}\hat{G}^2\tfrac{\hat{T}(\Omega)}{\hat{T}}dt$$

$$+ b_1(m-1)\hat{X}^{m-2}d\hat{X} + b_1\tfrac{1}{2}(m-1)(m-2)\hat{X}^{m-3}\hat{G}^2\tfrac{\hat{T}(\Omega)}{\hat{T}}dt$$

$$+\cdots$$

$$+ b_{m-1}d\hat{X}$$

$$= \left(b_0 m\hat{X}^{m-1} + b_1(m-1)\hat{X}^{m-2} + \cdots + b_{m-1}\right)d\hat{X}$$

$$+ \tfrac{1}{2}\left(b_0 m(m-1)\hat{X}^{m-2} + b_1(m-1)(m-2)\hat{X}^{m-3} + \cdots + 2b_{m-2}\right)\hat{G}^2\tfrac{\hat{T}(\Omega)}{\hat{T}}dt$$

$$= f'(\hat{X})d\hat{X} + \tfrac{1}{2}f''(\hat{X})\hat{G}^2\tfrac{\hat{T}(\Omega)}{\hat{T}}dt.$$

□

Theorem 1.9.74. *(iso-Itô's formula for iso-stochastic differential of the third kind) Suppose that $\hat{X}(\cdot)$ has an iso-stochastic differential of the third kind*

$$d\hat{X} = \hat{F}dt + \hat{G}d\frac{\frac{\tilde{W}^3}{\tilde{T}(\tilde{W}^3)}}{\ddot{T}},$$

for $\hat{F} \in \hat{\mathbb{L}}^1(0,T)$, $\hat{G} \in \hat{\mathbb{L}}^2(0,T)$. Assume $u : \mathbb{R} \times [0,T] \longrightarrow \mathbb{R}$ is continuous and

$$\frac{\partial u}{\partial t}, \quad \frac{\partial u}{\partial x}, \quad \frac{\partial^2 u}{\partial x^2}$$

exist and are continuous.

We set

$$\hat{Y}(t) := u(\hat{X}(t), t).$$

Then

$$d\hat{Y} = \tfrac{\partial u}{\partial t}dt + \tfrac{\partial u}{\partial x}d\hat{X} + \tfrac{1}{2}\tfrac{\partial^2 u}{\partial x^2}\hat{G}^2\tfrac{\hat{T}(\Omega)}{\hat{T}}dt$$

(36)

$$= \left(\tfrac{\partial u}{\partial t} + \tfrac{\partial u}{\partial x}\hat{F} + \tfrac{1}{2}\tfrac{\partial^2 u}{\partial x^2}\hat{G}^2\tfrac{\hat{T}(\Omega)}{\hat{T}}\right)dt + \tfrac{\partial u}{\partial x}\hat{G}d\frac{\frac{\tilde{W}^3}{\tilde{T}(\tilde{W}^3)}}{\hat{T}}.$$

Definition 1.9.75. *The formula* (36) *will be called iso-Itô's formula for iso-stochastic differential of the third kind.*

Proof. Suppose that

$$u(x,t) = f(x)g(t),$$

where f and g are polynomials. Then

$$d(u(\hat{X},t)) = d(f(\hat{X})g(t))$$

$$= g(t)df(\hat{X}) + f(\hat{X})dg(t)$$

$$= f(\hat{X})g'(t)dt + g(t)\left(f'(\hat{X})d\hat{X} + \tfrac{1}{2}f''(\hat{X})\hat{G}^2\tfrac{\hat{T}(\Omega)}{\hat{T}}dt\right)$$

$$= \tfrac{\partial u}{\partial t}dt + \tfrac{\partial u}{\partial x}d\hat{X} + \tfrac{1}{2}\tfrac{\partial^2 u}{\partial x^2}\hat{G}^2\tfrac{\hat{T}(\Omega)}{\hat{T}}dt$$

$$= \tfrac{\partial u}{\partial t}dt + \tfrac{\partial u}{\partial x}\left(\hat{F}dt + \hat{G}d\frac{\frac{\tilde{W}^3}{\hat{T}(\tilde{W}^3)}}{\hat{T}}\right) + \tfrac{1}{2}\tfrac{\partial^2 u}{\partial x^2}\hat{G}^2\tfrac{\hat{T}(\Omega)}{\hat{T}}dt$$

$$= \tfrac{\partial u}{\partial t}dt + \tfrac{\partial u}{\partial x}\hat{F}dt + \tfrac{\partial u}{\partial x}\hat{G}d\frac{\frac{\tilde{W}^3}{\hat{T}(\tilde{W}^3)}}{\hat{T}} + \tfrac{1}{2}\tfrac{\partial^2 u}{\partial x^2}\hat{G}^2\tfrac{\hat{T}(\Omega)}{\hat{T}}dt$$

$$= \left(\tfrac{\partial u}{\partial t} + \tfrac{\partial u}{\partial x}\hat{F} + \tfrac{1}{2}\tfrac{\partial^2 u}{\partial x^2}\hat{G}^2\tfrac{\hat{T}(\Omega)}{\hat{T}}\right)dt + \tfrac{\partial u}{\partial x}\hat{G}d\frac{\frac{\tilde{W}^3}{\hat{T}(\tilde{W}^3)}}{\hat{T}}.$$

Thus the iso-Itô's formula for iso-stochastic differential of the third kind is true as well for any function u having the form

$$u(x,t) = \sum_{i=1}^{m} f^i(x)g^i(t),$$

where f^i and g^i are polynomials.

Given u as in the iso-Itô's formula for iso-stochastic differential of the third kind, there exists a sequence of polynomials u^n such that

$$u^n \longrightarrow u, \quad \frac{\partial u^n}{\partial t} \longrightarrow \frac{\partial u}{\partial t}, \quad \frac{\partial u^n}{\partial x} \longrightarrow \frac{\partial u}{\partial x}, \quad \frac{\partial^2 u^n}{\partial x^2} \longrightarrow \frac{\partial^2 u}{\partial x^2}$$

uniformly on compact subsets of $\mathbb{R} \times [0,T]$. As in above, we have that for all $0 \leq r \leq T$

$$u^n(\hat{X}(r),r) = u^n(\hat{X}(0),0)$$

$$+ \int_0^r \left(\tfrac{\partial u^n}{\partial t} + \tfrac{\partial u^n}{\partial x}\hat{F} + \tfrac{1}{2}\tfrac{\partial^2 u^n}{\partial x^2}\hat{G}^2\tfrac{\hat{T}(\Omega)}{\hat{T}}\right)dt + \int_0^r \tfrac{\partial u^n}{\partial x}\hat{G}d\frac{\frac{\tilde{W}^3}{\hat{T}(\tilde{W}^3)}}{\hat{T}}$$

almost surely.

We may pass to limits as $n \longrightarrow \infty$ in this expression, thereby proving the iso-Itô's formula for iso-stochastic differential of the third kind. □

Example 1.9.76. *Let*

$$d\hat{X} = \sqrt{\frac{\hat{T}}{\hat{T}(\Omega)}}d\frac{\tilde{W}^3}{\hat{T}\tilde{T}(\tilde{W}^3)}.$$

Let also,

$$u(x,t) = e^{\sigma x + \left(\mu - \frac{\sigma^2}{2}\right)t}, \qquad \hat{Y} = u(\hat{X}, t),$$

where μ and σ are real constants.

Then

$$\frac{\partial u}{\partial t} = \left(\mu - \frac{\sigma^2}{t}\right) e^{\sigma x + \left(\mu - \frac{\sigma^2}{2}\right)t}, \qquad \frac{\partial u}{\partial x} = \sigma e^{\sigma x + \left(\mu - \frac{\sigma^2}{2}\right)t},$$

$$\frac{\partial^2 u}{\partial x^2} = \sigma^2 e^{\sigma x + \left(\mu - \frac{\sigma^2}{2}\right)t}.$$

Then

$$d\hat{Y} = \left(\mu - \frac{\sigma^2}{2}\right)\hat{Y}dt + \sigma\hat{Y}d\hat{X} + \frac{\sigma^2}{2}\hat{Y}\frac{\hat{T}}{\hat{T}(\Omega)}\frac{\hat{T}(\Omega)}{\hat{T}}dt$$

$$= \mu\hat{Y}dt + \sigma\hat{Y}\sqrt{\frac{\hat{T}}{\hat{T}(\Omega)}}d\frac{\tilde{W}^3}{\hat{T}\tilde{T}(\tilde{W}^3)},$$

also,

$$\hat{Y}(0) = Y_0,$$

i.e.,

$$\begin{cases} d\hat{Y} = \mu\hat{Y}dt + \sigma\sqrt{\frac{\hat{T}}{\hat{T}(\Omega)}}d\frac{\tilde{W}^3}{\hat{T}\tilde{T}(\tilde{W}^3)} \\ \hat{Y}(0) = Y_0, \end{cases}$$

which is an iso-stochastic differential equation

1.10 Hermite Polynomial and Iso-Wiener Iso-Processes

Definition 1.10.1. *For $n = 0, 1, 2, \ldots$, we define*

$$h_n(x,t) := \frac{(-1)^n}{n!} e^{\frac{x^2}{2t}} \frac{d^n}{dx^n}\left(e^{-\frac{x^2}{2t}}\right)$$

the nth Hermite polynomial.

Now we will give explicit expression of the Hermite polynomials $h_0(t)$, $h_1(t)$, $h_2(t)$, $h_3(t)$ and $h_4(t)$.

We have

$$h_0(x,t) = \frac{(-t)^0}{0!} e^{\frac{x^2}{2t}} e^{-\frac{x^2}{2t}}$$

$$= e^0$$

$$= 1,$$

$$h_1(x,t) = \frac{(-t)^1}{1!} e^{\frac{x^2}{2t}} \frac{d}{dx}\left(e^{-\frac{x^2}{2t}}\right)$$

$$= -te^{\frac{x^2}{2t}}\left(-\frac{2x}{2t}e^{-\frac{x^2}{2t}}\right)$$

$$= xe^{\frac{x^2}{2t}}e^{-\frac{x^2}{2t}}$$

$$= x,$$

$$\frac{d}{dx}\left(e^{-\frac{x^2}{2t}}\right) = -\frac{x}{t}e^{-\frac{x^2}{2t}},$$

$$h_2(x,t) = \frac{(-t)^2}{2!}e^{\frac{x^2}{2t}}\frac{d^2}{dx^2}\left(e^{-\frac{x^2}{2t}}\right)$$

$$= \frac{t^2}{2!}e^{\frac{x^2}{2t}}\frac{d}{dx}\left(\frac{d}{dx}\left(e^{-\frac{x^2}{2t}}\right)\right)$$

$$= \frac{t^2}{2!}e^{\frac{x^2}{2t}}\frac{d}{dx}\left(-\frac{x}{t}e^{-\frac{x^2}{2t}}\right)$$

$$= \frac{t^2}{2!}e^{\frac{x^2}{2t}}\left(-\frac{1}{t}e^{-\frac{x^2}{2t}} - \frac{x}{t}\frac{d}{dx}\left(e^{-\frac{x^2}{2t}}\right)\right)$$

$$= \frac{t^2}{2!}e^{\frac{x^2}{2t}}\left(-\frac{x}{t}e^{-\frac{x^2}{2t}} - \frac{x}{t}\left(-\frac{x}{t}\right)e^{-\frac{x^2}{2t}}\right)$$

$$= \frac{t^2}{2}e^{\frac{x^2}{2t}}e^{-\frac{x^2}{2t}}\left(-\frac{1}{t} + \frac{x^2}{t^2}\right)$$

$$= \frac{t^2}{2}\frac{-t+x^2}{t^2}$$

$$= \frac{x^2}{2} - \frac{t}{2},$$

$$\frac{d^2}{dx^2}\left(e^{-\frac{x^2}{2t}}\right) = e^{-\frac{x^2}{2t}}\left(-\frac{1}{t} + \frac{x^2}{t^2}\right),$$

$$h_3(x,t) = \frac{(-t)^3}{3!}e^{\frac{x^2}{2t}}\frac{d^3}{dx^3}\left(e^{-\frac{x^2}{2t}}\right)$$

$$= \frac{(-t)^3}{3!}e^{\frac{x^2}{2t}}\frac{d}{dx}\left(\frac{d^2}{dx^2}\left(e^{-\frac{x^2}{2t}}\right)\right)$$

$$= \frac{(-t)^3}{3!}e^{\frac{x^2}{2t}}\frac{d}{dx}\left(e^{-\frac{x^2}{2t}}\left(-\frac{1}{t} + \frac{x^2}{t^2}\right)\right)$$

$$= \frac{(-t)^3}{3!}e^{\frac{x^2}{2t}}\left(-\frac{2x}{2t}\left(-\frac{1}{t} + \frac{x^2}{t^2}\right)e^{-\frac{x^2}{2t}} + e^{-\frac{x^2}{2t}}\frac{2x}{t^2}\right)$$

$$= -\frac{t^3}{6}e^{\frac{x^2}{2t}}e^{-\frac{x^2}{2t}}\left(-\frac{x}{t}\left(-\frac{1}{t} + \frac{x^2}{t^2}\right) + \frac{2x}{t^2}\right)$$

$$= -\frac{t^3}{6}\left(\frac{x}{t^2} - \frac{x^3}{t^3} + 2\frac{x}{t^2}\right)$$

$$= -\frac{t^3}{6}\left(3\frac{x}{t^2} - \frac{x^3}{t^3}\right)$$

$$= -\frac{t^3}{6}\frac{3xt-x^3}{t^3}$$

$$= \frac{x^3-3xt}{6}$$

$$= \frac{x^3}{6} - \frac{1}{2}xt,$$

$$\frac{d^3}{dx^3}\left(e^{-\frac{x^2}{2t}}\right) = e^{-\frac{x^2}{2t}}\left(3\frac{x}{t^2} - \frac{x^3}{t^3}\right),$$

$$h_4(x,t) = \frac{(-t)^4}{4!}e^{\frac{x^2}{2t}}\frac{d^4}{dx^4}\left(e^{-\frac{x^2}{2t}}\right)$$

$$= \frac{t^4}{4!}e^{\frac{x^2}{2t}}\frac{d}{dx}\left(\frac{d^3}{dx^3}\left(e^{-\frac{x^2}{2t}}\right)\right)$$

$$= \frac{t^4}{24}e^{\frac{x^2}{2t}}\frac{d}{dx}\left(e^{-\frac{x^2}{2t}}\left(3\frac{x}{t^2} - \frac{x^3}{t^3}\right)\right)$$

$$= \frac{t^4}{24}e^{\frac{x^2}{2t}}\left(-\frac{2x}{2t}e^{-\frac{x^2}{2t}}\left(3\frac{x}{t^2} - \frac{x^3}{t^3}\right) + e^{-\frac{x^2}{2t}}\left(3\frac{1}{t^2} - 3\frac{x^2}{t^3}\right)\right)$$

$$= \frac{t^4}{24}e^{\frac{x^2}{2t}}e^{-\frac{x^2}{2t}}\left(-\frac{x}{t}\left(3\frac{x}{t^2} - \frac{x^3}{t^3}\right) + \frac{3}{t^2} - 3\frac{x^2}{t^3}\right)$$

$$= \frac{t^4}{24}\left(-3\frac{x^2}{t^3} + \frac{x^4}{t^4} + \frac{3}{t^2} - 3\frac{x^2}{t^3}\right)$$

$$= \frac{t^4}{24}\left(\frac{x^4}{t^4} - 6\frac{x^2}{t^3} + \frac{3}{t^2}\right)$$

$$= \frac{t^4}{24}\frac{x^4-6tx^2+3t^2}{t^4}$$

$$= \frac{x^4-6tx^2+3t^2}{24}$$

$$= \frac{x^4}{24} - \frac{tx^2}{4} + \frac{t^2}{8}.$$

Theorem 1.10.2. *(iso-stochastic calculus with Hermite polynomials) We have*

$$\int_0^t \sqrt{\frac{\hat{T}}{\hat{T}(\Omega)}}h_n\left(\frac{\tilde{W}^1}{\sqrt{\hat{T}\hat{T}(\Omega)}},s\right)d\frac{\tilde{W}^1}{\hat{T}} = h_{n+1}\left(\frac{\tilde{W}^1}{\sqrt{\hat{T}\hat{T}(\Omega)}},t\right) \qquad \text{for} \qquad t \geq 0,$$

$n = 0, 1, 2, \ldots$, *i.e.*,

$$dh_{n+1}\left(\frac{\tilde{W}^1}{\sqrt{\hat{T}\hat{T}(\Omega)}},t\right) = \sqrt{\frac{\hat{T}}{\hat{T}(\Omega)}}h_n\left(\frac{\tilde{W}^1}{\sqrt{\hat{T}\hat{T}(\Omega)}},t\right)d\frac{\tilde{W}^1}{\hat{T}}.$$

Proof. Here we will use the following formula.

$$\frac{d^n}{d\lambda^n}\left(e^{-\frac{(x-\lambda t)^2}{2t}}\right)\Big|_{\lambda=0} = (-t)^n\frac{d^n}{dx^n}\left(e^{-\frac{x^2}{2t}}\right).$$

Also,

$$\frac{d^n}{d\lambda^n}\left(e^{-\frac{(x-\lambda t)^2}{2t}}\right)=\frac{d^n}{d\lambda^n}\left(e^{-\frac{x^2-2x\lambda t+\lambda^2t^2}{2t}}\right)$$

$$=\frac{d^n}{d\lambda^n}\left(e^{-\frac{x^2}{2t}}e^{\frac{2x\lambda t-\lambda^2t^2}{2t}}\right)$$

$$=e^{-\frac{x^2}{2t}}\frac{d^n}{d\lambda^n}\left(e^{x\lambda-\frac{\lambda^2t}{2}}\right).$$

Hence,

$$\frac{d^n}{d\lambda^n}\left(e^{-\frac{(x-\lambda t)^2}{2t}}\right)\Big|_{\lambda=0}=e^{-\frac{x^2}{2t}}\frac{d^n}{d\lambda^n}\left(e^{x\lambda-\frac{\lambda^2t}{2}}\right)\Big|_{\lambda=0},$$

from where

$$e^{-\frac{x^2}{2t}}\frac{d^n}{d\lambda^n}\left(e^{x\lambda-\frac{\lambda^2t}{2}}\right)\Big|_{\lambda=0}=(-t)^n\frac{d^n}{dx^n}\left(e^{-\frac{x^2}{2t}}\right).$$

Therefore

$$\frac{d^n}{d\lambda^n}\left(e^{x\lambda-\frac{\lambda^2t}{2}}\right)\Big|_{\lambda=0}=(-t)^ne^{\frac{x^2}{2t}}\frac{d^n}{dx^n}\left(e^{-\frac{x^2}{2t}}\right)$$

$$=n!h_n(x,t).$$

From here,

$$e^{x\lambda-\frac{\lambda^2t}{2}}=\sum_{n=0}^{\infty}\frac{d^n}{d\lambda^n}\left(e^{x\lambda-\frac{\lambda^2t}{2}}\right)\Big|_{\lambda=0}\frac{\lambda^n}{n!}$$

$$=\sum_{n=0}^{\infty}n!h_n(x,t)\frac{\lambda^n}{n!}$$

$$=\sum_{n=0}^{\infty}\lambda^nh_n(x,t).$$

In this way we get that

$$\begin{aligned}\hat{Y}(t)&=e^{\lambda\frac{\tilde{W}^1}{\sqrt{\hat{T}\hat{T}(\Omega)}}-\frac{\lambda^2t}{2}}\\&=\sum_{n=0}^{\infty}\lambda^nh_n\left(\frac{\tilde{W}^1(t)}{\sqrt{\hat{T}\hat{T}(\Omega)}},t\right).\end{aligned}\tag{37}$$

We note that $\hat{Y}$ is a solution to the following problem

$$\begin{cases}d\hat{Y}=\lambda\hat{Y}\sqrt{\frac{\hat{T}}{\hat{T}(\Omega)}}d\frac{\tilde{W}^1}{\hat{T}}\\ \hat{Y}(0)=1.\end{cases}$$

Consequently,

$$\hat{Y}=1+\lambda\int_0^t\hat{Y}\sqrt{\frac{\hat{T}}{\hat{T}(\Omega)}}d\frac{\tilde{W}^1}{\hat{T}}$$

for all $t\geq 0$.

Now we put (37) in the last expression and we find

$$\sum_{n=0}^{\infty}\lambda^n h_n\left(\frac{\tilde{W}^1(t)}{\sqrt{\hat{T}\hat{T}(\Omega)}},t\right)=1+\lambda\int_0^t\sum_{n=0}^{\infty}\lambda^n\sqrt{\frac{\hat{T}}{\hat{T}(\Omega)}}h_n\left(\frac{\tilde{W}^1(s)}{\sqrt{\hat{T}\hat{T}(\Omega)}},s\right)d\frac{\tilde{W}^1}{\hat{T}}$$

$$=1+\sum_{n=0}^{\infty}\lambda^{n+1}\int_0^t\sqrt{\frac{\hat{T}}{\hat{T}(\Omega)}}h_n\left(\frac{\tilde{W}^1(s)}{\sqrt{\hat{T}\hat{T}(\Omega)}},s\right)d\frac{\tilde{W}^1}{\hat{T}}$$

$$=1+\sum_{n=1}^{\infty}\lambda^n\int_0^t h_{n-1}\sqrt{\frac{\hat{T}}{\hat{T}(\Omega)}}\left(\frac{\tilde{W}^1(s)}{\sqrt{\hat{T}\hat{T}(\Omega)}},s\right)d\frac{\tilde{W}^1}{\hat{T}}$$

or

$$h_0\left(\frac{\tilde{W}^1(t)}{\sqrt{\hat{T}\hat{T}(\Omega)}},t\right)+\lambda h_1\left(\frac{\tilde{W}^1(t)}{\sqrt{\hat{T}\hat{T}(\Omega)}},t\right)+\lambda^2 h_2\left(\frac{\tilde{W}^1(t)}{\sqrt{\hat{T}\hat{T}(\Omega)}},t\right)$$

$$+\cdots+\lambda^n h_n\left(\frac{\tilde{W}^1(t)}{\sqrt{\hat{T}\hat{T}(\Omega)}},t\right)+\cdots$$

$$=1+\lambda\int_0^t\sqrt{\frac{\hat{T}}{\hat{T}(\Omega)}}h_0\left(\frac{\tilde{W}^1(s)}{\hat{T}},s\right)d\frac{\tilde{W}^1}{\sqrt{\hat{T}\hat{T}(\Omega)}}+\lambda^2\int_0^t\sqrt{\frac{\hat{T}}{\hat{T}(\Omega)}}h_1\left(\frac{\tilde{W}^1(s)}{\sqrt{\hat{T}\hat{T}(\Omega)}},s\right)d\frac{\tilde{W}^1}{\hat{T}}$$

$$+\cdots+\lambda^n\int_0^t\sqrt{\frac{\hat{T}}{\hat{T}(\Omega)}}h_{n-1}\left(\frac{\tilde{W}^1(s)}{\sqrt{\hat{T}\hat{T}(\Omega)}},s\right)d\frac{\tilde{W}^1}{\hat{T}}+\cdots,$$

therefore

$$\int_0^t\sqrt{\frac{\hat{T}}{\hat{T}(\Omega)}}h_n\left(\frac{\tilde{W}^1(s)}{\sqrt{\hat{T}\hat{T}(\Omega)}},s\right)d\frac{\tilde{W}^1}{\hat{T}}=h_{n+1}\left(\frac{\tilde{W}^1(t)}{\sqrt{\hat{T}\hat{T}(\Omega)}},t\right)$$

for all $t\geq 0$ and $n=0,1,2,\ldots$. □

Theorem 1.10.3. *(iso-stochastic calculus with Hermite polynomials) We have*

$$\int_0^t\sqrt{\frac{\hat{T}}{\hat{T}(\Omega)}}h_n\left(\frac{\tilde{W}^2\tilde{T}(\tilde{W}^2)}{\sqrt{\hat{T}\hat{T}(\Omega)}},s\right)d\frac{\tilde{W}^2\tilde{T}(\tilde{W}^2)}{\hat{T}}=h_{n+1}\left(\frac{\tilde{W}^2\tilde{T}(\tilde{W}^2)}{\sqrt{\hat{T}\hat{T}(\Omega)}},t\right)\quad\text{for}\quad t\geq 0,$$

$n=0,1,2,\ldots$, *i.e.,*

$$dh_{n+1}\left(\frac{\tilde{W}^2\tilde{T}(\tilde{W}^2)}{\sqrt{\hat{T}\hat{T}(\Omega)}},t\right)=\sqrt{\frac{\hat{T}}{\hat{T}(\Omega)}}h_n\left(\frac{\tilde{W}^2\tilde{T}(\tilde{W}^2)}{\sqrt{\hat{T}\hat{T}(\Omega)}},t\right)d\frac{\tilde{W}^2\tilde{T}(\tilde{W}^2)}{\hat{T}}.$$

Proof. Here we will use that

$$\frac{d^n}{d\lambda^n}\left(e^{x\lambda-\frac{\lambda^2 t}{2}}\right)\Big|_{\lambda=0}=n!h_n(x,t).$$

From here,

$$e^{x\lambda-\frac{\lambda^2 t}{2}}=\sum_{n=0}^{\infty}\lambda^n h_n(x,t).$$

In this way we get that

$$\hat{Y}(t) = e^{\lambda\frac{\tilde{W}^2\tilde{T}(\tilde{W}^2)}{\sqrt{\hat{T}\hat{T}(\Omega)}} - \frac{\lambda^2 t}{2}}$$

(38)

$$= \sum_{n=0}^{\infty} \lambda^n h_n\left(\frac{\tilde{W}^2\tilde{T}(\tilde{W}^2)(t)}{\sqrt{\hat{T}\hat{T}(\Omega)}}, t\right).$$

We note that $\hat{Y}$ is a solution to the following problem

$$\begin{cases} d\hat{Y} = \lambda\hat{Y}\sqrt{\frac{\hat{T}}{\hat{T}(\Omega)}}d\frac{\tilde{W}^2\tilde{T}(\tilde{W}^2)}{\hat{T}} \\ \hat{Y}(0) = 1. \end{cases}$$

Consequently,

$$\hat{Y} = 1 + \lambda\int_0^t \hat{Y}\sqrt{\frac{\hat{T}}{\hat{T}(\Omega)}}d\frac{\tilde{W}^2\tilde{T}(\tilde{W}^2)}{\hat{T}}$$

for all $t \geq 0$.

Now we put (38) in the last expression and we find

$$\sum_{n=0}^{\infty} \lambda^n h_n\left(\frac{\tilde{W}^2\tilde{T}(\tilde{W}^2)(t)}{\sqrt{\hat{T}\hat{T}(\Omega)}}, t\right) = 1 + \lambda\int_0^t \sum_{n=0}^{\infty} \lambda^n \sqrt{\frac{\hat{T}}{\hat{T}(\Omega)}}h_n\left(\frac{\tilde{W}^2\tilde{T}(\tilde{W}^2)(s)}{\sqrt{\hat{T}\hat{T}(\Omega)}}, s\right)d\frac{\tilde{W}^2\tilde{T}(\tilde{W}^2)}{\hat{T}}$$

$$= 1 + \sum_{n=0}^{\infty} \lambda^{n+1}\int_0^t \sqrt{\frac{\hat{T}}{\hat{T}(\Omega)}}h_n\left(\frac{\tilde{W}^2\tilde{T}(\tilde{W}^2)(s)}{\sqrt{\hat{T}\hat{T}(\Omega)}}, s\right)d\frac{\tilde{W}^2\tilde{T}(\tilde{W}^2)}{\hat{T}}$$

$$= 1 + \sum_{n=1}^{\infty} \lambda^n \int_0^t h_{n-1}\sqrt{\frac{\hat{T}}{\hat{T}(\Omega)}}\left(\frac{\tilde{W}^2\tilde{T}(\tilde{W}^2)(s)}{\sqrt{\hat{T}\hat{T}(\Omega)}}, s\right)d\frac{\tilde{W}^2\tilde{T}(\tilde{W}^2)}{\hat{T}}$$

or

$$h_0\left(\frac{\tilde{W}^2\tilde{T}(\tilde{W}^2)(t)}{\sqrt{\hat{T}\hat{T}(\Omega)}}, t\right) + \lambda h_1\left(\frac{\tilde{W}^2\tilde{T}(\tilde{W}^2)(t)}{\sqrt{\hat{T}\hat{T}(\Omega)}}, t\right) + \lambda^2 h_2\left(\frac{\tilde{W}^2\tilde{T}(\tilde{W}^2)(t)}{\sqrt{\hat{T}\hat{T}(\Omega)}}, t\right)$$

$$+ \cdots + \lambda^n h_n\left(\frac{\tilde{W}^2\tilde{T}(\tilde{W}^2)(t)}{\sqrt{\hat{T}\hat{T}(\Omega)}}, t\right) + \cdots$$

$$= 1 + \lambda\int_0^t \sqrt{\frac{\hat{T}}{\hat{T}(\Omega)}}h_0\left(\frac{\tilde{W}^2\tilde{T}(\tilde{W}^2)(s)}{\hat{T}}, s\right)d\frac{\tilde{W}^2\tilde{T}(\tilde{W}^2)}{\sqrt{\hat{T}\hat{T}(\Omega)}} + \lambda^2\int_0^t \sqrt{\frac{\hat{T}}{\hat{T}(\Omega)}}h_1\left(\frac{\tilde{W}^2\tilde{T}(\tilde{W}^2)(s)}{\sqrt{\hat{T}\hat{T}(\Omega)}}, s\right)d\frac{\tilde{W}^2\tilde{T}(\tilde{W}^2)}{\hat{T}}$$

$$+ \cdots + \lambda^n\int_0^t \sqrt{\frac{\hat{T}}{\hat{T}(\Omega)}}h_{n-1}\left(\frac{\tilde{W}^2\tilde{T}(\tilde{W}^2)(s)}{\sqrt{\hat{T}\hat{T}(\Omega)}}, s\right)d\frac{\tilde{W}^2\tilde{T}(\tilde{W}^2)}{\hat{T}} + \cdots,$$

therefore

$$\int_0^t \sqrt{\frac{\hat{T}}{\hat{T}(\Omega)}}h_n\left(\frac{\tilde{W}^2\tilde{T}(\tilde{W}^2)(s)}{\sqrt{\hat{T}\hat{T}(\Omega)}}, s\right)d\frac{\tilde{W}^2\tilde{T}(\tilde{W}^2)}{\hat{T}} = h_{n+1}\left(\frac{\tilde{W}^2\tilde{T}(\tilde{W}^2)(t)}{\sqrt{\hat{T}\hat{T}(\Omega)}}, t\right)$$

for all $t \geq 0$ and $n = 0, 1, 2, \ldots$. □

Theorem 1.10.4. *(iso-stochastic calculus with Hermite polynomials) We have*

$$\int_0^t \sqrt{\frac{\hat{T}}{\hat{T}(\Omega)}} h_n\left(\frac{\frac{\tilde{W}^3}{\tilde{T}(\tilde{W}^3)}}{\sqrt{\hat{T}\hat{T}(\Omega)}}, s\right) d\frac{\frac{\tilde{W}^3}{\tilde{T}(\tilde{W}^3)}}{\hat{T}} = h_{n+1}\left(\frac{\frac{\tilde{W}^3}{\tilde{T}(\tilde{W}^3)}}{\sqrt{\hat{T}\hat{T}(\Omega)}}, t\right) \quad \text{for} \quad t \geq 0,$$

$n = 0, 1, 2, \ldots$, *i.e.,*

$$dh_{n+1}\left(\frac{\frac{\tilde{W}^3}{\tilde{T}(\tilde{W}^3)}}{\sqrt{\hat{T}\hat{T}(\Omega)}}, t\right) = \sqrt{\frac{\hat{T}}{\hat{T}(\Omega)}} h_n\left(\frac{\frac{\tilde{W}^3}{\tilde{T}(\tilde{W}^3)}}{\sqrt{\hat{T}\hat{T}(\Omega)}}, t\right) d\frac{\frac{\tilde{W}^3}{\tilde{T}(\tilde{W}^3)}}{\hat{T}}.$$

Proof. Here we will use that

$$\frac{d^n}{d\lambda^n}\left(e^{x\lambda - \frac{\lambda^2 t}{2}}\right)\Big|_{\lambda=0} = n! h_n(x,t).$$

From here,

$$e^{x\lambda - \frac{\lambda^2 t}{2}} = \sum_{n=0}^{\infty} \lambda^n h_n(x,t).$$

In this way we get that

$$\begin{aligned} \hat{Y}(t) &= e^{\lambda \frac{\frac{\tilde{W}^3}{\tilde{T}(\tilde{W}^3)}}{\sqrt{\hat{T}\hat{T}(\Omega)}} - \frac{\lambda^2 t}{2}} \\ &= \sum_{n=0}^{\infty} \lambda^n h_n\left(\frac{\frac{\tilde{W}^3}{\tilde{T}(\tilde{W}^3)}(t)}{\sqrt{\hat{T}\hat{T}(\Omega)}}, t\right). \end{aligned} \tag{39}$$

We note that $\hat{Y}$ is a solution to the following problem

$$\begin{cases} d\hat{Y} = \lambda \hat{Y} \sqrt{\frac{\hat{T}}{\hat{T}(\Omega)}} d\frac{\frac{\tilde{W}^3}{\tilde{T}(\tilde{W}^3)}}{\hat{T}} \\ \hat{Y}(0) = 1. \end{cases}$$

Consequently,

$$\hat{Y} = 1 + \lambda \int_0^t \hat{Y} \sqrt{\frac{\hat{T}}{\hat{T}(\Omega)}} d\frac{\frac{\tilde{W}^3}{\tilde{T}(\tilde{W}^3)}}{\hat{T}}$$

for all $t \geq 0$.

Now we put (39) in the last expression and we find

$$\begin{aligned} \sum_{n=0}^{\infty} \lambda^n h_n\left(\frac{\frac{\tilde{W}^3}{\tilde{T}(\tilde{W}^3)}(t)}{\sqrt{\hat{T}\hat{T}(\Omega)}}, t\right) &= 1 + \lambda \int_0^t \sum_{n=0}^{\infty} \lambda^n \sqrt{\frac{\hat{T}}{\hat{T}(\Omega)}} h_n\left(\frac{\frac{\tilde{W}^3}{\tilde{T}(\tilde{W}^3)}(s)}{\sqrt{\hat{T}\hat{T}(\Omega)}}, s\right) d\frac{\frac{\tilde{W}^3}{\tilde{T}(\tilde{W}^3)}}{\hat{T}} \\ &= 1 + \sum_{n=0}^{\infty} \lambda^{n+1} \int_0^t \sqrt{\frac{\hat{T}}{\hat{T}(\Omega)}} h_n\left(\frac{\frac{\tilde{W}^3}{\tilde{T}(\tilde{W}^3)}(s)}{\sqrt{\hat{T}\hat{T}(\Omega)}}, s\right) d\frac{\frac{\tilde{W}^3}{\tilde{T}(\tilde{W}^3)}}{\hat{T}} \\ &= 1 + \sum_{n=1}^{\infty} \lambda^n \int_0^t h_{n-1} \sqrt{\frac{\hat{T}}{\hat{T}(\Omega)}} \left(\frac{\frac{\tilde{W}^3}{\tilde{T}(\tilde{W}^3)}(s)}{\sqrt{\hat{T}\hat{T}(\Omega)}}, s\right) d\frac{\frac{\tilde{W}^3}{\tilde{T}(\tilde{W}^3)}}{\hat{T}} \end{aligned}$$

or

$$h_0\left(\frac{\frac{\tilde{W}^3}{\tilde{T}(\tilde{W}^3)}(t)}{\sqrt{\hat{T}\hat{T}(\Omega)}},t\right)+\lambda h_1\left(\frac{\frac{\tilde{W}^3}{\tilde{T}(\tilde{W}^3)}(t)}{\sqrt{\hat{T}\hat{T}(\Omega)}},t\right)+\lambda^2 h_2\left(\frac{\frac{\tilde{W}^3}{\tilde{T}(\tilde{W}^3)}(t)}{\sqrt{\hat{T}\hat{T}(\Omega)}},t\right)$$

$$+\cdots+\lambda^n h_n\left(\frac{\frac{\tilde{W}^3}{\tilde{T}(\tilde{W}^3)}(t)}{\sqrt{\hat{T}\hat{T}(\Omega)}},t\right)+\cdots$$

$$=1+\lambda\int_0^t\sqrt{\frac{\hat{T}}{\hat{T}(\Omega)}}h_0\left(\frac{\frac{\tilde{W}^3}{\tilde{T}(\tilde{W}^3)}(s)}{\hat{T}},s\right)d\frac{\frac{\tilde{W}^3}{\tilde{T}(\tilde{W}^3)}}{\sqrt{\hat{T}\hat{T}(\Omega)}}+\lambda^2\int_0^t\sqrt{\frac{\hat{T}}{\hat{T}(\Omega)}}h_1\left(\frac{\frac{\tilde{W}^3}{\tilde{T}(\tilde{W}^3)}(s)}{\sqrt{\hat{T}\hat{T}(\Omega)}},s\right)d\frac{\frac{\tilde{W}^3}{\tilde{T}(\tilde{W}^3)}}{\hat{T}}$$

$$+\cdots+\lambda^n\int_0^t\sqrt{\frac{\hat{T}}{\hat{T}(\Omega)}}h_{n-1}\left(\frac{\frac{\tilde{W}^3}{\tilde{T}(\tilde{W}^3)}(s)}{\sqrt{\hat{T}\hat{T}(\Omega)}},s\right)d\frac{\frac{\tilde{W}^3}{\tilde{T}(\tilde{W}^3)}}{\hat{T}}+\cdots,$$

therefore

$$\int_0^t\sqrt{\frac{\hat{T}}{\hat{T}(\Omega)}}h_n\left(\frac{\frac{\tilde{W}^3}{\tilde{T}(\tilde{W}^3)}(s)}{\sqrt{\hat{T}\hat{T}(\Omega)}},s\right)d\frac{\frac{\tilde{W}^3}{\tilde{T}(\tilde{W}^3)}}{\hat{T}}=h_{n+1}\left(\frac{\frac{\tilde{W}^3}{\tilde{T}(\tilde{W}^3)}(t)}{\sqrt{\hat{T}\hat{T}(\Omega)}},t\right)$$

for all $t \geq 0$ and $n = 0, 1, 2, \ldots$. □

Chapter 2

Iso-Stochastic Differential Equations, Existence and Uniqueness of Solutions

2.1 Definitions and Examples

Let

$$\hat{f}, \hat{g} : \mathbb{R} \times [0,T] \longrightarrow \mathbb{R},$$

$\hat{f} \in \hat{\mathbb{L}}^1(0,T)$, $\hat{g} \in \hat{\mathbb{L}}^2(0,T)$.

Definition 2.1.1. *We will say that a real-valued iso-stochastic process $\hat{X}(\cdot)$ is a solution of the Itô iso-stochastic differential equation of the first kind*

$$(1) \qquad \begin{cases} d\hat{X} = \hat{f}(\hat{X},t)dt + \hat{g}(\hat{X},t)d\frac{\tilde{W}^1}{\hat{T}} \\ \hat{X}(0) = \hat{X}_0 \end{cases}$$

for $0 \leq t \leq T$, provided

(i) *$\hat{X}(\cdot)$ is progressively measurable with respect to $\mathcal{F}(\cdot)$,*

(ii)

$$\hat{X}(t) = \hat{X}_0 + \int_0^t \hat{f}(\hat{X}(s),s)ds + \int_0^t \hat{g}(\hat{X}(s),s)d\frac{\tilde{W}^1}{\hat{T}} \qquad \text{a.s.}$$

for all $0 \leq t \leq T$.

Example 2.1.2. *Let*

$$\hat{X} = -\tfrac{1}{2}\int_0^t \hat{g}^2 ds + \int_0^t \hat{g}\sqrt{\tfrac{\hat{T}}{\hat{T}(\Omega)}}d\tfrac{\tilde{W}^1}{\hat{T}},$$

$$u(x,t) = e^x, \qquad \hat{Y} = e^{\hat{X}}.$$

Then

$$d\hat{X} = -\tfrac{1}{2}\hat{g}^2 dt + \hat{g}\sqrt{\tfrac{\hat{T}}{\hat{T}(\Omega)}}d\tfrac{\tilde{W}^1}{\hat{T}},$$

$$\tfrac{\partial u}{\partial x} = e^x, \qquad \tfrac{\partial^2 u}{\partial x^2} = e^x, \qquad \tfrac{\partial u}{\partial t} = 0.$$

From here,

$$\begin{aligned} d\hat{Y} &= \hat{Y}d\hat{X} + \tfrac{1}{2}\hat{Y}\hat{g}^2\tfrac{\hat{T}}{\hat{T}(\Omega)}\tfrac{\hat{T}(\Omega)}{\hat{T}}dt \\ &= \hat{Y}d\hat{X} + \tfrac{1}{2}\hat{Y}\hat{g}^2 dt \\ &= \hat{Y}\left(d\hat{X} + \tfrac{1}{2}\hat{g}^2 dt\right) \\ &= \hat{Y}\left(-\tfrac{1}{2}\hat{g}^2 dt + \hat{g}\sqrt{\tfrac{\hat{T}}{\hat{T}(\Omega)}}d\tfrac{\tilde{W}^1}{\hat{T}} + \tfrac{1}{2}\hat{g}^2 dt\right) \\ &= \hat{Y}\hat{g}\sqrt{\tfrac{\hat{T}}{\hat{T}(\Omega)}}d\tfrac{\tilde{W}^1}{\hat{T}}. \end{aligned}$$

Also, we have

$$\hat{X}(0) = 0, \qquad \hat{Y}(0) = 1.$$

In this way we obtain the Itô iso-stochastic differential equation of the first kind

$$\begin{cases} d\hat{Y} = \hat{Y}\hat{g}\sqrt{\tfrac{\hat{T}}{\hat{T}(\Omega)}}d\tfrac{\tilde{W}^1}{\hat{T}} \\ \hat{Y}(0) = 1. \end{cases}$$

Definition 2.1.3. *We will say that a real-valued iso-stochastic process $\hat{X}$ is a solution to the Itô iso-stochastic differential equation of the second kind*

$$(2) \qquad \begin{cases} d\hat{X} = \hat{f}(\hat{X},t)dt + \hat{g}(\hat{X},t)d\frac{\tilde{W}^2\tilde{T}(\tilde{W}^2)}{\hat{T}} \\ \hat{X}(0) = \hat{X}_0 \end{cases}$$

for $0 \le t \le T$, if

(i) *$\hat{X}$ is progressively measurable with respect to $\mathcal{F}(\cdot)$,*

(ii)

$$\hat{X}(t) = \hat{X}_0 + \int_0^t \hat{f}(\hat{X}(s),s)ds + \int_0^t \hat{g}(\hat{X}(s),s)d\frac{\tilde{W}^2\tilde{T}(\tilde{W}^2)}{\hat{T}}$$

for all $0 \le t \le T$.

Example 2.1.4. *Let*

$$\hat{X} = \int_0^t \left(\hat{f} - \tfrac{1}{2}\hat{g}^2\right)ds + \int_0^t \hat{g}\sqrt{\tfrac{\hat{T}}{\hat{T}(\Omega)}}d\tfrac{\tilde{W}^2\tilde{T}(\tilde{W}^2)}{\hat{T}},$$

$$\tfrac{\partial u}{\partial x} = \tfrac{\partial^2 u}{\partial x^2} = e^x, \qquad \tfrac{\partial u}{\partial t} = 0.$$

Hence,

$$d\hat{Y} = \hat{Y}d\hat{X} + \tfrac{1}{2}\hat{Y}\hat{g}^2\frac{\hat{T}}{\hat{T}(\Omega)}\frac{\hat{T}(\Omega)}{\hat{T}}dt$$

$$= \hat{Y}d\hat{X} + \tfrac{1}{2}\hat{Y}\hat{g}^2dt$$

$$= \hat{Y}\left(d\hat{X} + \tfrac{1}{2}\hat{g}^2dt\right)$$

$$= \hat{Y}\left(\left(\hat{f} - \tfrac{1}{2}\hat{g}^2\right)dt + \hat{g}\sqrt{\frac{\hat{T}}{\hat{T}(\Omega)}}d\frac{\tilde{W}^2\tilde{T}(\tilde{W}^2)}{\hat{T}} + \tfrac{1}{2}\hat{g}^2dt\right)$$

$$= \hat{Y}\hat{f}dt + \hat{Y}\hat{g}\sqrt{\frac{\hat{T}}{\hat{T}(\Omega)}}d\frac{\tilde{W}^2\tilde{T}(\tilde{W}^2)}{\hat{T}},$$

$$\hat{X}(0) = 0, \qquad \hat{Y}(0) = 1.$$

Consequently we get to the following Itô iso-differential equation of the second kind

$$\begin{cases} d\hat{Y} = \hat{Y}\hat{f}dt + \hat{Y}\hat{g}\sqrt{\frac{\hat{T}}{\hat{T}(\Omega)}}d\frac{\tilde{W}^2\tilde{T}(\tilde{W}^2)}{\hat{T}} \\ \hat{Y}(0) = 1. \end{cases}$$

Definition 2.1.5. *We will say that a real-valued iso-stochastic process $\hat{X}$ is a solution to the Itô iso-stochastic differential equation of the third kind*

$$(3) \qquad \begin{cases} d\hat{X} = \hat{f}(\hat{X},t)dt + \hat{g}(\hat{X},t)d\frac{\frac{\tilde{W}^3}{\tilde{T}(\tilde{W}^3)}}{\hat{T}} \\ \hat{X}(0) = \hat{X}_0 \end{cases}$$

for $0 \leq t \leq T$, *if*

(i) $\hat{X}$ *is progressively measurable with respect to* $\mathcal{F}(\cdot)$,

(ii)

$$\hat{X}(t) = \hat{X}_0 + \int_0^t \hat{f}(\hat{X}(s),s)ds + \int_0^t \hat{g}(\hat{X}(s),s)d\frac{\frac{\tilde{W}^3}{\tilde{T}(\tilde{W}^3)}}{\hat{T}}$$

for all $0 \leq t \leq T$.

Example 2.1.6. *Let*

$$\hat{X}(t) = (1-t)\int_0^t \frac{1}{1-s}d\frac{\frac{\tilde{W}^3}{\tilde{T}(\tilde{W}^3)}}{\hat{T}}, \qquad 0 \leq t < 1.$$

Then

$$\int_0^t \frac{1}{1-s}d\frac{\frac{\tilde{W}^3}{\tilde{T}(\tilde{W}^3)}}{\hat{T}} = \frac{1}{1-t}\hat{X}(t)$$

and

$$d\hat{X} = -\int_0^t \frac{1}{1-s} d\frac{\frac{\tilde{W}^3}{\tilde{T}(\tilde{W}^3)}}{\hat{T}} + (1-t)\frac{1}{1-t} d\frac{\frac{\tilde{W}^3}{\tilde{T}(\tilde{W}^3)}}{\hat{T}}$$

$$= -\int_0^t \frac{1}{1-s} d\frac{\frac{\tilde{W}^3}{\tilde{T}(\tilde{W}^3)}}{\hat{T}} + d\frac{\frac{\tilde{W}^3}{\tilde{T}(\tilde{W}^3)}}{\hat{T}}$$

$$= -\frac{1}{1-t}\hat{X} + d\frac{\frac{\tilde{W}^3}{\tilde{T}(\tilde{W}^3)}}{\hat{T}},$$

$$\hat{X}(0) = 0.$$

Thus, we obtain the following Itô iso-stochastic differential equation of the third kind

$$\begin{cases} d\hat{X} = -\frac{1}{1-t}\hat{X}dt + d\frac{\frac{\tilde{W}^3}{\tilde{T}(\tilde{W}^3)}}{\hat{T}}, & t \in [0,1) \\ \hat{X}(0) = 0. \end{cases}$$

Example 2.1.7. *Let b and σ be real constants. Let also,*

$$\hat{X}(t) = e^{-bt}\hat{X}_0 + \sigma \int_0^t e^{-b(t-s)} d\frac{\tilde{W}^2\tilde{T}(\tilde{W}^2)}{\hat{T}}, \qquad t \geq 0.$$

Then

$$d\hat{X} = -be^{-bt}\hat{X}dt - b\sigma \int_0^t e^{-b(t-s)} d\frac{\tilde{W}^2\tilde{T}(\tilde{W}^2)}{\hat{T}} dt + \sigma e^{-b(t-t)} d\frac{\tilde{W}^2\tilde{T}(\tilde{W}^2)}{\hat{T}}$$

$$= -b\left(e^{-bt}\hat{X}_0 + \sigma \int_0^t e^{-b(t-s)} s d\frac{\tilde{W}^2\tilde{T}(\tilde{W}^2)}{\hat{T}}\right) dt + \sigma d\frac{\tilde{W}^2\tilde{T}(\tilde{W}^2)}{\hat{T}}$$

$$= -b\hat{X}dt + \sigma d\frac{\tilde{W}^2\tilde{T}(\tilde{W}^2)}{\hat{T}}.$$

Also,

$$\hat{X}(0) = \hat{X}_0.$$

Therefore we obtain the following Itô iso-stochastic differential equation of the second kind

$$\begin{cases} d\hat{X} = -b\hat{X}dt + \sigma d\frac{\tilde{W}^2\tilde{T}(\tilde{W}^2)}{\hat{T}}, & t \geq 0, \\ \hat{X}(0) = \hat{X}_0. \end{cases}$$

2.2 A General Existence and Uniqueness Theorem

Before to formulate and prove a general existence and uniqueness theorem we will formulate and prove the Gronwall's lemma.

Lemma 2.2.1. *(Gronwall's lemma) Let ϕ and f be nonnegative, continuous functions defined for $0 \leq t \leq T$, and let C_0 denotes a constant. If*

$$\phi(t) \leq C_0 + \int_0^t f(s)\phi(s)ds \qquad \text{for} \quad \text{all} \quad 0 \leq t \leq T,$$

then

$$\phi(t) \le C_0 e^{\int_0^t f(s)ds} \quad \text{for} \quad \text{all} \quad 0 \le t \le T.$$

Proof. Let

$$\Phi(t) := C_0 + \int_0^t f(s)\phi(s)ds, \qquad 0 \le t \le T.$$

Then

(4) $\phi(t) \le \Phi(t), \qquad 0 \le t \le T.$

Also,

$$\Phi'(t) = \left(C_0 + \int_0^t f(s)\phi(s)ds\right)'$$

$$= f(t)\phi(t) \qquad ((4))$$

$$\le f(t)\Phi(t), \qquad 0 \le t \le T,$$

i.e.,

(5) $\Phi'(t) \le f(t)\Phi(t), \qquad 0 \le t \le T.$

Also,

$$\left(e^{-\int_0^t f(s)ds}\Phi(t)\right)' = \left(e^{-\int_0^t f(s)ds}\right)'\Phi(t) + e^{-\int_0^t f(s)ds}\Phi'(t)$$

$$= -f(t)e^{-\int_0^t f(s)ds}\Phi(t) + e^{-\int_0^t f(s)ds}\Phi'(t) \qquad ((5))$$

$$\le -f(t)e^{-\int_0^t f(s)ds}\Phi(t) + e^{-\int_0^t f(s)ds}f(t)\Phi(t)$$

$$= 0, \qquad 0 \le t \le T.$$

Consequently the function

$$e^{-\int_0^t f(s)ds}\Phi(t)$$

is a non-increasing function in $[0,T]$.

Therefore

$$e^{-\int_0^t f(s)ds}\Phi(t) \le e^{-\int_0^0 f(s)ds}\Phi(0)$$

$$= \Phi(0)$$

$$= C_0, \qquad 0 \le t \le T,$$

whereupon

$$\Phi(t) \le C_0 e^{\int_0^t f(s)ds}, \qquad 0 \le t \le T.$$

Hence and (4) we conclude that

$$\phi(t) \le C_0 e^{\int_0^t f(s)ds}, \qquad 0 \le t \le T.$$

□

Theorem 2.2.2. *Suppose that $\hat{f}$, $\hat{g} : \mathbb{R} \times [0,T]$ are continuous and satisfy the following conditions*

$$|\hat{f}(x,t)| \le L(1+|x|), \qquad |\hat{g}(x,t)| \le L(1+|x|),$$

$$|\hat{f}(x,t) - \hat{f}(y,t)| \le L|x-y|, \qquad |\hat{g}(x,t) - \hat{g}(y,t)| \le L|x-y|$$

for all $t \in [0,T]$, for all x, $y \in \mathbb{R}$, for some positive constant L. Let also, $\hat{T} \ge N$ on Ω and U, $P(A) \le P(B)$, $\hat{T}(A) \ge \hat{T}(B)$ for every A, $B \in U$, $A \subset B$, $\hat{E}(|\hat{X}_0|^{\hat{2}}) < \infty$, $\hat{X}_0$ is independent of $\left(\frac{\tilde{W}^1(0)}{\hat{T}}\right)^+$. Then the problem

$$(1) \qquad \begin{aligned} d\hat{X} &= \hat{f}(\hat{X},t)dt + \hat{g}(\hat{X},t)d\tfrac{\tilde{W}^1}{\hat{T}}, \qquad 0 \le t \le T, \\ \hat{X}(0) &= \hat{X}_0 \end{aligned}$$

has unique solution $\hat{X} \in \hat{\mathbb{L}}^{\hat{2}}(0,T)$.

Proof. 1. Existence. We define

$$\begin{cases} \hat{X}^0(t) := \hat{X}_0 \\ \hat{X}^{n+1}(t) := \hat{X}_0 + \int_0^t \hat{f}(\hat{X}^n(s),s)ds + \int_0^t \hat{g}(\hat{X}^n(s),s)d\frac{\tilde{W}^1}{\hat{T}} \end{cases}$$

for $n = 0,1,2,\ldots$.

Let also,

$$d^n(t) := \hat{E}\left(|\hat{X}^{n+1}(t) - \hat{X}^n(t)|^{\hat{2}}\right).$$

Then

$$\begin{aligned}
&\hat{E}\left(|\hat{X}^1(t) - \hat{X}^0(t)|^{\hat{2}}\right) \\
&= \hat{E}\left(\hat{T}|\hat{X}^1(t) - \hat{X}^0(t)|^2\right) \\
&= \hat{E}\left(\hat{T}\left|\hat{X}_0 + \int_0^t \hat{f}(\hat{X}^0(s),s)ds + \int_0^t \hat{g}(\hat{X}^0(s),s)d\frac{\tilde{W}^1}{\hat{T}} - \hat{X}_0\right|^2\right) \\
&= \hat{E}\left(\hat{T}\left|\int_0^t \hat{f}(\hat{X}^0(s),s)ds + \int_0^t \hat{g}(\hat{X}^0(s),s)d\frac{\tilde{W}^1}{\hat{T}}\right|^2\right) \\
&(a+b)^2 \le 2(a^2+b^2) \\
&\le \hat{E}\left(2\hat{T}\left|\int_0^t \hat{f}(\hat{X}_0,s)ds\right|^2 + 2\hat{T}\left|\int_0^t \hat{g}(\hat{X}_0,s)d\frac{\tilde{W}^1}{\hat{T}}\right|^2\right) \\
&= \hat{E}\left(2\hat{T}\left|\int_0^t \hat{f}(\hat{X}_0,s)ds\right|^2\right) + \hat{E}\left(2\hat{T}\left|\int_0^t \hat{g}(\hat{X}_0,s)d\frac{\tilde{W}^1}{\hat{T}}\right|^2\right) \\
&\le 2\hat{E}\left(\hat{T}\left(\int_0^t (L+|\hat{X}_0|)ds\right)^2\right) + 2\hat{E}\left(\left(\int_0^t \hat{g}(\hat{X}_0,s)d\frac{\tilde{W}^1}{\hat{T}}\right)^{\hat{2}}\right)
\end{aligned}$$

$$= 2\hat{E}\left(\hat{T}(L+|\hat{X}_0|)^2 t^2\right) + 2\hat{E}\left(\left(\int_0^t \hat{g}(\hat{X}_0,s)d\frac{\tilde{W}^1}{\hat{T}}\right)^{\hat{2}}\right)$$

$$\leq 2\hat{E}\left(\hat{T}(L+|\hat{X}_0|)^2\right)Tt + 2\hat{E}\left(\left(\hat{\int}_0^t \tilde{T}\hat{g}(\hat{X}_0,s)\frac{\tilde{T}\tilde{T}}{\tilde{T}\tilde{T}}d\frac{\tilde{W}^1}{\hat{T}}\right)^{\hat{2}}\right)$$

$$= 2\hat{E}\left(\hat{T}(L+|\hat{x}_0|)^2\right)Tt + 2\hat{E}\left(\left(\hat{\int}_0^t \frac{\hat{g}(\hat{X}_0,s)}{\tilde{T}}\hat{\times} d\frac{\hat{W}^1}{\hat{T}}\right)^{\hat{2}}\right)$$

$$= 2\hat{E}\left(\hat{T}(L+|\hat{x}_0|)^2\right)Tt + 2\hat{E}\left(\int_0^t \frac{\hat{g}^2(\hat{X}_0,s)}{\tilde{T}^2}\frac{\tilde{T}^2}{\hat{T}}ds\right)$$

$$= 2\hat{E}\left(\hat{T}(L+|\hat{x}_0|)^2\right)Tt + 2\hat{E}\left(\frac{1}{\hat{T}}\int_0^t \hat{g}^2(\hat{X}_0,s)ds\right)$$

$$= 2\hat{E}\left(\hat{T}(L+|\hat{x}_0|)^2\right)Tt + 2\hat{E}\left(\frac{1}{\hat{T}}\int_0^t (L+|\hat{X}_0|)^2 ds\right)$$

$$= 2\hat{E}\left(\hat{T}(L+|\hat{x}_0|)^2\right)Tt + 2\hat{E}\left(\frac{1}{\hat{T}}(L+|\hat{X}_0|)^2 t\right)$$

$$= 2\hat{E}\left(\hat{T}(L+|\hat{x}_0|)^2\right)Tt + 2\hat{E}\left(\frac{(L+|\hat{X}_0|)^2}{\hat{T}}\right)t$$

$$= \left(2\hat{E}\left(\hat{T}(L+|\hat{x}_0|)^2\right)T + 2\hat{E}\left(\frac{(L+|\hat{X}_0|)^2}{\hat{T}}\right)\right)t,$$

i.e.,

$$d^0(t) \leq \left(2\hat{E}\left(\hat{T}(L+|\hat{x}_0|)^2\right)T + 2\hat{E}\left(\frac{(L+|\hat{X}_0|)^2}{\hat{T}}\right)\right)t.$$

Let

$$M := \max\left\{2\hat{E}\left(\hat{T}(L+|\hat{x}_0|)^2\right)T + 2\hat{E}\left(\frac{(L+|\hat{X}_0|)^2}{\hat{T}}\right), 2TL^2 + 2\frac{L^2}{N^2}\right\}.$$

Therefore

$$d^0(t) \leq Mt.$$

Now we consider $d^1(t)$. For it we have

$$d^1(t) = \hat{E}\left(|\hat{X}^2(t) - \hat{X}^1(t)|^{\hat{2}}\right)$$

$$= \hat{E}\left(\hat{T}|\hat{X}^2(t) - \hat{X}^1(t)|^2\right)$$

$$= \hat{E}\Big(\hat{T}\Big|\hat{X}_0 + \int_0^t \hat{f}(\hat{X}^1(s),s)ds + \int_0^t \hat{g}(\hat{X}^1(s),s)d\frac{\tilde{W}^1}{\hat{T}}$$

$$-\hat{X}_0 - \int_0^t \hat{f}(\hat{X}^0(s),s)ds - \int_0^t \hat{g}(\hat{X}^0(s),s)d\frac{\tilde{W}^1}{\hat{T}}\Big|^2\Big)$$

$$= \hat{E}\left(\hat{T}\left|\int_0^t(\hat{f}(\hat{X}^1(s),s) - \hat{f}(\hat{X}_0,s))ds + \int_0^t(\hat{g}(\hat{X}^1(s),s) - \hat{g}(\hat{X}^0(s),s))\frac{d\tilde{W}^1}{\hat{T}}\right|^2\right)$$

$$(a+b)^2 \leq 2a^2+2b^2$$

$$\leq \hat{E}\left(2\hat{T}\left|\int_0^t(\hat{f}(\hat{X}^1(s),s)-\hat{f}(\hat{X}_0,s))ds\right|^2+2\hat{T}\left(\int_0^t(\hat{g}(\hat{X}^1(s),s)-\hat{g}(\hat{X}^0,s))d\frac{\tilde{W}^1}{\hat{T}}\right)^2\right)$$

$$=\hat{E}\left(2\hat{T}\left|\int_0^t(\hat{f}(\hat{X}^1(s),s)-\hat{f}(\hat{X}_0,s))ds\right|^2\right)$$

$$+\hat{E}\left(2\hat{T}\left(\int_0^t(\hat{g}(\hat{X}^1(s),s)-\hat{g}(\hat{X}^0,s))d\frac{\tilde{W}^1}{\hat{T}}\right)^2\right)$$

$$=\hat{E}\left(2\hat{T}\left|\int_0^t(\hat{f}(\hat{X}^1(s),s)-\hat{f}(\hat{X}_0,s))ds\right|^2\right)$$

$$+\hat{E}\left(2\left(\int_0^t(\hat{g}(\hat{X}^1(s),s)-\hat{g}(\hat{X}^0,s))d\frac{\tilde{W}^1}{\hat{T}}\right)^{\hat{2}}\right)$$

$$\leq 2\hat{E}\left(L^2\hat{T}\left(|\hat{X}^1(s)-\hat{X}^0|ds\right)^2\right)$$

$$+2\hat{E}\left(\left(\int_0^t(\hat{g}(\hat{X}^1(s),s)-\hat{g}(\hat{X}^0,s))d\frac{\tilde{W}^1}{\hat{T}}\right)^{\hat{2}}\right)$$

$$\leq 2L^2\hat{E}\left(\hat{T}t\int_0^t|\hat{X}^1(s)-\hat{X}^0|^2ds\right)$$

$$+2\hat{E}\left(\left(\int_0^t\frac{1}{\hat{T}}\tilde{T}(\hat{g}(\hat{X}^1(s),s)-\hat{g}(\hat{X}_0,s))\frac{\tilde{T}\tilde{T}}{\tilde{T}\tilde{T}}d\frac{\tilde{W}^1}{\hat{T}}\right)^{\hat{2}}\right)$$

$$\leq 2TL^2\hat{E}\left(\int_0^t\hat{T}|\hat{X}^1(s)-\hat{X}_0|^2ds\right)+2\hat{E}\left(\left(\hat{\int}_0^t\frac{\hat{g}(\hat{X}^1(s),s)-\hat{g}(\hat{X}_0,s)}{\hat{T}}\hat{\times}d\frac{\hat{\tilde{W}}^1}{\hat{T}}\right)^{\hat{2}}\right)$$

$$=2T\hat{E}\left(\int_0^t|\hat{X}^1(s)-\hat{X}^0|^2ds\right)+2\hat{E}\left(\int_0^t\frac{(\hat{g}(\hat{X}^1(s),s)-\hat{g}(\hat{X}^0,s))^2}{\tilde{T}^2}\frac{\tilde{T}^2}{\hat{T}}ds\right)$$

$$=2TL^2\int_0^t\hat{E}\left(|\hat{X}^1(s)-\hat{X}^0|^{\hat{2}}\right)ds+2\hat{E}\left(\int_0^t\frac{L^2(\hat{X}^1(s)-\hat{X}^0)^2}{\hat{T}}ds\right)$$

$$=2TL^2\int_0^t d^0(s)ds+2L^2\hat{E}\left(\int_0^t\frac{\hat{T}(\hat{X}^1(s)-\hat{X}^0)^2}{\hat{T}^2}ds\right)$$

$$\leq 2TL^2\int_0^t d^0(s)ds+2\frac{L^2}{N^2}\hat{E}\left(\int_0^t(\hat{X}^1(s)-\hat{X}^0)^{\hat{2}}ds\right)$$

$$=2TL^2\int_0^t d^0(s)ds+2\frac{L^2}{N^2}\int_0^t\hat{E}\left((\hat{X}^1(s)-\hat{X}^0)^{\hat{2}}\right)ds$$

$$=2TL^2\int_0^t d^0(s)ds+2\frac{L^2}{N^2}\int_0^t d^0(s)ds$$

$$=\left(2TL^2+2\frac{L^2}{N^2}\right)\int_0^t d^0(s)ds$$

$$\leq \left(2TL^2+2\tfrac{L^2}{N^2}\right)M\int_0^t sds = \left(2TL^2+2\tfrac{L^2}{N^2}\right)M\tfrac{t^2}{2}$$

$$\leq \tfrac{(MT)^2}{2},$$

i.e.,

$$d^1(t) \leq \frac{(Mt)^2}{2}.$$

We suppose that

$$d^n(t) \leq \frac{(Mt)^{n+1}}{(n+1)!}$$

for some $n \in \mathbb{N}$.

We will prove that

$$d^{n+1}(t) \leq \frac{(Mt)^{n+2}}{(n+2)!}$$

We have

$$d^{n+1}(t) = \hat{E}\left(|\hat{X}^{n+2}(t)-\hat{X}^{n+1}(t)|^{\hat{2}}\right)$$

$$= \hat{E}\left(\hat{T}|\hat{X}^{n+2}(t)-\hat{X}^{n+1}(t)|^2\right)$$

$$= \hat{E}\left(\hat{T}\left|\hat{X}_0+\int_0^t \hat{f}(\hat{X}^{n+1}(s),s)ds+\int_0^t \hat{g}(\hat{X}^{n+1}(s),s)d\tfrac{\tilde{W}^1}{\hat{T}}\right.\right.$$

$$\left.\left.-\hat{X}_0-\int_0^t \hat{f}(\hat{X}^n(s),s)ds-\int_0^t \hat{g}(\hat{X}^n(s),s)d\tfrac{\tilde{W}^1}{\hat{T}}\right|^2\right)$$

$$= \hat{E}\left(\hat{T}\left|\int_0^t(\hat{f}(\hat{X}^{n+1}(s),s)-\hat{f}(\hat{X}^n(s),s))ds+\int_0^t(\hat{g}(\hat{X}^{n+1}(s),s)-\hat{g}(\hat{X}^n(s),s))\tfrac{d\tilde{W}^1}{\hat{T}}\right|^2\right)$$

$$(a+b)^2 \leq 2a^2+2b^2$$

$$\leq \hat{E}\left(2\hat{T}\left|\int_0^t(\hat{f}(\hat{X}^{n+1}(s),s)-\hat{f}(\hat{X}^n(s),s))ds\right|^2\right.$$

$$\left.+2\hat{T}\left(\int_0^t(\hat{g}(\hat{X}^{n+1}(s),s)-\hat{g}(\hat{X}^n(s),s))d\tfrac{\tilde{W}^1}{\hat{T}}\right)^2\right)$$

$$= \hat{E}\left(2\hat{T}\left|\int_0^t(\hat{f}(\hat{X}^{n+1}(s),s)-\hat{f}(\hat{X}^n(s),s))ds\right|^2\right)$$

$$+\hat{E}\left(2\hat{T}\left(\int_0^t(\hat{g}(\hat{X}^{n+1}(s),s)-\hat{g}(\hat{X}^n(s),s))d\frac{\tilde{W}^1}{\hat{T}}\right)^2\right)$$

$$=\hat{E}\left(2\hat{T}\left|\int_0^t(\hat{f}(\hat{X}^{n+1}(s),s)-\hat{f}(\hat{X}^n(s),s))ds\right|^2\right)$$

$$+\hat{E}\left(2\left(\int_0^t(\hat{g}(\hat{X}^{n+1}(s),s)-\hat{g}(\hat{X}^n(s),s))d\frac{\tilde{W}^1}{\hat{T}}\right)^{\hat{2}}\right)$$

$$\leq 2\hat{E}\left(L^2\hat{T}\left(|\hat{X}^{n+1}(s)-\hat{X}^n(s)|ds\right)^2\right)$$

$$+2\hat{E}\left(\left(\int_0^t(\hat{g}(\hat{X}^{n+1}(s),s)-\hat{g}(\hat{X}^n(s),s))d\frac{\tilde{W}^1}{\hat{T}}\right)^{\hat{2}}\right)$$

$$\leq 2L^2\hat{E}\left(\hat{T}t\int_0^t|\hat{X}^{n+1}(s)-\hat{X}^n(s)|^2ds\right)$$

$$+2\hat{E}\left(\left(\int_0^t\frac{1}{\tilde{T}}\tilde{T}(\hat{g}(\hat{X}^{n+1}(s),s)-\hat{g}(\hat{X}^n(s),s))\frac{\tilde{T}\tilde{T}}{\tilde{T}\tilde{T}}d\frac{\tilde{W}^1}{\hat{T}}\right)^{\hat{2}}\right)$$

$$\leq 2TL^2\hat{E}\left(\int_0^t\hat{T}|\hat{X}^{n+1}(s)-\hat{X}^n(s)|^2ds\right)$$

$$+2\hat{E}\left(\left(\hat{\int}_0^t\frac{\hat{g}(\hat{X}^{n+1}(s),s)-\hat{g}(\hat{X}^n(s),s)}{\tilde{T}}\hat{\times}d\frac{\hat{\tilde{W}}^1}{\hat{T}}\right)^{\hat{2}}\right)$$

$$=2T\hat{E}\left(\int_0^t|\hat{X}^{n+1}(s)-\hat{X}^n(s)|^{\hat{2}}ds\right)+2\hat{E}\left(\int_0^t\frac{(\hat{g}(\hat{X}^{n+1}(s),s)-\hat{g}(\hat{X}^n(s),s))^2}{\tilde{T}^2}\frac{\tilde{T}^2}{\hat{T}}ds\right)$$

$$=2TL^2\int_0^t\hat{E}\left(|\hat{X}^{n+1}(s)-\hat{X}^n(s)|^{\hat{2}}\right)ds+2\hat{E}\left(\int_0^t\frac{L^2(\hat{X}^{n+1}(s)-\hat{X}^n(s))^2}{\hat{T}}ds\right)$$

$$=2TL^2\int_0^t d^n(s)ds+2L^2\hat{E}\left(\int_0^t\frac{\hat{T}(\hat{X}^{n+1}(s)-\hat{X}^n(s))^2}{\hat{T}^2}ds\right)$$

$$\leq 2TL^2\int_0^t d^n(s)ds+2\frac{L^2}{N^2}\hat{E}\left(\int_0^t(\hat{X}^{n+1}(s)-\hat{X}^n(s))^{\hat{2}}ds\right)$$

$$=2TL^2\int_0^t d^n(s)ds+2\frac{L^2}{N^2}\int_0^t\hat{E}\left((\hat{X}^{n+1}(s)-\hat{X}^n(s))^{\hat{2}}\right)ds$$

$$=2TL^2\int_0^t d^n(s)ds+2\frac{L^2}{N^2}\int_0^t d^n(s)ds$$

$$=\left(2TL^2+2\frac{L^2}{N^2}\right)\int_0^t d^n(s)ds$$

$$\leq\left(2TL^2+2\frac{L^2}{N^2}\right)\frac{M^{n+1}}{(n+1)!}\int_0^t s^{n+1}ds$$

$$=\left(2TL^2+2\frac{L^2}{N^2}\right)M^{n+1}\frac{t^{n+2}}{(n+2)!}$$

$$\leq \tfrac{(MT)^{n+2}}{(n+2)!}.$$

Consequently

$$d^n(t) \leq \frac{(Mt)^{n+1}}{(n+1)!}$$

for every $n \in \mathbb{N}$.

Hence,

$$\hat{E}\left(\max_{0\leq t\leq T}|\hat{X}^{n+1}(t)-\hat{X}^n(t)|^{\hat{2}}\right) \leq \frac{M^{n+1}T^{n+1}}{(n+1)!}.$$

From the iso-Chebyshev,s inequality we get

$$P\left(\max_{0\leq t\leq T}|\hat{X}^{n+1}(t)-\hat{X}^n(t)|>\tfrac{1}{2^n}\right) \leq \tfrac{1}{\left(\frac{1}{2^n}\right)^2}\hat{E}\left(\max_{0\leq t\leq T}|\hat{X}^{n+1}(t)-\hat{X}^n(t)|^2\right)$$

$$+\tfrac{1}{\left(\frac{1}{2^n}\right)^2}\int_\Omega \max_{0\leq t\leq T}|\hat{X}^{n+1}(t)-\hat{X}^n(t)|^2\tfrac{\hat{T}'(\omega)}{\hat{T}(\omega)}P(\omega)d\omega$$

$$\leq C2^{2n}\hat{E}\left(\max_{0\leq t\leq T}|\hat{X}^{n+1}(t)-\hat{X}^n(t)|^2\right)$$

$$= C2^{2n}\hat{E}\left(\tfrac{1}{\hat{T}}\max_{0\leq t\leq T}|\hat{X}^{n+1}(t)-\hat{X}^n(t)|^{\hat{2}}\right)$$

$$\leq \tfrac{C}{N}2^{2n}\hat{E}\left(\max_{0\leq t\leq T}|\hat{X}^{n+1}(t) \quad \hat{X}^n(t)|^{\hat{2}}\right)$$

$$\leq \tfrac{C}{N}2^{2n}\tfrac{(MT)^{n+1}}{(n+1)!}$$

$$\leq \tfrac{C}{N}\tfrac{(4MT)^{n+1}}{(n+1)!}$$

for some positive constant C.

From here and iso-Borel-Cantelli lemma it follows that

$$\hat{P}\left(\max_{0\leq t\leq T}|\hat{X}^{n+1}(t)-\hat{X}^n(t)|>\frac{1}{2^n}\text{i.o.}\right)=0.$$

Therefore

$$\hat{X}^n=\hat{X}^0+\sum_{j=0}^{n-1}\left(\hat{X}^{j+1}-\hat{X}^j\right)$$

converges uniformly on $[0,T]$ for almost every ω to a process $\hat{X}(\cdot)$.

We pas to limits in the definition of $\hat{X}^{n+1}$ and we get

$$\hat{X}(t)=\hat{X}_0+\int_0^t \hat{f}(\hat{X}(s),s)ds+\int_0^t \hat{g}(\hat{X}(s),s)d\frac{\tilde{W}^1}{\hat{T}}, \quad 0\leq t\leq T.$$

Now we will prove that $\hat{X}\in\hat{\mathbb{L}}^{\hat{2}}(0,T)$.

We have

$$|\hat{X}^{n+1}|^{\hat{2}} = \hat{T}|\hat{X}^{n+1}|^2$$

$$= \hat{T}\left|\hat{X}_0 + \int_0^t \hat{f}(\hat{X}^n(s),s)ds + \int_0^t \hat{g}(\hat{X}^n(s),s)d\frac{\tilde{W}^1}{\hat{T}}\right|^2$$

$$\leq C_1\hat{T}\left(|\hat{X}_0|^2 + \left(\int_0^t |\hat{f}(\hat{X}^n(s),s)|ds\right)^2 + \left(\int_0^t |\hat{g}(\hat{X}^n(s),s)|d\frac{\tilde{W}^1}{\hat{T}}\right)^2\right)$$

$$\leq C_1\hat{T}|\hat{X}_0|^2 + C_1 T\hat{T}\int_0^t |\hat{f}(\hat{X}^n(s),s)|^2 ds + C_1\left(\int_0^t |\hat{g}(\hat{X}^n(s),s)|d\frac{\tilde{W}^1}{\hat{T}}\right)^{\hat{2}}$$

$$\leq C_1|\hat{X}_0|^{\hat{2}} + C_1 T\hat{T}\int_0^t L^2(1+|\hat{X}^n|)^2 ds + C_1\left(\int_0^t \frac{\tilde{T}}{\tilde{T}}|\hat{g}(\hat{X}^n(s),s)|\frac{T\tilde{T}}{\tilde{T}T}d\frac{\tilde{W}^1}{\hat{T}}\right)^{\hat{2}}$$

$$\leq C_1|\hat{X}_0|^{\hat{2}} + C_1 T\hat{T}L^2\int_0^t (1+|\hat{X}^0|)^2 ds + C_1\left(\hat{\int}_0^t \frac{|\hat{g}(\hat{X}^n(s),s)|}{\tilde{T}}\hat{\times}\hat{d}\frac{\tilde{W}^1}{\hat{T}}\right)^{\hat{2}}$$

$$\leq C_2|\hat{X}_0|^{\hat{2}} + C_2\hat{T}\int_0^t |\hat{X}^n|^2 ds + C_2\left(\hat{\int}_0^t \frac{|\hat{g}(\hat{X}^n(s),s)|}{\tilde{T}}\hat{\times}\hat{d}\frac{\tilde{W}^1}{\hat{T}}\right)^{\hat{2}}$$

$$\leq C_2|\hat{X}_0|^{\hat{2}} + C_2\int_0^t |\hat{X}^n|^{\hat{2}} ds + C_2\left(\hat{\int}_0^t \frac{|\hat{g}(\hat{X}^n(s),s)|}{\tilde{T}}\hat{\times}\hat{d}\frac{\tilde{W}^1}{\hat{T}}\right)^{\hat{2}},$$

i.e.,

$$|\hat{X}^{n+1}|^{\hat{2}} \leq C_2|\hat{X}_0|^{\hat{2}} + C_2\int_0^t |\hat{X}^n|^{\hat{2}} ds + C_2\left(\hat{\int}_0^t \frac{|\hat{g}(\hat{X}^n(s),s)|}{\tilde{T}}\hat{\times}\hat{d}\frac{\tilde{W}^1}{\hat{T}}\right)^{\hat{2}}.$$

Here C_1 and C_2 are positive constants.

From here,

$$\hat{E}\left(|\hat{X}^{n+1}|^{\hat{2}}\right) \leq C_2\hat{E}\left(|\hat{X}_0|^{\hat{2}}\right) + C_2\int_0^t \hat{E}\left(|\hat{X}^n|^{\hat{2}}\right)ds + C_2\hat{E}\left(\left(\hat{\int}_0^t \frac{|\hat{g}(\hat{X}^n(s),s)|}{\tilde{T}}\hat{\times}\hat{d}\frac{\tilde{W}^1}{\hat{T}}\right)^{\hat{2}}\right)$$

$$\leq C_2\hat{E}\left(|\hat{X}_0|^{\hat{2}}\right) + C_2\int_0^t \hat{E}\left(|\hat{X}^n|^{\hat{2}}\right)ds + C_2\hat{E}\left(\int_0^t \frac{|\hat{g}(\hat{X}^n(s),s)|^2}{\tilde{T}^2}\frac{\tilde{T}^2}{\hat{T}}ds\right)$$

$$\leq C_2\hat{E}\left(|\hat{X}_0|^{\hat{2}}\right) + C_2\int_0^t \hat{E}\left(|\hat{X}^n|^{\hat{2}}\right)ds + C_2\hat{E}\left(\int_0^t \frac{|\hat{g}(\hat{X}^n(s),s)|^2}{\hat{T}}ds\right)$$

$$\leq C_3\hat{E}\left(|\hat{X}_0|^{\hat{2}}\right) + C_3\int_0^t \hat{E}\left(|\hat{X}^n|^{\hat{2}}\right)ds + C_3L^2\hat{E}\left(\int_0^t \frac{|\hat{X}^n(s)|^2}{\hat{T}}\right)$$

$$\leq C_3\hat{E}\left(|\hat{X}_0|^{\hat{2}}\right) + C_3\int_0^t \hat{E}\left(|\hat{X}^n|^{\hat{2}}\right)ds + C_3L^2\hat{E}\left(\int_0^t \frac{|\hat{X}^n(s)|^{\hat{2}}}{\hat{T}^2}ds\right)$$

$$\leq C_3\hat{E}\left(|\hat{X}_0|^{\hat{2}}\right) + C_3\int_0^t \hat{E}\left(|\hat{X}^n|^{\hat{2}}\right)ds + C_3\frac{L^2}{N^2}\int_0^t \hat{E}\left(|\hat{X}^n|^{\hat{2}}\right)ds$$

$$\leq C_4\hat{E}\left(|\hat{X}_0|^{\hat{2}}\right) + C_4\int_0^t \hat{E}\left(|\hat{X}^n(s)|^{\hat{2}}\right)ds,$$

i.e.,

$$(6)\qquad \hat{E}\left(|\hat{X}^{n+1}|^{\hat{2}}\right) \le C_4\hat{E}\left(|\hat{X}_0|^{\hat{2}}\right) + C_4\int_0^t \hat{E}\left(|\hat{X}^n(s)|^{\hat{2}}\right)ds,$$

where C_3 and C_4 are positive constants and $C_4 > 1$.

For $n = 0$, using (6), we get

$$\hat{E}\left(|\hat{X}^1|^{\hat{2}}\right) \le C_4\hat{E}\left(|\hat{X}_0|^{\hat{2}}\right) + C_4\int_0^t \hat{E}\left(|\hat{X}_0|^{\hat{2}}\right)ds$$

$$= C_4(1+t)\hat{E}\left(|\hat{X}_0|^{\hat{2}}\right).$$

From here and (6), for $n = 1$, we find

$$\hat{E}\left(|\hat{X}^2|^{\hat{2}}\right) \le C_4\hat{E}\left(|\hat{X}_0|^{\hat{2}}\right) + C_4\int_0^t \hat{E}\left(|\hat{X}^1|^{\hat{2}}\right)ds$$

$$\le C_4\hat{E}\left(|\hat{X}_0|^{\hat{2}}\right) + C_4^2\int_0^t(1+s)\hat{E}\left(|\hat{X}_0|^{\hat{2}}\right)ds$$

$$= \left(C_4 + C_4^2\frac{(1+t)^2}{2} - \frac{C_4}{2}\right)\hat{E}\left(|\hat{X}_0|^{\hat{2}}\right)$$

$$\le \left(C_4 + C_4(1+t) + C_4^2\frac{(1+t)^2}{2}\right)\hat{E}\left(|\hat{X}_0|^{\hat{2}}\right).$$

We suppose that

$$(7)\qquad \hat{E}\left(|\hat{X}_0^n|^{\hat{2}}\right) \le \left(C_4 + C_4(1+t) + C_4^2\frac{(1+t)^2}{2} + \cdots + C_4^n\frac{(1+t)^n}{n!}\right)\hat{E}\left(|\hat{X}_0|^{\hat{2}}\right)$$

for some $n \in \mathbb{N}$.

We will prove that

$$\hat{E}\left(|\hat{X}_0^{n+1}|^{\hat{2}}\right)$$

$$\le \left(C_4 + C_4(1+t) + C_4^2\frac{(1+t)^2}{2} + \cdots + C_4^n\frac{(1+t)^n}{n!} + C_4^{n+1}\frac{(1+t)^{n+1}}{(n+1)!}\right)\hat{E}\left(|\hat{X}_0|^{\hat{2}}\right).$$

Indeed, from (6), using (7), we find that

$$\hat{E}\left(|\hat{X}^{n+1}|^{\hat{2}}\right) \le C_4\hat{E}\left(|\hat{X}_0|^{\hat{2}}\right) + C_4\int_0^t \hat{E}\left(|\hat{X}^n(s)|^{\hat{2}}\right)ds$$

$$\le C_4\hat{E}\left(|\hat{X}_0|^{\hat{2}}\right)$$

$$+C_4\int_0^t\left(C_4 + C_4(1+s) + C_4^2\frac{(1+s)^2}{2!} + \cdots + C_4^n\frac{(1+s)^n}{n!}\right)ds\hat{E}\left(|\hat{X}_0|^{\hat{2}}\right)$$

$$\le \left(C_4 + C_4(1+t) + C_4\frac{(1+t)^2}{2!} + \cdots + C_4^{n+1}\frac{(1+t)^{n+1}}{(n+1)!}\right)\hat{E}\left(|\hat{X}_0|^{\hat{2}}\right).$$

Consequently (7) is valid for every $n \in \mathbb{N}$. From it we find

$$\hat{E}\left(|\hat{X}^n|^{\hat{2}}\right) \leq \left(C_4 + e^{C_4(1+t)}\right)\hat{E}\left(|\hat{X}_0|^{\hat{2}}\right),$$

whereupon, when $n \longrightarrow \infty$,

$$\hat{E}\left(|\hat{X}|^{\hat{2}}\right) \leq \left(C_4 + e^{C_4(1+t)}\right)\hat{E}\left(|\hat{X}_0|^{\hat{2}}\right),$$

i.e.,

$$\hat{X} \in \hat{\mathbb{L}}^{\hat{2}}(0,T).$$

2. Uniqueness. Now we suppose that the problem (1) has two solutions $\hat{X}$ and $\hat{Y}$. Then

$$\begin{cases} \hat{X}(t) = \hat{X}_0 + \int_0^t \hat{f}(\hat{X}(s),s)ds + \int_0^t \hat{g}(\hat{X}(s),s)d\frac{\tilde{W}^1}{\hat{T}} \\ \hat{Y}(t) = \hat{X}_0 + \int_0^t \hat{f}(\hat{Y}(s),s)ds + \int_0^t \hat{g}(\hat{Y}(s),s)d\frac{\tilde{W}^1}{\hat{T}}, \qquad 0 \leq t \leq T, \end{cases}$$

whereupon

$$\hat{X}(t) - \hat{Y}(t) = \int_0^t \left(\hat{f}(\hat{X}(s),s) - \hat{f}(\hat{Y}(s),s)\right)ds + \int_0^t \left(\hat{g}(\hat{X}(s),s) - \hat{g}(\hat{Y}(s),s)\right)d\frac{\tilde{W}^1}{\hat{T}}$$

and

$$\hat{E}\left(|\hat{X}(t) - \hat{Y}(t)|^{\hat{2}}\right) = \hat{E}\left(\hat{T}|\hat{X}(t) - \hat{Y}(t)|^2\right)$$

$$\hat{E}\left(\hat{T}\left(\int_0^t \left(\hat{f}(\hat{X}(s),s) - \hat{f}(\hat{Y}(s),s)\right)ds + \int_0^t \left(\hat{g}(\hat{X}(s),s) - \hat{g}(\hat{Y}(s),s)\right)d\frac{\tilde{W}^1}{\hat{T}}\right)^2\right)$$

$$\leq 2\hat{E}\left(\hat{T}\left(\int_0^t(\hat{f}(\hat{X}(s),s) - \hat{f}(\hat{Y}(s),s))\right)^2\right)$$

$$+2\hat{E}\left(\hat{T}\left(\int_0^t(\hat{g}(\hat{X}(s),s) - \hat{g}(\hat{Y}(s),s))d\frac{\tilde{W}^1}{\hat{T}}\right)^2\right)$$

$$\leq 2\hat{E}\left(\hat{T}t\int_0^t(\hat{f}(\hat{X}(s),s)-\hat{f}(\hat{Y}(s),s))^2ds\right)$$

$$+2\hat{E}\left(\hat{T}\left(\int_0^t\tilde{T}(s)\frac{1}{\tilde{T}(s)}\frac{\hat{g}(\hat{X}(s),s)-\hat{g}(\hat{Y}(s),s)}{\tilde{T}^2(s)}\tilde{T}^2(s)d\frac{\tilde{W}^1}{\hat{T}}\right)^2\right)$$

$$\leq 2TL^2\hat{E}\left(\hat{T}\int_0^t|\hat{X}(s)-\hat{Y}(s)|^2ds\right)+2\hat{E}\left(\left(\hat{\int}_0^t\frac{\hat{g}(\hat{X}(s),s)-\hat{g}(\hat{Y}(s),s)}{\tilde{T}}d\frac{\tilde{W}^1}{\hat{T}}\right)^{\hat{2}}\right)$$

$$=2TL^2\hat{E}\left(\int_0^t|\hat{X}(s)-\hat{Y}(s)|^{\hat{2}}ds\right)+2\hat{E}\left(\int_0^t\frac{(\hat{g}(\hat{X}(s),s)-\hat{g}(\hat{Y}(s),s))^2}{\hat{T}}ds\right)$$

$$\leq 2TL^2\int_0^t\hat{E}\left(|\hat{X}(s)-\hat{Y}(s)|^{\hat{2}}\right)ds+2L^2\hat{E}\left(\int_0^t\frac{|\hat{X}(s)-\hat{Y}(s)|^2}{\hat{T}}ds\right)$$

$$=2TL^2\int_0^t\hat{E}(|\hat{X}(s)-\hat{Y}(s)|^{\hat{2}})ds+2L^2\hat{E}\left(\int_0^t\frac{|\hat{X}(s)-\hat{Y}(s)|^{\hat{2}}}{\hat{T}}ds\right)$$

$$\leq 2TL^2\int_0^t\hat{E}(|\hat{X}(s)-\hat{Y}(s)|^{\hat{2}})ds+2\frac{L^2}{N^2}\int_0^t\hat{E}(|\hat{X}(s)-\hat{Y}(s)|^{\hat{2}})ds$$

$$=2L^2\left(T+\frac{1}{N^2}\right)\int_0^t\hat{E}(|\hat{X}(s)-\hat{Y}(s)|^{\hat{2}})ds,$$

i.e.,

$$\int_0^t\hat{E}(|\hat{X}(s)-\hat{Y}(s)|^{\hat{2}})ds\leq 2L^2\left(T+\frac{1}{N^2}\right)\int_0^t\hat{E}(|\hat{X}(s)-\hat{Y}(s)|^{\hat{2}})ds.$$

Hence and Gronwall's lemma it follows that

$$\hat{E}(|\hat{X}(t)-\hat{Y}(t)|^{\hat{2}})=0 \qquad \text{for} \qquad t\in[0,T].$$

Thus

$$\hat{X}(t)=\hat{Y}(t) \qquad \text{a.s.}$$

for all $0\leq t\leq T$.

So

$$\hat{X}(r)=\hat{Y}(r)$$

for all rational $0\leq r\leq T$, except for some set of probability zero. As $\hat{X}$ and $\hat{Y}$ have continuous sample paths almost surely,

$$\hat{P}\left(\max_{0\leq t\leq T}|\hat{X}(t)-\hat{Y}(t)|>0\right)=0.$$

□

Theorem 2.2.3. *Suppose that $\hat{f}$, $\hat{g}:\mathbb{R}\times[0,T]$ are continuous and satisfy the following conditions*

$$|\hat{f}(x,t)|\leq L(1+|x|), \qquad |\hat{g}(x,t)|\leq L(1+|x|),$$

$$|\hat{f}(x,t)-\hat{f}(y,t)|\leq L|x-y|, \qquad |\hat{g}(x,t)-\hat{g}(y,t)|\leq L|x-y|$$

for all $t \in [0,T]$, for all x, $y \in \mathbb{R}$, for some positive constant L. Let also, $\hat{T} \geq N$ on Ω and U, $P(A) \leq P(B)$, $\hat{T}(A) \geq \hat{T}(B)$ for every A, $B \in U$, $A \subset B$, $\hat{E}(|\hat{X}_0|^{\hat{2}}) < \infty$, $\hat{X}_0$ is independent of $\left(\frac{\tilde{W}^2\tilde{T}(\tilde{W}^2)(0)}{\hat{T}}\right)^+$. Then the problem

$$(2) \qquad \begin{aligned} d\hat{X} &= \hat{f}(\hat{X},t)dt + \hat{g}(\hat{X},t)d\frac{\tilde{W}^2\tilde{T}(\tilde{W}^2)}{\hat{T}}, \qquad 0 \leq t \leq T, \\ \hat{X}(0) &= \hat{X}_0 \end{aligned}$$

has unique solution $\hat{X} \in \hat{\mathbb{L}}^{\hat{2}}(0,T)$.

Proof. 1. Existence. We define

$$\begin{cases} \hat{X}^0(t) := \hat{X}_0 \\ \hat{X}^{n+1}(t) := \hat{X}_0 + \int_0^t \hat{f}(\hat{X}^n(s),s)ds + \int_0^t \hat{g}(\hat{X}^n(s),s)d\frac{\tilde{W}^2\tilde{T}(\tilde{W}^2)}{\hat{T}} \end{cases}$$

for $n = 0,1,2,\ldots$.

Let also,

$$d^n(t) := \hat{E}\left(|\hat{X}^{n+1}(t) - \hat{X}^n(t)|^{\hat{2}}\right).$$

Then

$$\begin{aligned}
&\hat{E}\left(|\hat{X}^1(t) - \hat{X}^0(t)|^{\hat{2}}\right) \\
&= \hat{E}\left(\hat{T}|\hat{X}^1(t) - \hat{X}^0(t)|^2\right) \\
&= \hat{E}\left(\hat{T}\left|\hat{X}_0 + \int_0^t \hat{f}(\hat{X}^0(s),s)ds + \int_0^t \hat{g}(\hat{X}^0(s),s)d\frac{\tilde{W}^2\tilde{T}(\tilde{W}^2)}{\hat{T}} - \hat{X}_0\right|^2\right) \\
&= \hat{E}\left(\hat{T}\left|\int_0^t \hat{f}(\hat{X}^0(s),s)ds + \int_0^t \hat{g}(\hat{X}^0(s),s)d\frac{\tilde{W}^2\tilde{T}(\tilde{W}^2)}{\hat{T}}\right|^2\right) \\
&(a+b)^2 \leq 2(a^2+b^2) \\
&\leq \hat{E}\left(2\hat{T}\left|\int_0^t \hat{f}(\hat{X}_0,s)ds\right|^2 + 2\hat{T}\left|\int_0^t \hat{g}(\hat{X}_0,s)d\frac{\tilde{W}^2\tilde{T}(\tilde{W}^2)}{\hat{T}}\right|^2\right) \\
&= \hat{E}\left(2\hat{T}\left|\int_0^t \hat{f}(\hat{X}_0,s)ds\right|^2\right) + \hat{E}\left(2\hat{T}\left|\int_0^t \hat{g}(\hat{X}_0,s)d\frac{\tilde{W}^2\tilde{T}(\tilde{W}^2)}{\hat{T}}\right|^2\right) \\
&\leq 2\hat{E}\left(\hat{T}\left(\int_0^t (L+|\hat{X}_0|)ds\right)^2\right) + 2\hat{E}\left(\left(\int_0^t \hat{g}(\hat{X}_0,s)d\frac{\tilde{W}^2\tilde{T}(\tilde{W}^2)}{\hat{T}}\right)^{\hat{2}}\right)
\end{aligned}$$

$$= 2\hat{E}\left(\hat{T}(L+|\hat{X}_0|)^2 t^2\right) + 2\hat{E}\left(\left(\int_0^t \hat{g}(\hat{X}_0,s)d\frac{\tilde{W}^2\tilde{T}(\tilde{W}^2)}{\hat{T}}\right)^{\hat{2}}\right)$$

$$\leq 2\hat{E}\left(\hat{T}(L+|\hat{X}_0|)^2\right)Tt + 2\hat{E}\left(\left(\hat{\int}_0^t \tilde{T}\hat{g}(\hat{X}_0,s)\frac{\tilde{T}\tilde{T}}{\tilde{T}\tilde{T}}d\frac{\tilde{W}^2\tilde{T}(\tilde{W}^2)}{\hat{T}}\right)^{\hat{2}}\right)$$

$$= 2\hat{E}\left(\hat{T}(L+|\hat{x}_0|)^2\right)Tt + 2\hat{E}\left(\left(\hat{\int}_0^t \frac{\hat{g}(\hat{X}_0,s)}{\tilde{T}}\hat{\times}\hat{d}\frac{\tilde{W}^2\tilde{T}(\tilde{W}^2)}{\hat{T}}\right)^{\hat{2}}\right)$$

$$= 2\hat{E}\left(\hat{T}(L+|\hat{x}_0|)^2\right)Tt + 2\hat{E}\left(\int_0^t \frac{\hat{g}^2(\hat{X}_0,s)}{\tilde{T}^2}\frac{\tilde{T}^2}{\hat{T}}ds\right)$$

$$= 2\hat{E}\left(\hat{T}(L+|\hat{x}_0|)^2\right)Tt + 2\hat{E}\left(\frac{1}{\hat{T}}\int_0^t \hat{g}^2(\hat{X}_0,s)ds\right)$$

$$= 2\hat{E}\left(\hat{T}(L+|\hat{x}_0|)^2\right)Tt + 2\hat{E}\left(\frac{1}{\hat{T}}\int_0^t (L+|\hat{X}_0|)^2 ds\right)$$

$$= 2\hat{E}\left(\hat{T}(L+|\hat{x}_0|)^2\right)Tt + 2\hat{E}\left(\frac{1}{\hat{T}}(L+|\hat{X}_0|)^2 t\right)$$

$$= 2\hat{E}\left(\hat{T}(L+|\hat{x}_0|)^2\right)Tt + 2\hat{E}\left(\frac{(L+|\hat{X}_0|)^2}{\hat{T}}\right)t$$

$$= \left(2\hat{E}\left(\hat{T}(L+|\hat{x}_0|)^2\right)T + 2\hat{E}\left(\frac{(L+|\hat{X}_0|)^2}{\hat{T}}\right)\right)t,$$

i.e.,

$$d^0(t) \leq \left(2\hat{E}\left(\hat{T}(L+|\hat{x}_0|)^2\right)T + 2\hat{E}\left(\frac{(L+|\hat{X}_0|)^2}{\hat{T}}\right)\right)t.$$

Let

$$M := \max\left\{2\hat{E}\left(\hat{T}(L+|\hat{x}_0|)^2\right)T + 2\hat{E}\left(\frac{(L+|\hat{X}_0|)^2}{\hat{T}}\right), 2TL^2 + 2\frac{L^2}{N^2}\right\}.$$

Therefore

$$d^0(t) \leq Mt.$$

Now we consider $d^1(t)$. For it we have

$$d^1(t) = \hat{E}\left(|\hat{X}^2(t) - \hat{X}^1(t)|^{\hat{2}}\right)$$

$$= \hat{E}\left(\hat{T}|\hat{X}^2(t) - \hat{X}^1(t)|^2\right)$$

$$= \hat{E}\left(\hat{T}\left|\hat{X}_0 + \int_0^t \hat{f}(\hat{X}^1(s),s)ds + \int_0^t \hat{g}(\hat{X}^1(s),s)d\frac{\tilde{W}^2\tilde{T}(\tilde{W}^2)}{\hat{T}}\right.\right.$$

$$\left.\left. -\hat{X}_0 - \int_0^t \hat{f}(\hat{X}^0(s),s)ds - \int_0^t \hat{g}(\hat{X}^0(s),s)d\frac{\tilde{W}^2\tilde{T}(\tilde{W}^2)}{\hat{T}}\right|^2\right)$$

$$= \hat{E}\left(\hat{T}\left|\int_0^t (\hat{f}(\hat{X}^1(s),s) - \hat{f}(\hat{X}_0,s))ds + \int_0^t (\hat{g}(\hat{X}^1(s),s) - \hat{g}(\hat{X}^0(s),s))\frac{d\tilde{W}^2\tilde{T}(\tilde{W}^2)}{\hat{T}}\right|^2\right)$$

$$
\begin{aligned}
&(a+b)^2 \le 2a^2+2b^2\\
&\le \hat{E}\left(2\hat{T}\left|\int_0^t(\hat{f}(\hat{X}^1(s),s)-\hat{f}(\hat{X}_0,s))ds\right|^2+2\hat{T}\left(\int_0^t(\hat{g}(\hat{X}^1(s),s)-\hat{g}(\hat{X}^0,s))d\frac{\tilde{W}^2\tilde{T}(\tilde{W}^2)}{\hat{T}}\right)^2\right)\\
&=\hat{E}\left(2\hat{T}\left|\int_0^t(\hat{f}(\hat{X}^1(s),s)-\hat{f}(\hat{X}_0,s))ds\right|^2\right)+\hat{E}\left(2\hat{T}\left(\int_0^t(\hat{g}(\hat{X}^1(s),s)-\hat{g}(\hat{X}^0,s))d\frac{\tilde{W}^2\tilde{T}(\tilde{W}^2)}{\hat{T}}\right)^2\right)\\
&=\hat{E}\left(2\hat{T}\left|\int_0^t(\hat{f}(\hat{X}^1(s),s)-\hat{f}(\hat{X}_0,s))ds\right|^2\right)+\hat{E}\left(2\left(\int_0^t(\hat{g}(\hat{X}^1(s),s)-\hat{g}(\hat{X}^0,s))d\frac{\tilde{W}^2\tilde{T}(\tilde{W}^2)}{\hat{T}}\right)^{\hat{2}}\right)\\
&\le 2\hat{E}\left(L^2\hat{T}\left(|\hat{X}^1(s)-\hat{X}^0|ds\right)^2\right)+2\hat{E}\left(\left(\int_0^t(\hat{g}(\hat{X}^1(s),s)-\hat{g}(\hat{X}^0,s))d\frac{\tilde{W}^2\tilde{T}(\tilde{W}^2)}{\hat{T}}\right)^{\hat{2}}\right)\\
&\le 2L^2\hat{E}\left(\hat{T}t\int_0^t|\hat{X}^1(s)-\hat{X}^0|^2ds\right)+2\hat{E}\left(\left(\int_0^t\frac{1}{\tilde{T}}\tilde{T}(\hat{g}(\hat{X}^1(s),s)-\hat{g}(\hat{X}_0,s))\frac{\tilde{T}\tilde{T}}{\tilde{T}\tilde{T}}d\frac{\tilde{W}^2\tilde{T}(\tilde{W}^2)}{\hat{T}}\right)^{\hat{2}}\right)\\
&\le 2TL^2\hat{E}\left(\int_0^t\hat{T}|\hat{X}^1(s)-\hat{X}_0|^2ds\right)+2\hat{E}\left(\left(\hat{\int}_0^t\frac{\hat{g}(\hat{X}^1(s),s)-\hat{g}(\hat{X}_0,s)}{\tilde{T}}\hat{\times}\hat{d}\frac{\tilde{W}^2\tilde{T}(\tilde{W}^2)}{\hat{T}}\right)^{\hat{2}}\right)\\
&=2T\hat{E}\left(\int_0^t|\hat{X}^1(s)-\hat{X}^0|^{\hat{2}}ds\right)+2\hat{E}\left(\int_0^t\frac{(\hat{g}(\hat{X}^1(s),s)-\hat{g}(\hat{X}^0,s))^2}{\tilde{T}^2}\frac{\tilde{T}^2}{\hat{T}}ds\right)\\
&=2TL^2\int_0^t\hat{E}\left(|\hat{X}^1(s)-\hat{X}^0|^{\hat{2}}\right)ds+2\hat{E}\left(\int_0^t\frac{L^2(\hat{X}^1(s)-\hat{X}^0)^2}{\hat{T}}ds\right)\\
&=2TL^2\int_0^t d^0(s)ds+2L^2\hat{E}\left(\int_0^t\frac{\hat{T}(\hat{X}^1(s)-\hat{X}^0)^2}{\hat{T}^2}ds\right)\\
&\le 2TL^2\int_0^t d^0(s)ds+2\frac{L^2}{N^2}\hat{E}\left(\int_0^t(\hat{X}^1(s)-\hat{X}^0)^{\hat{2}}ds\right)\\
&=2TL^2\int_0^t d^0(s)ds+2\frac{L^2}{N^2}\int_0^t\hat{E}\left((\hat{X}^1(s)-\hat{X}^0)^{\hat{2}}\right)ds\\
&=2TL^2\int_0^t d^0(s)ds+2\frac{L^2}{N^2}\int_0^t d^0(s)ds\\
&=\left(2TL^2+2\frac{L^2}{N^2}\right)\int_0^t d^0(s)ds\\
&\le\left(2TL^2+2\frac{L^2}{N^2}\right)M\int_0^t sds\\
&=\left(2TL^2+2\frac{L^2}{N^2}\right)M\frac{t^2}{2}\\
&\le\frac{(MT)^2}{2},
\end{aligned}
$$

i.e.,

$$d^1(t) \le \frac{(Mt)^2}{2}.$$

We suppose that

$$d^n(t) \le \frac{(Mt)^{n+1}}{(n+1)!}$$

for some $n \in \mathbb{N}$.

We will prove that

$$d^{n+1}(t) \le \frac{(Mt)^{n+2}}{(n+2)!}$$

$$d^{n+1}(t) = \hat{E}\left(|\hat{X}^{n+2}(t) - \hat{X}^{n+1}(t)|^{\hat{2}}\right)$$

$$= \hat{E}\left(\hat{T}|\hat{X}^{n+2}(t) - \hat{X}^{n+1}(t)|^2\right)$$

$$= \hat{E}\left(\hat{T}\middle|\hat{X}_0 + \int_0^t \hat{f}(\hat{X}^{n+1}(s),s)ds + \int_0^t \hat{g}(\hat{X}^{n+1}(s),s)d\frac{\tilde{W}^2\tilde{T}(\tilde{W}^2)}{\hat{T}}\right.$$

$$\left.-\hat{X}_0 - \int_0^t \hat{f}(\hat{X}^n(s),s)ds - \int_0^t \hat{g}(\hat{X}^n(s),s)d\frac{\tilde{W}^2\tilde{T}(\tilde{W}^2)}{\hat{T}}\right|^2\Bigg)$$

$$= \hat{E}\left(\hat{T}\left|\int_0^t(\hat{f}(\hat{X}^{n+1}(s),s) - \hat{f}(\hat{X}^n(s),s))ds + \int_0^t(\hat{g}(\hat{X}^{n+1}(s),s) - \hat{g}(\hat{X}^n(s),s))\frac{d\tilde{W}^2\tilde{T}(\tilde{W}^2)}{\hat{T}}\right|^2\right)$$

$$(a+b)^2 \le 2a^2 + 2b^2$$

$$\le \hat{E}\Bigg(2\hat{T}\left|\int_0^t(\hat{f}(\hat{X}^{n+1}(s),s) - \hat{f}(\hat{X}^n(s),s))ds\right|^2$$

$$+2\hat{T}\left(\int_0^t(\hat{g}(\hat{X}^{n+1}(s),s) - \hat{g}(\hat{X}^n(s),s))d\frac{\tilde{W}^2\tilde{T}(\tilde{W}^2)}{\hat{T}}\right)^2\Bigg)$$

$$= \hat{E}\left(2\hat{T}\left|\int_0^t(\hat{f}(\hat{X}^{n+1}(s),s) - \hat{f}(\hat{X}^n(s),s))ds\right|^2\right)$$

$$+\hat{E}\left(2\hat{T}\left(\int_0^t(\hat{g}(\hat{X}^{n+1}(s),s) - \hat{g}(\hat{X}^n(s),s))d\frac{\tilde{W}^2\tilde{T}(\tilde{W}^2)}{\hat{T}}\right)^2\right)$$

$$= \hat{E}\left(2\hat{T}\left|\int_0^t(\hat{f}(\hat{X}^{n+1}(s),s) - \hat{f}(\hat{X}^n(s),s))ds\right|^2\right)$$

$$+\hat{E}\left(2\left(\int_0^t(\hat{g}(\hat{X}^{n+1}(s),s) - \hat{g}(\hat{X}^n(s),s))d\frac{\tilde{W}^2\tilde{T}(\tilde{W}^2)}{\hat{T}}\right)^{\hat{2}}\right)$$

$$\le 2\hat{E}\left(L^2\hat{T}\left(|\hat{X}^{n+1}(s) - \hat{X}^n(s)|ds\right)^2\right)$$

$$+2\hat{E}\left(\left(\int_0^t(\hat{g}(\hat{X}^{n+1}(s),s) - \hat{g}(\hat{X}^n(s),s))d\frac{\tilde{W}^2\tilde{T}(\tilde{W}^2)}{\hat{T}}\right)^{\hat{2}}\right)$$

$$\leq 2L^2\hat{E}\left(\hat{T}t\int_0^t|\hat{X}^{n+1}(s)-\hat{X}^n(s)|^2ds\right)$$

$$+2\hat{E}\left(\left(\int_0^t\frac{1}{\tilde{T}}\tilde{T}(\hat{g}(\hat{X}^{n+1}(s),s)-\hat{g}(\hat{X}^n(s),s))\frac{\tilde{T}\tilde{T}}{\tilde{T}\tilde{T}}d\frac{\tilde{W}^2\tilde{T}(\tilde{W}^2)}{\hat{T}}\right)^{\hat{2}}\right)$$

$$\leq 2TL^2\hat{E}\left(\int_0^t\hat{T}|\hat{X}^{n+1}(s)-\hat{X}^n(s)|^2ds\right)$$

$$+2\hat{E}\left(\left(\hat{\int}_0^t\frac{\hat{g}(\hat{X}^{n+1}(s),s)-\hat{g}(\hat{X}^n(s),s)}{\tilde{T}}\hat{\times}\hat{d}\frac{\tilde{W}^2\tilde{T}(\tilde{W}^2)}{\hat{T}}\right)^{\hat{2}}\right)$$

$$=2T\hat{E}\left(\int_0^t|\hat{X}^{n+1}(s)-\hat{X}^n(s)|^{\hat{2}}ds\right)+2\hat{E}\left(\int_0^t\frac{(\hat{g}(\hat{X}^{n+1}(s),s)-\hat{g}(\hat{X}^n(s),s))^2}{\tilde{T}^2}\frac{\tilde{T}^2}{\hat{T}}ds\right)$$

$$=2TL^2\int_0^t\hat{E}\left(|\hat{X}^{n+1}(s)-\hat{X}^n(s)|^{\hat{2}}\right)ds+2\hat{E}\left(\int_0^t\frac{L^2(\hat{X}^{n+1}(s)-\hat{X}^n(s))^2}{\hat{T}}ds\right)$$

$$=2TL^2\int_0^t d^n(s)ds+2L^2\hat{E}\left(\int_0^t\frac{\hat{T}(\hat{X}^{n+1}(s)-\hat{X}^n(s))^2}{\hat{T}^2}ds\right)$$

$$\leq 2TL^2\int_0^t d^n(s)ds+2\frac{L^2}{N^2}\hat{E}\left(\int_0^t(\hat{X}^{n+1}(s)-\hat{X}^n(s))^{\hat{2}}ds\right)$$

$$=2TL^2\int_0^t d^n(s)ds+2\frac{L^2}{N^2}\int_0^t\hat{E}\left((\hat{X}^{n+1}(s)-\hat{X}^n(s))^{\hat{2}}\right)ds$$

$$=2TL^2\int_0^t d^n(s)ds+2\frac{L^2}{N^2}\int_0^t d^n(s)ds$$

$$=\left(2TL^2+2\frac{L^2}{N^2}\right)\int_0^t d^n(s)ds$$

$$\leq\left(2TL^2+2\frac{L^2}{N^2}\right)\frac{M^{n+1}}{(n+1)!}\int_0^t s^{n+1}ds$$

$$=\left(2TL^2+2\frac{L^2}{N^2}\right)M^{n+1}\frac{t^{n+2}}{(n+2)!}$$

$$\leq\frac{(MT)^{n+2}}{(n+2)!}.$$

Consequently

$$d^n(t)\leq\frac{(Mt)^{n+1}}{(n+1)!}$$

for every $n\in\mathbb{N}$.

Hence,

$$\hat{E}\left(\max_{0\leq t\leq T}|\hat{X}^{n+1}(t)-\hat{X}^n(t)|^{\hat{2}}\right)\leq\frac{M^{n+1}T^{n+1}}{(n+1)!}.$$

From the iso-Chebyshev,s inequality we get

$$P\left(\max_{0\leq t\leq T}|\hat{X}^{n+1}(t)-\hat{X}^{n}(t)|>\frac{1}{2^n}\right)\leq \frac{1}{\left(\frac{1}{2^n}\right)^2}\hat{E}\left(\max_{0\leq t\leq T}|\hat{X}^{n+1}(t)-\hat{X}^{n}(t)|^2\right)$$

$$+\frac{1}{\left(\frac{1}{2^n}\right)^2}\int_\Omega \max_{0\leq t\leq T}|\hat{X}^{n+1}(t)-\hat{X}^{n}(t)|^2\frac{\hat{T}'(\omega)}{\hat{T}(\omega)}P(\omega)d\omega$$

$$\leq C2^{2n}\hat{E}\left(\max_{0\leq t\leq T}|\hat{X}^{n+1}(t)-\hat{X}^{n}(t)|^2\right)$$

$$=C2^{2n}\hat{E}\left(\frac{1}{\hat{T}}\max_{0\leq t\leq T}|\hat{X}^{n+1}(t)-\hat{X}^{n}(t)|^{\hat{2}}\right)$$

$$\leq \frac{C}{N}2^{2n}\hat{E}\left(\max_{0\leq t\leq T}|\hat{X}^{n+1}(t)-\hat{X}^{n}(t)|^{\hat{2}}\right)$$

$$\leq \frac{C}{N}2^{2n}\frac{(MT)^{n+1}}{(n+1)!}$$

$$\leq \frac{C}{N}\frac{(4MT)^{n+1}}{(n+1)!}$$

for some positive constant C.

From here and iso-Borel-Cantelli lemma it follows that

$$\hat{P}\left(\max_{0\leq t\leq T}|\hat{X}^{n+1}(t)-\hat{X}^{n}(t)|>\frac{1}{2^n}\text{i.o.}\right)=0.$$

Therefore

$$\hat{X}^n=\hat{X}^0+\sum_{j=0}^{n-1}\left(\hat{X}^{j+1}-\hat{X}^j\right)$$

converges uniformly on $[0,T]$ for almost every ω to a process $\hat{X}(\cdot)$.

We pas to limits in the definition of $\hat{X}^{n+1}$ and we get

$$\hat{X}(t)=\hat{X}_0+\int_0^t \hat{f}(\hat{X}(s),s)ds+\int_0^t \hat{g}(\hat{X}(s),s)d\frac{\tilde{W}^2\tilde{T}(\tilde{W}^2)}{\hat{T}},\quad 0\leq t\leq T.$$

Now we will prove that $\hat{X}\in\hat{\mathbb{L}}^{\hat{2}}(0,T)$.

We have

$$|\hat{X}^{n+1}|^{\hat{2}}=\hat{T}|\hat{X}^{n+1}|^2$$

$$=\hat{T}\left|\hat{X}_0+\int_0^t\hat{f}(\hat{X}^n(s),s)ds+\int_0^t\hat{g}(\hat{X}^n(s),s)d\frac{\tilde{W}^2\tilde{T}(\tilde{W}^2)}{\hat{T}}\right|^2$$

$$\leq C_5\hat{T}\left(|\hat{X}_0|^2+\left(\int_0^t|\hat{f}(\hat{X}^n(s),s)|ds\right)^2+\left(\int_0^t|\hat{g}(\hat{X}^n(s),s)|d\frac{\tilde{W}^2\tilde{T}(\tilde{W}^2)}{\hat{T}}\right)^2\right)$$

$$\leq C_5\hat{T}|\hat{X}_0|^2+C_5T\hat{T}\int_0^t|\hat{f}(\hat{X}^n(s),s)|^2ds+C_5\left(\int_0^t|\hat{g}(\hat{X}^n(s),s)|d\frac{\tilde{W}^2\tilde{T}(\tilde{W}^2)}{\hat{T}}\right)^{\hat{2}}$$

$$\leq C_5|\hat{X}_0|^{\hat{2}} + C_5 T\hat{T}\int_0^t L^2(1+|\hat{X}^n|)^2 ds + C_5\left(\int_0^t \frac{\tilde{T}}{\hat{T}}|\hat{g}(\hat{X}^n(s),s)|\frac{\tilde{T}\tilde{T}}{\tilde{T}\tilde{T}} d\frac{\tilde{W}^2\tilde{T}(\tilde{W}^2)}{\hat{T}}\right)^{\hat{2}}$$

$$\leq C_5|\hat{X}_0|^{\hat{2}} + C_5 T\hat{T}L^2\int_0^t (1+|\hat{X}^0|)^2 ds + C_5\left(\hat{\int}_0^t \frac{|\hat{g}(\hat{X}^n(s),s)|}{\tilde{T}}\hat{\times}\hat{d}\frac{\tilde{W}^2\tilde{T}(\tilde{W}^2)}{\hat{T}}\right)^{\hat{2}}$$

$$\leq C_6|\hat{X}_0|^{\hat{2}} + C_6\hat{T}\int_0^t |\hat{X}^n|^2 ds + C_6\left(\hat{\int}_0^t \frac{|\hat{g}(\hat{X}^n(s),s)|}{\tilde{T}}\hat{\times}\hat{d}\frac{\tilde{W}^2\tilde{T}(\tilde{W}^2)}{\hat{T}}\right)^{\hat{2}}$$

$$\leq C_6|\hat{X}_0|^{\hat{2}} + C_6\int_0^t |\hat{X}^n|^{\hat{2}} ds + C_6\left(\hat{\int}_0^t \frac{|\hat{g}(\hat{X}^n(s),s)|}{\tilde{T}}\hat{\times}\hat{d}\frac{\tilde{W}^2\tilde{T}(\tilde{W}^2)}{\hat{T}}\right)^{\hat{2}},$$

i.e.,

$$|\hat{X}^{n+1}|^{\hat{2}} \leq C_6|\hat{X}_0|^{\hat{2}} + C_6\int_0^t |\hat{X}^n|^{\hat{2}} ds + C_6\left(\hat{\int}_0^t \frac{|\hat{g}(\hat{X}^n(s),s)|}{\tilde{T}}\hat{\times}\hat{d}\frac{\tilde{W}^2\tilde{T}(\tilde{W}^2)}{\hat{T}}\right)^{\hat{2}}.$$

Here C_5 and C_6 are positive constants.

From here,

$$\hat{E}\left(|\hat{X}^{n+1}|^{\hat{2}}\right) \leq C_6\hat{E}\left(|\hat{X}_0|^{\hat{2}}\right) + C_6\int_0^t \hat{E}\left(|\hat{X}^n|^{\hat{2}}\right) ds + C_6\hat{E}\left(\left(\hat{\int}_0^t \frac{|\hat{g}(\hat{X}^n(s),s)|}{\tilde{T}}\hat{\times}\hat{d}\frac{\tilde{W}^2\tilde{T}(\tilde{W}^2)}{\hat{T}}\right)^{\hat{2}}\right)$$

$$\leq C_6\hat{E}\left(|\hat{X}_0|^{\hat{2}}\right) + C_6\int_0^t \hat{E}\left(|\hat{X}^n|^{\hat{2}}\right) ds + C_6\hat{E}\left(\int_0^t \frac{|\hat{g}(\hat{X}^n(s),s)|^2}{\tilde{T}^2}\frac{\tilde{T}^2}{\hat{T}} ds\right)$$

$$\leq C_6\hat{E}\left(|\hat{X}_0|^{\hat{2}}\right) + C_6\int_0^t \hat{E}\left(|\hat{X}^n|^{\hat{2}}\right) ds + C_6\hat{E}\left(\int_0^t \frac{|\hat{g}(\hat{X}^n(s),s)|^2}{\hat{T}} ds\right)$$

$$\leq C_7\hat{E}\left(|\hat{X}_0|^{\hat{2}}\right) + C_7\int_0^t \hat{E}\left(|\hat{X}^n|^{\hat{2}}\right) ds + C_7 L^2\hat{E}\left(\int_0^t \frac{|\hat{X}^n(s)|^2}{\hat{T}}\right)$$

$$\leq C_7\hat{E}\left(|\hat{X}_0|^{\hat{2}}\right) + C_7\int_0^t \hat{E}\left(|\hat{X}^n|^{\hat{2}}\right) ds + C_7 L^2\hat{E}\left(\int_0^t \frac{|\hat{X}^n(s)|^{\hat{2}}}{\hat{T}^2} ds\right)$$

$$\leq C_7\hat{E}\left(|\hat{X}_0|^{\hat{2}}\right) + C_7\int_0^t \hat{E}\left(|\hat{X}^n|^{\hat{2}}\right) ds + C_7\frac{L^2}{N^2}\int_0^t \hat{E}\left(|\hat{X}^n|^{\hat{2}}\right) ds$$

$$\leq C_8\hat{E}\left(|\hat{X}_0|^{\hat{2}}\right) + C_8\int_0^t \hat{E}\left(|\hat{X}^n(s)|^{\hat{2}}\right) ds,$$

i.e.,

$$\hat{E}\left(|\hat{X}^{n+1}|^{\hat{2}}\right) \leq C_8\hat{E}\left(|\hat{X}_0|^{\hat{2}}\right) + C_8\int_0^t \hat{E}\left(|\hat{X}^n(s)|^{\hat{2}}\right) ds, \tag{8}$$

where C_7 and C_8 are positive constants and $C_8 > 1$.

For $n = 0$, using (8), we get

$$\hat{E}\left(|\hat{X}^1|^{\hat{2}}\right) \leq C_8\hat{E}\left(|\hat{X}_0|^{\hat{2}}\right) + C_8\int_0^t \hat{E}\left(|\hat{X}_0|^{\hat{2}}\right) ds$$

$$= C_8(1+t)\hat{E}\left(|\hat{X}_0|^{\hat{2}}\right).$$

From here and (8), for $n = 1$, we find

$$\hat{E}\left(|\hat{X}^2|^{\hat{2}}\right) \leq C_8\hat{E}\left(|\hat{X}_0|^{\hat{2}}\right) + C_8\int_0^t \hat{E}\left(|\hat{X}^1|^{\hat{2}}\right) ds$$

$$\leq C_8\hat{E}\left(|\hat{X}_0|^{\hat{2}}\right) + C_8^2\int_0^t (1+s)\hat{E}\left(|\hat{X}_0|^{\hat{2}}\right) ds$$

$$= \left(C_8 + C_8^2\frac{(1+t)^2}{2} - \frac{C_8}{2}\right)\hat{E}\left(|\hat{X}_0|^{\hat{2}}\right)$$

$$\leq \left(C_8 + C_8(1+t) + C_8^2\frac{(1+t)^2}{2}\right)\hat{E}\left(|\hat{X}_0|^{\hat{2}}\right).$$

We suppose that

$$\hat{E}\left(|\hat{X}_0^n|^{\hat{2}}\right) \leq \left(C_8 + C_8(1+t) + C_8^2\frac{(1+t)^2}{2} + \cdots + C_8^n\frac{(1+t)^n}{n!}\right)\hat{E}\left(|\hat{X}_0|^{\hat{2}}\right) \tag{9}$$

for some $n \in \mathbb{N}$.

We will prove that

$$\hat{E}\left(|\hat{X}_0^{n+1}|^{\hat{2}}\right)$$

$$\leq \left(C_8 + C_8(1+t) + C_8^2\frac{(1+t)^2}{2} + \cdots + C_8^n\frac{(1+t)^n}{n!} + C_8^{n+1}\frac{(1+t)^{n+1}}{(n+1)!}\right)\hat{E}\left(|\hat{X}_0|^{\hat{2}}\right).$$

Really, from (8), using (9), we have that

$$\hat{E}\left(|\hat{X}^{n+1}|^{\hat{2}}\right) \leq C_8\hat{E}\left(|\hat{X}_0|^{\hat{2}}\right) + C_8\int_0^t \hat{E}\left(|\hat{X}^n(s)|^{\hat{2}}\right) ds$$

$$\leq C_8\hat{E}\left(|\hat{X}_0|^{\hat{2}}\right)$$

$$+C_8\int_0^t \left(C_8 + C_8(1+s) + C_8^2\frac{(1+s)^2}{2!} + \cdots + C_8^n\frac{(1+s)^n}{n!}\right) ds\hat{E}\left(|\hat{X}_0|^{\hat{2}}\right)$$

$$\leq \left(C_8 + C_8(1+t) + C_8\frac{(1+t)^2}{2!} + \cdots + C_8^{n+1}\frac{(1+t)^{n+1}}{(n+1)!}\right)\hat{E}\left(|\hat{X}_0|^{\hat{2}}\right).$$

Consequently (9) is valid for every $n \in \mathbb{N}$. From it we find

$$\hat{E}\left(|\hat{X}^n|^{\hat{2}}\right) \leq \left(C_8 + e^{C_8(1+t)}\right)\hat{E}\left(|\hat{X}_0|^{\hat{2}}\right),$$

whereupon, when $n \longrightarrow \infty$,

$$\hat{E}\left(|\hat{X}|^{\hat{2}}\right) \leq \left(C_8 + e^{C_8(1+t)}\right)\hat{E}\left(|\hat{X}_0|^{\hat{2}}\right),$$

i.e.,

$$\hat{X} \in \hat{\mathbb{L}}^{\hat{2}}(0,T).$$

2. Uniqueness. Now we suppose that the problem (2) has two solutions $\hat{X}$ and $\hat{Y}$. Then

$$\begin{cases} \hat{X}(t) = \hat{X}_0 + \int_0^t \hat{f}(\hat{X}(s),s)ds + \int_0^t \hat{g}(\hat{X}(s),s)d\frac{\tilde{W}^2\tilde{T}(\tilde{W}^2)}{\hat{T}} \\ \hat{Y}(t) = \hat{X}_0 + \int_0^t \hat{f}(\hat{Y}(s),s)ds + \int_0^t \hat{g}(\hat{Y}(s),s)d\frac{\tilde{W}^2\tilde{T}(\tilde{W}^2)}{\hat{T}}, \qquad 0 \le t \le T, \end{cases}$$

whereupon

$$\hat{X}(t) - \hat{Y}(t) = \int_0^t \left(\hat{f}(\hat{X}(s),s) - \hat{f}(\hat{Y}(s),s)\right)ds + \int_0^t \left(\hat{g}(\hat{X}(s),s) - \hat{g}(\hat{Y}(s),s)\right)d\frac{\tilde{W}^2\tilde{T}(\tilde{W}^2)}{\hat{T}}$$

and

$$\hat{E}\left(|\hat{X}(t) - \hat{Y}(t)|^{\hat{2}}\right) = \hat{E}\left(\hat{T}|\hat{X}(t) - \hat{Y}(t)|^2\right)$$

$$\hat{E}\left(\hat{T}\left(\int_0^t \left(\hat{f}(\hat{X}(s),s) - \hat{f}(\hat{Y}(s),s)\right)ds + \int_0^t \left(\hat{g}(\hat{X}(s),s) - \hat{g}(\hat{Y}(s),s)\right)d\frac{\tilde{W}^2\tilde{T}(\tilde{W}^2)}{\hat{T}}\right)^2\right)$$

$$\le 2\hat{E}\left(\hat{T}\left(\int_0^t(\hat{f}(\hat{X}(s),s) - \hat{f}(\hat{Y}(s),s))\right)^2\right)$$

$$+2\hat{E}\left(\hat{T}\left(\int_0^t(\hat{g}(\hat{X}(s),s) - \hat{g}(\hat{Y}(s),s))d\frac{\tilde{W}^2\tilde{T}(\tilde{W}^2)}{\hat{T}}\right)^2\right)$$

$$\le 2\hat{E}\left(\hat{T}t\int_0^t(\hat{f}(\hat{X}(s),s) - \hat{f}(\hat{Y}(s),s))^2 ds\right)$$

$$+2\hat{E}\left(\hat{T}\left(\int_0^t \tilde{T}(s)\frac{1}{\tilde{T}(s)}\frac{\hat{g}(\hat{X}(s),s) - \hat{g}(\hat{Y}(s),s)}{\tilde{T}^2(s)}\tilde{T}^2(s)d\frac{\tilde{W}^2\tilde{T}(\tilde{W}^2)}{\hat{T}}\right)^2\right)$$

$$\le 2TL^2\hat{E}\left(\hat{T}\int_0^t|\hat{X}(s) - \hat{Y}(s)|^2 ds\right) + 2\hat{E}\left(\left(\hat{\int}_0^t \frac{\hat{g}(\hat{X}(s),s) - \hat{g}(\hat{Y}(s),s)}{\tilde{T}}d\frac{\tilde{W}^2\tilde{T}(\tilde{W}^2)}{\hat{T}}\right)^{\hat{2}}\right)$$

$$= 2TL^2\hat{E}\left(\int_0^t|\hat{X}(s) - \hat{Y}(s)|^2 ds\right) + 2\hat{E}\left(\int_0^t \frac{(\hat{g}(\hat{X}(s),s) - \hat{g}(\hat{Y}(s),s))^2}{\hat{T}}ds\right)$$

$$\le 2TL^2\int_0^t \hat{E}\left(|\hat{X}(s) - \hat{Y}(s)|^2\right)ds + 2L^2\hat{E}\left(\int_0^t \frac{|\hat{X}(s) - \hat{Y}(s)|^2}{\hat{T}}ds\right)$$

$$= 2TL^2\int_0^t \hat{E}(|\hat{X}(s) - \hat{Y}(s)|^{\hat{2}})ds + 2L^2\hat{E}\left(\int_0^t \frac{|\hat{X}(s) - \hat{Y}(s)|^{\hat{2}}}{\hat{T}}ds\right)$$

$$\le 2TL^2\int_0^t \hat{E}(|\hat{X}(s) - \hat{Y}(s)|^{\hat{2}})ds + 2\frac{L^2}{N^2}\int_0^t \hat{E}(|\hat{X}(s) - \hat{Y}(s)|^{\hat{2}})ds$$

$$= 2L^2\left(T + \frac{1}{N^2}\right)\int_0^t \hat{E}(|\hat{X}(s) - \hat{Y}(s)|^{\hat{2}})ds,$$

i.e.,

$$\int_0^t \hat{E}(|\hat{X}(s) - \hat{Y}(s)|^{\hat{2}})ds \le 2L^2\left(T + \frac{1}{N^2}\right)\int_0^t \hat{E}(|\hat{X}(s) - \hat{Y}(s)|^{\hat{2}})ds.$$

Hence and Gronwall's lemma it follows that

$$\hat{E}(|\hat{X}(t)-\hat{Y}(t)|^{\hat{2}})=0 \qquad \text{for} \qquad t\in[0,T].$$

Thus

$$\hat{X}(t)=\hat{Y}(t) \qquad \text{a.s.}$$

for all $0\leq t\leq T$.

Thus

$$\hat{X}(r)=\hat{Y}(r)$$

for all rational $0\leq r\leq T$, except for some set of probability zero. As $\hat{X}$ and $\hat{Y}$ have continuous sample paths almost surely,

$$\hat{P}\left(\max_{0\leq t\leq T}|\hat{X}(t)-\hat{Y}(t)|>0\right)=0.$$

□

Theorem 2.2.4. *Suppose that $\hat{f},\ \hat{g}:\mathbb{R}\times[0,T]$ are continuous and satisfy the following conditions*

$$|\hat{f}(x,t)|\leq L(1+|x|), \qquad |\hat{g}(x,t)|\leq L(1+|x|),$$

$$|\hat{f}(x,t)-\hat{f}(y,t)|\leq L|x-y|, \qquad |\hat{g}(x,t)-\hat{g}(y,t)|\leq L|x-y|$$

for all $t\in[0,T]$, for all $x,\ y\in\mathbb{R}$, for some positive constant L. Let also, $\hat{T}\geq N$ on Ω and U, $P(A)\leq P(B)$, $\hat{T}(A)\geq\hat{T}(B)$ for every $A,\ B\in U$, $A\subset B$, $\hat{E}(|\hat{X}_0|^{\hat{2}})<\infty$, $\hat{X}_0$ is independent of $\left(\frac{\frac{\tilde{W}^3}{\hat{T}(\tilde{W}^3)}(0)}{\hat{T}}\right)^+$. Then the problem

$$(3)\qquad \begin{aligned} &d\hat{X}=\hat{f}(\hat{X},t)dt+\hat{g}(\hat{X},t)d\frac{\frac{\tilde{W}^3}{\hat{T}(\tilde{W}^3)}}{\hat{T}}, \qquad 0\leq t\leq T,\\ &\hat{X}(0)=\hat{X}_0 \end{aligned}$$

has unique solution $\hat{X}\in\hat{\mathbb{L}}^{\hat{2}}(0,T)$.

Proof. 1. Existence. We define

$$\begin{cases} \hat{X}^0(t):=\hat{X}_0\\ \hat{X}^{n+1}(t):=\hat{X}_0+\int_0^t\hat{f}(\hat{X}^n(s),s)ds+\int_0^t\hat{g}(\hat{X}^n(s),s)d\frac{\frac{\tilde{W}^3}{\hat{T}(\tilde{W}^3)}}{\hat{T}}\end{cases}$$

for $n=0,1,2,\ldots$.

Let also,

$$d^n(t):=\hat{E}\left(|\hat{X}^{n+1}(t)-\hat{X}^n(t)|^{\hat{2}}\right).$$

Then

$$\hat{E}\left(|\hat{X}^1(t)-\hat{X}^0(t)|^{\hat{2}}\right)$$

$$=\hat{E}\left(\hat{T}|\hat{X}^1(t)-\hat{X}^0(t)|^2\right)$$

$$=\hat{E}\left(\hat{T}\left|\hat{X}_0+\int_0^t\hat{f}(\hat{X}^0(s),s)ds+\int_0^t\hat{g}(\hat{X}^0(s),s)d\frac{\frac{\tilde{W}^3}{\tilde{T}(\tilde{W}^3)}}{\hat{T}}-\hat{X}_0\right|^2\right)$$

$$=\hat{E}\left(\hat{T}\left|\int_0^t\hat{f}(\hat{X}^0(s),s)ds+\int_0^t\hat{g}(\hat{X}^0(s),s)d\frac{\frac{\tilde{W}^3}{\tilde{T}(\tilde{W}^3)}}{\hat{T}}\right|^2\right)$$

$$(a+b)^2\leq 2(a^2+b^2)$$

$$\leq\hat{E}\left(2\hat{T}\left|\int_0^t\hat{f}(\hat{X}_0,s)ds\right|^2+2\hat{T}\left|\int_0^t\hat{g}(\hat{X}_0,s)d\frac{\frac{\tilde{W}^3}{\tilde{T}(\tilde{W}^3)}}{\hat{T}}\right|^2\right)$$

$$=\hat{E}\left(2\hat{T}\left|\int_0^t\hat{f}(\hat{X}_0,s)ds\right|^2\right)+\hat{E}\left(2\hat{T}\left|\int_0^t\hat{g}(\hat{X}_0,s)d\frac{\frac{\tilde{W}^3}{\tilde{T}(\tilde{W}^3)}}{\hat{T}}\right|^2\right)$$

$$\leq 2\hat{E}\left(\hat{T}\left(\int_0^t(L+|\hat{X}_0|)ds\right)^2\right)+2\hat{E}\left(\left(\int_0^t\hat{g}(\hat{X}_0,s)d\frac{\frac{\tilde{W}^3}{\tilde{T}(\tilde{W}^3)}}{\hat{T}}\right)^{\hat{2}}\right)$$

$$=2\hat{E}\left(\hat{T}(L+|\hat{X}_0|)^2t^2\right)+2\hat{E}\left(\left(\int_0^t\hat{g}(\hat{X}_0,s)d\frac{\frac{\tilde{W}^3}{\tilde{T}(\tilde{W}^3)}}{\hat{T}}\right)^{\hat{2}}\right)$$

$$\leq 2\hat{E}\left(\hat{T}(L+|\hat{X}_0|)^2\right)Tt+2\hat{E}\left(\left(\hat{\int}_0^t\tilde{T}\hat{g}(\hat{X}_0,s)\frac{\tilde{T}\tilde{T}}{\tilde{T}\tilde{T}}d\frac{\frac{\tilde{W}^3}{\tilde{T}(\tilde{W}^3)}}{\hat{T}}\right)^{\hat{2}}\right)$$

$$=2\hat{E}\left(\hat{T}(L+|\hat{x}_0|)^2\right)Tt+2\hat{E}\left(\left(\hat{\int}_0^t\frac{\hat{g}(\hat{X}_0,s)}{\tilde{T}}\hat{\times}\hat{d}\frac{\frac{\tilde{W}^3}{\tilde{T}(\tilde{W}^3)}}{\hat{T}}\right)^{\hat{2}}\right)$$

$$=2\hat{E}\left(\hat{T}(L+|\hat{x}_0|)^2\right)Tt+2\hat{E}\left(\int_0^t\frac{\hat{g}^2(\hat{X}_0,s)}{\tilde{T}^2}\frac{\tilde{T}^2}{\hat{T}}ds\right)$$

$$=2\hat{E}\left(\hat{T}(L+|\hat{x}_0|)^2\right)Tt+2\hat{E}\left(\frac{1}{\hat{T}}\int_0^t\hat{g}^2(\hat{X}_0,s)ds\right)$$

$$=2\hat{E}\left(\hat{T}(L+|\hat{x}_0|)^2\right)Tt+2\hat{E}\left(\frac{1}{\hat{T}}\int_0^t(L+|\hat{X}_0|)^2ds\right)$$

$$=2\hat{E}\left(\hat{T}(L+|\hat{x}_0|)^2\right)Tt+2\hat{E}\left(\frac{1}{\hat{T}}(L+|\hat{X}_0|)^2t\right)$$

$$= 2\hat{E}\left(\hat{T}(L+|\hat{x}_0|)^2\right)Tt + 2\hat{E}\left(\frac{(L+|\hat{X}_0|)^2}{\hat{T}}\right)t$$

$$= \left(2\hat{E}\left(\hat{T}(L+|\hat{x}_0|)^2\right)T + 2\hat{E}\left(\frac{(L+|\hat{X}_0|)^2}{\hat{T}}\right)\right)t,$$

i.e.,

$$d^0(t) \leq \left(2\hat{E}\left(\hat{T}(L+|\hat{x}_0|)^2\right)T + 2\hat{E}\left(\frac{(L+|\hat{X}_0|)^2}{\hat{T}}\right)\right)t.$$

Let

$$M := \max\left\{2\hat{E}\left(\hat{T}(L+|\hat{x}_0|)^2\right)T + 2\hat{E}\left(\frac{(L+|\hat{X}_0|)^2}{\hat{T}}\right), 2TL^2 + 2\frac{L^2}{N^2}\right\}.$$

Therefore

$$d^0(t) \leq Mt.$$

Now we consider $d^1(t)$. For it we have

$$d^1(t) - \hat{E}\left(|\hat{X}^2(t) - \hat{X}^1(t)|^{\hat{2}}\right)$$

$$= \hat{E}\left(\hat{T}|\hat{X}^2(t) - \hat{X}^1(t)|^2\right)$$

$$= \hat{E}\Bigg(\hat{T}\Bigg|\hat{X}_0 + \int_0^t \hat{f}(\hat{X}^1(s),s)ds + \int_0^t \hat{g}(\hat{X}^1(s),s)d\frac{\frac{\tilde{W}^3}{\tilde{T}(\tilde{W}^3)}}{\hat{T}}$$

$$-\hat{X}_0 - \int_0^t \hat{f}(\hat{X}^0(s),s)ds - \int_0^t \hat{g}(\hat{X}^0(s),s)d\frac{\frac{\tilde{W}^3}{\tilde{T}(\tilde{W}^3)}}{\hat{T}}\Bigg|^2\Bigg)$$

$$= \hat{E}\left(\hat{T}\left|\int_0^t(\hat{f}(\hat{X}^1(s),s) - \hat{f}(\hat{X}_0,s))ds + \int_0^t(\hat{g}(\hat{X}^1(s),s) - \hat{g}(\hat{X}^0(s),s))\frac{d\frac{\tilde{W}^3}{\tilde{T}(\tilde{W}^3)}}{\hat{T}}\right|^2\right)$$

$$(a+b)^2 \leq 2a^2 + 2b^2$$

$$\leq \hat{E}\left(2\hat{T}\left|\int_0^t(\hat{f}(\hat{X}^1(s),s) - \hat{f}(\hat{X}_0,s))ds\right|^2 + 2\hat{T}\left(\int_0^t(\hat{g}(\hat{X}^1(s),s) - \hat{g}(\hat{X}^0,s))d\frac{\frac{\tilde{W}^3}{\tilde{T}(\tilde{W}^3)}}{\hat{T}}\right)^2\right)$$

$$= \hat{E}\left(2\hat{T}\left|\int_0^t(\hat{f}(\hat{X}^1(s),s)-\hat{f}(\hat{X}_0,s))ds\right|^2\right)+\hat{E}\left(2\hat{T}\left(\int_0^t(\hat{g}(\hat{X}^1(s),s)-\hat{g}(\hat{X}^0,s))d\ \frac{\frac{\tilde{W}^3}{\tilde{T}(\tilde{W}^3)}}{\hat{T}}\right)^2\right)$$

$$= \hat{E}\left(2\hat{T}\left|\int_0^t(\hat{f}(\hat{X}^1(s),s)-\hat{f}(\hat{X}_0,s))ds\right|^2\right)+\hat{E}\left(2\left(\int_0^t(\hat{g}(\hat{X}^1(s),s)-\hat{g}(\hat{X}^0,s))d\ \frac{\frac{\tilde{W}^3}{\tilde{T}(\tilde{W}^3)}}{\hat{T}}\right)^{\hat{2}}\right)$$

$$\leq 2\hat{E}\left(L^2\hat{T}\left(|\hat{X}^1(s)-\hat{X}^0|ds\right)^2\right)+2\hat{E}\left(\left(\int_0^t(\hat{g}(\hat{X}^1(s),s)-\hat{g}(\hat{X}^0,s))d\ \frac{\frac{\tilde{W}^3}{\tilde{T}(\tilde{W}^3)}}{\hat{T}}\right)^{\hat{2}}\right)$$

$$\leq 2L^2\hat{E}\left(\hat{T}t\int_0^t|\hat{X}^1(s)-\hat{X}^0|^2ds\right)+2\hat{E}\left(\left(\int_0^t \frac{1}{\tilde{T}}\tilde{T}(\hat{g}(\hat{X}^1(s),s)-\hat{g}(\hat{X}_0,s))\frac{\tilde{T}\tilde{T}}{\tilde{T}\tilde{T}}d\frac{\frac{\tilde{W}^3}{\tilde{T}(\tilde{W}^3)}}{\hat{T}}\right)^{\hat{2}}\right)$$

$$\leq 2TL^2\hat{E}\left(\int_0^t\hat{T}|\hat{X}^1(s)-\hat{X}_0|^2ds\right)+2\hat{E}\left(\left(\hat{\int}_0^t\frac{\hat{g}(\hat{X}^1(s),s)-\hat{g}(\hat{X}_0,s)}{\tilde{T}}\hat{\times}\hat{d}\frac{\frac{\tilde{W}^3}{\tilde{T}(\tilde{W}^3)}}{\hat{T}}\right)^{\hat{2}}\right)$$

$$= 2T\hat{E}\left(\int_0^t|\hat{X}^1(s)-\hat{X}^0|^{\hat{2}}ds\right)+2\hat{E}\left(\int_0^t\frac{(\hat{g}(\hat{X}^1(s),s)-\hat{g}(\hat{X}^0,s))^2}{\tilde{T}^2}\frac{\tilde{T}^2}{\hat{T}}ds\right)$$

$$= 2TL^2\int_0^t\hat{E}\left(|\hat{X}^1(s)-\hat{X}^0|^{\hat{2}}\right)ds+2\hat{E}\left(\int_0^t\frac{L^2(\hat{X}^1(s)-\hat{X}^0)^2}{\hat{T}}ds\right)$$

$$= 2TL^2\int_0^t d^0(s)ds+2L^2\hat{E}\left(\int_0^t\frac{\hat{T}(\hat{X}^1(s)-\hat{X}^0)^2}{\hat{T}^2}ds\right)$$

$$\leq 2TL^2\int_0^t d^0(s)ds+2\frac{L^2}{N^2}\hat{E}\left(\int_0^t(\hat{X}^1(s)-\hat{X}^0)^{\hat{2}}ds\right)$$

$$= 2TL^2\int_0^t d^0(s)ds+2\frac{L^2}{N^2}\int_0^t\hat{E}\left((\hat{X}^1(s)-\hat{X}^0)^{\hat{2}}\right)ds$$

$$= 2TL^2\int_0^t d^0(s)ds+2\frac{L^2}{N^2}\int_0^t d^0(s)ds$$

$$= \left(2TL^2+2\frac{L^2}{N^2}\right)\int_0^t d^0(s)ds$$

$$\leq \left(2TL^2+2\frac{L^2}{N^2}\right)M\int_0^t sds$$

$$= \left(2TL^2+2\frac{L^2}{N^2}\right)M\frac{t^2}{2}$$

$$\leq \frac{(MT)^2}{2},$$

i.e.,

$$d^1(t) \leq \frac{(Mt)^2}{2}.$$

We suppose that

$$d^n(t) \leq \frac{(Mt)^{n+1}}{(n+1)!}$$

for some $n \in \mathbb{N}$.

We will prove that

$$d^{n+1}(t) \leq \frac{(Mt)^{n+2}}{(n+2)!}$$

Really,

$$d^{n+1}(t) = \hat{E}\left(|\hat{X}^{n+2}(t) - \hat{X}^{n+1}(t)|^{\hat{2}}\right)$$

$$= \hat{E}\left(\hat{T}|\hat{X}^{n+2}(t) - \hat{X}^{n+1}(t)|^2\right)$$

$$= \hat{E}\left(\hat{T}\middle|\hat{X}_0 + \int_0^t \hat{f}(\hat{X}^{n+1}(s),s)ds + \int_0^t \hat{g}(\hat{X}^{n+1}(s),s)d\frac{\frac{\tilde{W}^3}{\tilde{T}(\tilde{W}^3)}}{\hat{T}}\right.$$

$$\left.-\hat{X}_0 - \int_0^t \hat{f}(\hat{X}^n(s),s)ds - \int_0^t \hat{g}(\hat{X}^n(s),s)d\frac{\frac{\tilde{W}^3}{\tilde{T}(\tilde{W}^3)}}{\hat{T}}\right|^2\Bigg)$$

$$= \hat{E}\left(\hat{T}\left|\int_0^t(\hat{f}(\hat{X}^{n+1}(s),s) - \hat{f}(\hat{X}^n(s),s))ds + \int_0^t(\hat{g}(\hat{X}^{n+1}(s),s) - \hat{g}(\hat{X}^n(s),s))\frac{d\frac{\tilde{W}^3}{\tilde{T}(\tilde{W}^3)}}{\hat{T}}\right|^2\right)$$

$$(a+b)^2 \leq 2a^2 + 2b^2$$

$$\leq \hat{E}\left(2\hat{T}\left|\int_0^t(\hat{f}(\hat{X}^{n+1}(s),s) - \hat{f}(\hat{X}^n(s),s))ds\right|^2\right.$$

$$\left.+2\hat{T}\left(\int_0^t(\hat{g}(\hat{X}^{n+1}(s),s) - \hat{g}(\hat{X}^n(s),s))d\frac{\frac{\tilde{W}^3}{\tilde{T}(\tilde{W}^3)}}{\hat{T}}\right)^2\right)$$

$$= \hat{E}\left(2\hat{T}\left|\int_0^t(\hat{f}(\hat{X}^{n+1}(s),s) - \hat{f}(\hat{X}^n(s),s))ds\right|^2\right)$$

$$+\hat{E}\left(2\hat{T}\left(\int_0^t(\hat{g}(\hat{X}^{n+1}(s),s) - \hat{g}(\hat{X}^n(s),s))d\frac{\frac{\tilde{W}^3}{\tilde{T}(\tilde{W}^3)}}{\hat{T}}\right)^2\right)$$

$$= \hat{E}\left(2\hat{T}\left|\int_0^t(\hat{f}(\hat{X}^{n+1}(s),s) - \hat{f}(\hat{X}^n(s),s))ds\right|^2\right)$$

$$+\hat{E}\left(2\left(\int_0^t(\hat{g}(\hat{X}^{n+1}(s),s) - \hat{g}(\hat{X}^n(s),s))d\frac{\frac{\tilde{W}^3}{\tilde{T}(\tilde{W}^3)}}{\hat{T}}\right)^{\hat{2}}\right)$$

$$\leq 2\hat{E}\left(L^2\hat{T}\left(|\hat{X}^{n+1}(s)-\hat{X}^n(s)|ds\right)^2\right)$$

$$+2\hat{E}\left(\left(\int_0^t(\hat{g}(\hat{X}^{n+1}(s),s)-\hat{g}(\hat{X}^n(s),s))d\frac{\frac{\tilde{W}^3}{\tilde{T}(\tilde{W}^3)}}{\hat{T}}\right)^{\hat{2}}\right)$$

$$\leq 2L^2\hat{E}\left(\hat{T}t\int_0^t|\hat{X}^{n+1}(s)-\hat{X}^n(s)|^2ds\right)$$

$$+2\hat{E}\left(\left(\int_0^t\frac{1}{\tilde{T}}\tilde{T}(\hat{g}(\hat{X}^{n+1}(s),s)-\hat{g}(\hat{X}^n(s),s))\frac{\tilde{T}\tilde{T}}{\tilde{T}\tilde{T}}d\frac{\frac{\tilde{W}^3}{\tilde{T}(\tilde{W}^3)}}{\hat{T}}\right)^{\hat{2}}\right)$$

$$\leq 2TL^2\hat{E}\left(\int_0^t\hat{T}|\hat{X}^{n+1}(s)-\hat{X}^n(s)|^2ds\right)$$

$$+2\hat{E}\left(\left(\hat{\int}_0^t\frac{\hat{g}(\hat{X}^{n+1}(s),s)-\hat{g}(\hat{X}^n(s),s)}{\tilde{T}}\hat{\times}\hat{d}\frac{\frac{\tilde{W}^3}{\tilde{T}(\tilde{W}^3)}}{\hat{T}}\right)^{\hat{2}}\right)$$

$$=2T\hat{E}\left(\int_0^t|\hat{X}^{n+1}(s)-\hat{X}^n(s)|^{\hat{2}}ds\right)+2\hat{E}\left(\int_0^t\frac{(\hat{g}(\hat{X}^{n+1}(s),s)-\hat{g}(\hat{X}^n(s),s))^2}{\tilde{T}^2}\frac{\tilde{T}^2}{\hat{T}}ds\right)$$

$$=2TL^2\int_0^t\hat{E}\left(|\hat{X}^{n+1}(s)-\hat{X}^n(s)|^{\hat{2}}\right)ds+2\hat{E}\left(\int_0^t\frac{L^2(\hat{X}^{n+1}(s)-\hat{X}^n(s))^2}{\hat{T}}ds\right)$$

$$=2TL^2\int_0^t d^n(s)ds+2L^2\hat{E}\left(\int_0^t\frac{\hat{T}(\hat{X}^{n+1}(s)-\hat{X}^n(s))^2}{\hat{T}^2}ds\right)$$

$$\leq 2TL^2\int_0^t d^n(s)ds+2\frac{L^2}{N^2}\hat{E}\left(\int_0^t(\hat{X}^{n+1}(s)-\hat{X}^n(s))^{\hat{2}}ds\right)$$

$$=2TL^2\int_0^t d^n(s)ds+2\frac{L^2}{N^2}\int_0^t\hat{E}\left((\hat{X}^{n+1}(s)-\hat{X}^n(s))^{\hat{2}}\right)ds$$

$$=2TL^2\int_0^t d^n(s)ds+2\frac{L^2}{N^2}\int_0^t d^n(s)ds$$

$$=\left(2TL^2+2\frac{L^2}{N^2}\right)\int_0^t d^n(s)ds$$

$$\leq\left(2TL^2+2\frac{L^2}{N^2}\right)\frac{M^{n+1}}{(n+1)!}\int_0^t s^{n+1}ds$$

$$=\left(2TL^2+2\frac{L^2}{N^2}\right)M^{n+1}\frac{t^{n+2}}{(n+2)!}$$

$$\leq\frac{(MT)^{n+2}}{(n+2)!}.$$

Consequently

$$d^n(t)\leq\frac{(Mt)^{n+1}}{(n+1)!}$$

for every $n \in \mathbb{N}$.

Hence,

$$\hat{E}\left(\max_{0\le t\le T}|\hat{X}^{n+1}(t)-\hat{X}^{n}(t)|^{\hat{2}}\right)\le\frac{M^{n+1}T^{n+1}}{(n+1)!}.$$

From the iso-Chebyshev,s inequality we get

$$P\left(\max_{0\le t\le T}|\hat{X}^{n+1}(t)-\hat{X}^{n}(t)|>\tfrac{1}{2^n}\right)\le\frac{1}{\left(\frac{1}{2^n}\right)^2}\hat{E}\left(\max_{0\le t\le T}|\hat{X}^{n+1}(t)-\hat{X}^{n}(t)|^2\right)$$

$$+\frac{1}{\left(\frac{1}{2^n}\right)^2}\int_\Omega \max_{0\le t\le T}|\hat{X}^{n+1}(t)-\hat{X}^{n}(t)|^2\frac{\hat{T}'(\omega)}{\hat{T}(\omega)}P(\omega)d\omega$$

$$\le C2^{2n}\hat{E}\left(\max_{0\le t\le T}|\hat{X}^{n+1}(t)-\hat{X}^{n}(t)|^2\right)$$

$$=C2^{2n}\hat{E}\left(\tfrac{1}{\hat{T}}\max_{0\le t\le T}|\hat{X}^{n+1}(t)-\hat{X}^{n}(t)|^{\hat{2}}\right)$$

$$\le\tfrac{C}{N}2^{2n}\hat{E}\left(\max_{0\le t\le T}|\hat{X}^{n+1}(t)-\hat{X}^{n}(t)|^{\hat{2}}\right)$$

$$\le\tfrac{C}{N}2^{2n}\frac{(MT)^{n+1}}{(n+1)!}$$

$$\le\tfrac{C}{N}\frac{(4MT)^{n+1}}{(n+1)!}$$

for some positive constant C.

From here and iso-Borel-Cantelli lemma it follows that

$$\hat{P}\left(\max_{0\le t\le T}|\hat{X}^{n+1}(t)-\hat{X}^{n}(t)|>\frac{1}{2^n}\text{i.o.}\right)=0.$$

Therefore

$$\hat{X}^n=\hat{X}^0+\sum_{j=0}^{n-1}\left(\hat{X}^{j+1}-\hat{X}^j\right)$$

converges uniformly on $[0,T]$ for almost every ω to a process $\hat{X}(\cdot)$.

We pas to limits in the definition of $\hat{X}^{n+1}$ and we get

$$\hat{X}(t)=\hat{X}_0+\int_0^t\hat{f}(\hat{X}(s),s)ds+\int_0^t\hat{g}(\hat{X}(s),s)d\frac{\frac{\tilde{W}^3}{\tilde{T}(\tilde{W}^3)}}{\hat{T}},\quad 0\le t\le T.$$

Now we will prove that $\hat{X}\in\hat{\mathbb{L}}^{\hat{2}}(0,T)$.

We have

$$
\begin{aligned}
|\hat{X}^{n+1}|^{\hat{2}} &= \hat{T}|\hat{X}^{n+1}|^{2}\\
&= \hat{T}\left|\hat{X}_0 + \int_0^t \hat{f}(\hat{X}^n(s),s)ds + \int_0^t \hat{g}(\hat{X}^n(s),s)d\frac{\frac{\tilde{W}^3}{\tilde{T}(\tilde{W}^3)}}{\hat{T}}\right|^2\\
&\leq C_9\hat{T}\left(|\hat{X}_0|^2 + \left(\int_0^t |\hat{f}(\hat{X}^n(s),s)|ds\right)^2 + \left(\int_0^t |\hat{g}(\hat{X}^n(s),s)|d\frac{\frac{\tilde{W}^3}{\tilde{T}(\tilde{W}^3)}}{\hat{T}}\right)^2\right)\\
&\leq C_9\hat{T}|\hat{X}_0|^2 + C_9T\hat{T}\int_0^t |\hat{f}(\hat{X}^n(s),s)|^2 ds + C_9\left(\int_0^t |\hat{g}(\hat{X}^n(s),s)|d\frac{\frac{\tilde{W}^3}{\tilde{T}(\tilde{W}^3)}}{\hat{T}}\right)^{\hat{2}}\\
&\leq C_9|\hat{X}_0|^{\hat{2}} + C_9T\hat{T}\int_0^t L^2(1+|\hat{X}^n|)^2 ds + C_9\left(\int_0^t \frac{\tilde{T}}{\tilde{T}}|\hat{g}(\hat{X}^n(s),s)|\frac{\tilde{T}\tilde{T}}{\tilde{T}\tilde{T}}d\frac{\frac{\tilde{W}^3}{\tilde{T}(\tilde{W}^3)}}{\hat{T}}\right)^{\hat{2}}\\
&\leq C_9|\hat{X}_0|^{\hat{2}} + C_9T\hat{T}L^2\int_0^t (1+|\hat{X}^0|)^2 ds + C_9\left(\hat{\int}_0^t \frac{|\hat{g}(\hat{X}^n(s),s)|}{\tilde{T}}\hat{\times}\hat{d}\frac{\frac{\tilde{W}^3}{\tilde{T}(\tilde{W}^3)}}{\hat{T}}\right)^{\hat{2}}\\
&\leq C_{10}|\hat{X}_0|^{\hat{2}} + C_{10}\hat{T}\int_0^t |\hat{X}^n|^2 ds + C_{10}\left(\hat{\int}_0^t \frac{|\hat{g}(\hat{X}^n(s),s)|}{\tilde{T}}\hat{\times}\hat{d}\frac{\frac{\tilde{W}^3}{\tilde{T}(\tilde{W}^3)}}{\hat{T}}\right)^{\hat{2}}\\
&\leq C_{10}|\hat{X}_0|^{\hat{2}} + C_{10}\int_0^t |\hat{X}^n|^{\hat{2}} ds + C_{10}\left(\hat{\int}_0^t \frac{|\hat{g}(\hat{X}^n(s),s)|}{\tilde{T}}\hat{\times}\hat{d}\frac{\frac{\tilde{W}^3}{\tilde{T}(\tilde{W}^3)}}{\hat{T}}\right)^{\hat{2}},
\end{aligned}
$$

i.e.,

$$
|\hat{X}^{n+1}|^{\hat{2}} \leq C_{10}|\hat{X}_0|^{\hat{2}} + C_{10}\int_0^t |\hat{X}^n|^{\hat{2}} ds + C_{10}\left(\hat{\int}_0^t \frac{|\hat{g}(\hat{X}^n(s),s)|}{\tilde{T}}\hat{\times}\hat{d}\frac{\frac{\tilde{W}^3}{\tilde{T}(\tilde{W}^3)}}{\hat{T}}\right)^{\hat{2}}.
$$

Here C_9 and C_{10} are positive constants.

From here,

$$
\begin{aligned}
\hat{E}\left(|\hat{X}^{n+1}|^{\hat{2}}\right) &\leq C_{10}\hat{E}\left(|\hat{X}_0|^{\hat{2}}\right) + C_{10}\int_0^t \hat{E}\left(|\hat{X}^n|^{\hat{2}}\right)ds + C_{10}\hat{E}\left(\left(\hat{\int}_0^t \frac{|\hat{g}(\hat{X}^n(s),s)|}{\tilde{T}}\hat{\times}\hat{d}\frac{\frac{\tilde{W}^3}{\tilde{T}(\tilde{W}^3)}}{\hat{T}}\right)^{\hat{2}}\right)\\
&\leq C_{10}\hat{E}\left(|\hat{X}_0|^{\hat{2}}\right) + C_{10}\int_0^t \hat{E}\left(|\hat{X}^n|^{\hat{2}}\right)ds + C_{10}\hat{E}\left(\int_0^t \frac{|\hat{g}(\hat{X}^n(s),s)|^2}{\tilde{T}^2}\frac{\tilde{T}^2}{\hat{T}}ds\right)\\
&\leq C_{10}\hat{E}\left(|\hat{X}_0|^{\hat{2}}\right) + C_{10}\int_0^t \hat{E}\left(|\hat{X}^n|^{\hat{2}}\right)ds + C_{10}\hat{E}\left(\int_0^t \frac{|\hat{g}(\hat{X}^n(s),s)|^2}{\hat{T}}ds\right)
\end{aligned}
$$

$$\leq C_{11}\hat{E}\left(|\hat{X}_0|^{\hat{2}}\right)+C_{11}\int_0^t \hat{E}\left(|\hat{X}^n|^{\hat{2}}\right)ds+C_{11}L^2\hat{E}\left(\int_0^t \frac{|\hat{X}^n(s)|^2}{\hat{T}}\right)$$

$$\leq C_{11}\hat{E}\left(|\hat{X}_0|^{\hat{2}}\right)+C_{11}\int_0^t \hat{E}\left(|\hat{X}^n|^{\hat{2}}\right)ds+C_{11}L^2\hat{E}\left(\int_0^t \frac{|\hat{X}^n(s)|^{\hat{2}}}{\hat{T}^2}ds\right)$$

$$\leq C_{11}\hat{E}\left(|\hat{X}_0|^{\hat{2}}\right)+C_{11}\int_0^t \hat{E}\left(|\hat{X}^n|^{\hat{2}}\right)ds+C_{11}\frac{L^2}{N^2}\int_0^t \hat{E}\left(|\hat{X}^n|^{\hat{2}}\right)ds$$

$$\leq C_{12}\hat{E}\left(|\hat{X}_0|^{\hat{2}}\right)+C_{12}\int_0^t \hat{E}\left(|\hat{X}^n(s)|^{\hat{2}}\right)ds,$$

i.e.,

$$\hat{E}\left(|\hat{X}^{n+1}|^{\hat{2}}\right)\leq C_{12}\hat{E}\left(|\hat{X}_0|^{\hat{2}}\right)+C_{12}\int_0^t \hat{E}\left(|\hat{X}^n(s)|^{\hat{2}}\right)ds, \tag{10}$$

where C_{11} and C_{12} are positive constants and $C_{12}>1$.

For $n=0$, using (10), we get

$$\hat{E}\left(|\hat{X}^1|^{\hat{2}}\right)\leq C_{12}\hat{E}\left(|\hat{X}_0|^{\hat{2}}\right)+C_{12}\int_0^t \hat{E}\left(|\hat{X}_0|^{\hat{2}}\right)ds$$

$$=C_{12}(1+t)\hat{E}\left(|\hat{X}_0|^{\hat{2}}\right).$$

From here and (10), for $n=1$, we find

$$\hat{E}\left(|\hat{X}^2|^{\hat{2}}\right)\leq C_{12}\hat{E}\left(|\hat{X}_0|^{\hat{2}}\right)+C_{12}\int_0^t \hat{E}\left(|\hat{X}^1|^{\hat{2}}\right)ds$$

$$\leq C_{12}\hat{E}\left(|\hat{X}_0|^{\hat{2}}\right)+C_{12}^2\int_0^t (1+s)\hat{E}\left(|\hat{X}_0|^{\hat{2}}\right)ds$$

$$=\left(C_{12}+C_{12}^2\frac{(1+t)^2}{2}-\frac{C_{12}}{2}\right)\hat{E}\left(|\hat{X}_0|^{\hat{2}}\right)$$

$$\leq\left(C_{12}+C_{12}(1+t)+C_{12}^2\frac{(1+t)^2}{2}\right)\hat{E}\left(|\hat{X}_0|^{\hat{2}}\right).$$

We suppose that

$$\hat{E}\left(|\hat{X}_0^n|^{\hat{2}}\right)\leq\left(C_{12}+C_{12}(1+t)+C_{12}^2\frac{(1+t)^2}{2}+\cdots+C_{12}^n\frac{(1+t)^n}{n!}\right)\hat{E}\left(|\hat{X}_0|^{\hat{2}}\right) \tag{11}$$

for some $n\in\mathbb{N}$.

We will prove that

$$\hat{E}\left(|\hat{X}_0^{n+1}|^{\hat{2}}\right)$$

$$\leq\left(C_{12}+C_{12}(1+t)+C_{12}^2\frac{(1+t)^2}{2}+\cdots+C_{12}^n\frac{(1+t)^n}{n!}+C_{12}^{n+1}\frac{(1+t)^{n+1}}{(n+1)!}\right)\hat{E}\left(|\hat{X}_0|^{\hat{2}}\right).$$

Indeed, from (10), using (11), we find that

$$\hat{E}\left(|\hat{X}^{n+1}|^{\hat{2}}\right) \leq C_{12}\hat{E}\left(|\hat{X}_0|^{\hat{2}}\right) + C_{12}\int_0^t \hat{E}\left(|\hat{X}^n(s)|^{\hat{2}}\right)ds$$

$$\leq C_{12}\hat{E}\left(|\hat{X}_0|^{\hat{2}}\right)$$

$$+C_{12}\int_0^t \left(C_{12}+C_{12}(1+s)+C_{12}^2\frac{(1+s)^2}{2!}+\cdots+C_{12}^n\frac{(1+s)^n}{n!}\right)ds\hat{E}\left(|\hat{X}_0|^{\hat{2}}\right)$$

$$\leq \left(C_{12}+C_{12}(1+t)+C_{12}\frac{(1+t)^2}{2!}+\cdots+C_{12}^{n+1}\frac{(1+t)^{n+1}}{(n+1)!}\right)\hat{E}\left(|\hat{X}_0|^{\hat{2}}\right).$$

Consequently (11) is valid for every $n \in \mathbb{N}$. From it we find

$$\hat{E}\left(|\hat{X}^n|^{\hat{2}}\right) \leq \left(C_{12}+e^{C_{12}(1+t)}\right)\hat{E}\left(|\hat{X}_0|^{\hat{2}}\right),$$

whereupon, when $n \longrightarrow \infty$,

$$\hat{E}\left(|\hat{X}|^{\hat{2}}\right) \leq \left(C_{12}+e^{C_{12}(1+t)}\right)\hat{E}\left(|\hat{X}_0|^{\hat{2}}\right),$$

i.e.,

$$\hat{X} \in \hat{\mathbb{L}}^{\hat{2}}(0,T).$$

2. Uniqueness. Now we suppose that the problem (3) has two solutions $\hat{X}$ and $\hat{Y}$. Then

$$\begin{cases} \hat{X}(t) = \hat{X}_0 + \int_0^t \hat{f}(\hat{X}(s),s)ds + \int_0^t \hat{g}(\hat{X}(s),s)d\frac{\frac{\tilde{W}^3}{\tilde{T}(\tilde{W}^3)}}{\hat{T}} \\ \hat{Y}(t) = \hat{X}_0 + \int_0^t \hat{f}(\hat{Y}(s),s)ds + \int_0^t \hat{g}(\hat{Y}(s),s)d\frac{\frac{\tilde{W}^3}{\tilde{T}(\tilde{W}^3)}}{\hat{T}}, \qquad 0 \leq t \leq T, \end{cases}$$

whereupon

$$\hat{X}(t)-\hat{Y}(t) = \int_0^t \left(\hat{f}(\hat{X}(s),s)-\hat{f}(\hat{Y}(s),s)\right)ds + \int_0^t \left(\hat{g}(\hat{X}(s),s)-\hat{g}(\hat{Y}(s),s)\right)d\frac{\frac{\tilde{W}^3}{\tilde{T}(\tilde{W}^3)}}{\hat{T}}$$

and

$$\hat{E}\left(|\hat{X}(t)-\hat{Y}(t)|^{\hat{2}}\right)=\hat{E}\left(\hat{T}|\hat{X}(t)-\hat{Y}(t)|^{2}\right)$$

$$\hat{E}\left(\hat{T}\left(\int_0^t\left(\hat{f}(\hat{X}(s),s)-\hat{f}(\hat{Y}(s),s)\right)ds+\int_0^t\left(\hat{g}(\hat{X}(s),s)-\hat{g}(\hat{Y}(s),s)\right)d\frac{\frac{\tilde{W}^3}{\tilde{T}(\tilde{W}^3)}}{\hat{T}}\right)^2\right)$$

$$\leq 2\hat{E}\left(\hat{T}\left(\int_0^t(\hat{f}(\hat{X}(s),s)-\hat{f}(\hat{Y}(s),s))\right)^2\right)$$

$$+2\hat{E}\left(\hat{T}\left(\int_0^t(\hat{g}(\hat{X}(s),s)-\hat{g}(\hat{Y}(s),s))d\frac{\frac{\tilde{W}^3}{\tilde{T}(\tilde{W}^3)}}{\hat{T}}\right)^2\right)$$

$$\leq 2\hat{E}\left(\hat{T}t\int_0^t(\hat{f}(\hat{X}(s),s)-\hat{f}(\hat{Y}(s),s))^2ds\right)$$

$$+2\hat{E}\left(\hat{T}\left(\int_0^t\tilde{T}(s)\frac{1}{\tilde{T}(s)}\frac{\hat{g}(\hat{X}(s),s)-\hat{g}(\hat{Y}(s),s)}{\tilde{T}^2(s)}\tilde{T}^2(s)d\frac{\frac{\tilde{W}^3}{\tilde{T}(\tilde{W}^3)}}{\hat{T}}\right)^2\right)$$

$$\leq 2TL^2\hat{E}\left(\hat{T}\int_0^t|\hat{X}(s)-\hat{Y}(s)|^2ds\right)+2\hat{E}\left(\left(\hat{\int}_0^t\frac{\hat{g}(\hat{X}(s),s)-\hat{g}(\hat{Y}(s),s)}{\tilde{T}}d\frac{\frac{\tilde{W}^3}{\tilde{T}(\tilde{W}^3)}}{\hat{T}}\right)^{\hat{2}}\right)$$

$$=2TL^2\hat{E}\left(\int_0^t|\hat{X}(s)-\hat{Y}(s)|^{\hat{2}}ds\right)+2\hat{E}\left(\int_0^t\frac{(\hat{g}(\hat{X}(s),s)-\hat{g}(\hat{Y}(s),s))^2}{\hat{T}}ds\right)$$

$$\leq 2TL^2\int_0^t\hat{E}\left(|\hat{X}(s)-\hat{Y}(s)|^{\hat{2}}\right)ds+2L^2\hat{E}\left(\int_0^t\frac{|\hat{X}(s)-\hat{Y}(s)|^2}{\hat{T}}ds\right)$$

$$=2TL^2\int_0^t\hat{E}(|\hat{X}(s)-\hat{Y}(s)|^{\hat{2}})ds+2L^2\hat{E}\left(\int_0^t\frac{|\hat{X}(s)-\hat{Y}(s)|^{\hat{2}}}{\hat{T}}ds\right)$$

$$\leq 2TL^2\int_0^t\hat{E}(|\hat{X}(s)-\hat{Y}(s)|^{\hat{2}})ds+2\frac{L^2}{N^2}\int_0^t\hat{E}(|\hat{X}(s)-\hat{Y}(s)|^{\hat{2}})ds$$

$$=2L^2\left(T+\frac{1}{N^2}\right)\int_0^t\hat{E}(|\hat{X}(s)-\hat{Y}(s)|^{\hat{2}})ds,$$

i.e.,

$$\int_0^t\hat{E}(|\hat{X}(s)-\hat{Y}(s)|^{\hat{2}})ds\leq 2L^2\left(T+\frac{1}{N^2}\right)\int_0^t\hat{E}(|\hat{X}(s)-\hat{Y}(s)|^{\hat{2}})ds.$$

Hence and Gronwall's lemma it follows that

$$\hat{E}(|\hat{X}(t)-\hat{Y}(t)|^{\hat{2}})=0 \qquad \text{for} \qquad t\in[0,T].$$

Thus

$$\hat{X}(t)=\hat{Y}(t) \qquad \text{a.s.}$$

for all $0\leq t\leq T$.

In this way

$$\hat{X}(r) = \hat{Y}(r)$$

for all rational $0 \le r \le T$, except for some set of probability zero. As $\hat{X}$ and $\hat{Y}$ have continuous sample paths almost surely,

$$\hat{P}\left(\max_{0\le t\le T} |\hat{X}(t) - \hat{Y}(t)| > 0\right) = 0.$$

□

Example 2.2.5. *Let us suppose that $b : \mathbb{R} \longrightarrow \mathbb{R}$ is C^1 with $|b'| \le L$ for some positive constant L, and let us consider the iso-stochastic differential equation*

$$\begin{cases} d\hat{X} = b(\hat{X})dt + d\frac{\tilde{W}^1}{\hat{T}} \\ \hat{X}(0) = x, \qquad 0 \le t \le T, \end{cases}$$

where $x \in \mathbb{R}$.

We define

$$(12) \qquad \begin{aligned} &\hat{X}^0(t) = x, \\ &\hat{X}^{n+1}(t) = x + \textstyle\int_0^t b(\hat{X}^n(s))ds + \frac{\tilde{W}^1(t)}{\hat{T}}, \qquad 0 \le t \le T, \end{aligned}$$

for $n = 0, 1, 2, \ldots$. Next we write

$$D^n(t) = |\hat{X}^{n+1}(t) - \hat{X}^n(t)|, \qquad n = 0, 1, 2, \ldots.$$

Then

$$\begin{aligned} D^0(t) &= |\hat{X}^1(t) - \hat{X}^0(t)| \\ &= \left|x + \textstyle\int_0^t b(\hat{X}^0(s))ds + \frac{\tilde{W}^1(t)}{\hat{T}} - x\right| \\ &= \left|\textstyle\int_0^t b(x)ds + \frac{\tilde{W}^1(t)}{\hat{T}}\right| \\ &\le C, \qquad 0 \le t \le T, \end{aligned}$$

where C is a positive constant. Such constants exists for the function b, which is C^1, and for a given sample continuous path of the iso-Brownian iso-motion $\tilde{W}^1$.

Also, we have

$$D^1(t) = |\hat{X}^2(t) - \hat{X}^1(t)|$$

$$= \left| x + \int_0^T b(\hat{X}^1(s))ds + \frac{\tilde{W}^1(t)}{\hat{T}} - x - \int_0^t b(\hat{X}^0(s))ds - \frac{\tilde{W}^1(t)}{\hat{T}} \right|$$

$$= \left| \int_0^t b(\hat{X}^1(s))ds - \int_0^t b(\hat{X}^0(s))ds \right|$$

$$= \left| \int_0^t \left(b(\hat{X}^1(s)) - b(\hat{X}^0(s)) \right) ds \right|$$

$$\leq \int_0^t \left| b(\hat{X}^1(s)) - b(\hat{X}^0(s)) \right| ds$$

$$\leq L \int_0^t |\hat{X}^1(s) - \hat{X}^0(s)| ds$$

$$= L \int_0^t D^0(s) ds$$

$$\leq CLt.$$

We suppose that

$$(13) \qquad D^n(t) \leq C\frac{(Lt)^n}{n!}, \qquad 0 \leq t \leq T.$$

Really,

$$D^{n+1}(t) = |\hat{X}^{n+1}(t) - \hat{X}^n(t)|$$

$$= \left| x + \int_0^t b(\hat{X}^{n+1}(s))ds + \frac{\tilde{W}^1(t)}{\hat{T}} - x - \int_0^t b(\hat{X}^n(s))ds - \frac{\tilde{W}^1(t)}{\hat{T}} \right|$$

$$= \left| \int_0^t b(\hat{X}^{n+1}(s))ds - \int_0^x b(\hat{X}^n(s))ds \right|$$

$$= \left| \int_0^t (b(\hat{X}^{n+1}(s)) - b(\hat{X}^n(s)))ds \right|$$

$$\leq \int_0^t |b(\hat{X}^{n+1}(s)) - b(\hat{X}^n(s))| ds$$

$$\leq L \int_0^t |\hat{X}^{n+1}(s) - \hat{X}^n(s)| ds$$

$$= L \int_0^t D^n(s) ds$$

$$\leq CL \int_0^t \frac{(Ls)^n}{(n+1)!}$$

$$= C\frac{(Lt)^{n+1}}{(n+1)!}.$$

Consequently (13) *holds for every* $n \in \mathbb{N}$.

From here, for $m \geq n$,

$$\max_{0 \leq t \leq T} |\hat{X}^m(t) - \hat{X}^n(t)| \leq C \sum_{k=n}^{\infty} \frac{(LT)^k}{k!}$$

$$\longrightarrow_{n \longrightarrow \infty} 0.$$

Thus for almost every $\omega \in \Omega$, $\hat{X}^n(\cdot)$ *converges uniformly for* $0 \le t \le T$ *to a limit process* $\hat{X}(\cdot)$. *Hence and* (12) *we get that*

$$\hat{X}(t) = x + \int_0^t b(\hat{X}(s))ds + \frac{\tilde{W}^1(t)}{\hat{T}}, \qquad 0 \le t \le T.$$

Problem 2.2.6. *Let b is* C^1 *with* $|b'| \le L$ *for some positive constant L. Prove that*

$$\hat{X}(t) = x + \int_0^t b(\hat{X}(s))ds + \frac{\tilde{W}^2(t)\tilde{T}(\tilde{W}^2(t))}{\hat{T}}, \qquad 0 \le t \le T,$$

is a solution to the iso-stochastic differential equation

$$\begin{cases} d\hat{X} = b(\hat{X})dt + d\frac{\tilde{W}^2\tilde{T}(\tilde{W}^2)}{\hat{T}} \\ \hat{X}(0) = x, \qquad 0 \le t \le T. \end{cases}$$

Problem 2.2.7. *Let b is* C^1 *with* $|b'| \le L$ *for some positive constant L. Prove that*

$$\hat{X}(t) = x + \int_0^t b(\hat{X}(s))ds + \frac{\frac{\tilde{W}^3(t)}{\tilde{T}(\tilde{W}^3(t))}}{\hat{T}}, \qquad 0 \le t \le T,$$

is a solution to the iso-stochastic differential equation

$$\begin{cases} d\hat{X} = b(\hat{X})dt + d\frac{\frac{\tilde{W}^3}{\tilde{T}(\tilde{W}^3)}}{\hat{T}} \\ \hat{X}(0) = x, \qquad 0 \le t \le T. \end{cases}$$

Problem 2.2.8. *Find a solution to the iso-stochastic differential equation*

$$\begin{cases} d\hat{X} = tdt + d\frac{\tilde{W}^1}{\hat{T}} \\ \hat{X}(0) = 1, \qquad 0 \le t \le T. \end{cases}$$

Answer.

$$\hat{X}(t) = 1 + \frac{t^2}{2} + \frac{\tilde{W}^1(t)}{\hat{T}}, \qquad 0 \le t \le T.$$

Problem 2.2.9. *Find a solution to the iso-stochastic differential equation*

$$\begin{cases} d\hat{X} = \cos(t)dt + d\frac{\tilde{W}^2\tilde{T}(\tilde{W}^2)}{\hat{T}} \\ \hat{X}(0) = 0, \qquad 0 \le t \le T. \end{cases}$$

Answer.

$$\hat{X}(t) = \sin(t) + \frac{\tilde{W}^2(t)\tilde{T}(\tilde{W}^2)}{\hat{T}}, \qquad 0 \le t \le T.$$

Problem 2.2.10. *Find a solution to the iso-stochastic differential equation*

$$\begin{cases} d\hat{X} = e^t dt + 3d\frac{\frac{\tilde{W}^3}{\hat{T}(\tilde{W}^3)}}{\hat{T}} \\ \hat{X}(0) = 2, \qquad 0 \le t \le T. \end{cases}$$

Answer.

$$\hat{X}(t) = 2 + e^t + 3\frac{\frac{\tilde{W}^3}{\hat{T}(\tilde{W}^3)}}{\hat{T}}, \qquad 0 \le t \le T.$$

2.3 Solving Iso-Stochastic Differential Equations by Changing Variables

Given a general iso-stochastic differential equation of the following form

$$(14) \qquad \begin{cases} d\hat{X} = f(\hat{X})dt + g(\hat{X})d\frac{\tilde{W}^1}{\hat{T}} \\ \hat{X}(0) = x, \qquad 0 \le t \le T, \end{cases}$$

where x is a given real number.

Let us solve

$$(15) \qquad \begin{cases} d\hat{Y} = h(\hat{Y})dt + d\frac{\tilde{W}^1}{\hat{T}} \\ \hat{Y}(0) = y, \qquad 0 \le t \le T, \end{cases}$$

where h will be selected later. We will try to find a function u so that

$$\hat{X} = u(\hat{Y}),$$

which solves (14).

We have

$$d\hat{X} = u'(\hat{Y})d\hat{Y} + \frac{1}{2}u''(\hat{Y})\frac{\hat{T}(\Omega)}{\hat{T}}dt.$$

Hence and (15) we get

$$\begin{aligned} d\hat{X} &= u'(\hat{Y})\left(h(\hat{Y})dt + d\frac{\tilde{W}^1}{\hat{T}}\right) + \frac{1}{2}u''(\hat{Y})\frac{\hat{T}(\Omega)}{\hat{T}}dt \\ &= \left(u'(\hat{Y})h(\hat{Y}) + \frac{1}{2}u''(\hat{Y})\frac{\hat{T}(\Omega)}{\hat{T}}\right)dt + u'(\hat{Y})d\frac{\tilde{W}^1}{\hat{T}}. \end{aligned}$$

From here and (14) we find

$$\left(u'(\hat{Y})h(\hat{Y}) + \frac{1}{2}u''(\hat{Y})\frac{\hat{T}(\Omega)}{\hat{T}}\right)dt + u'(\hat{Y})d\frac{\tilde{W}^1}{\hat{T}} = f(\hat{X})dt + g(\hat{X})d\frac{\tilde{W}^1}{\hat{T}}.$$

Therefore

$$\begin{cases} u'(\hat{Y}) = g(\hat{X}) = g(u(\hat{Y})) \\ u'(\hat{Y})h(\hat{Y}) + \frac{1}{2}u''(\hat{Y})\frac{\hat{T}(\Omega)}{\hat{T}} = f(u(\hat{Y})) \\ u(y) = x. \end{cases}$$

Problem 2.3.1. *Let $\hat{X}$ be a solution to the iso-stochastic differential equation*

$$\begin{cases} d\hat{X} = f(\hat{X})dt + g(\hat{X})d\frac{\tilde{W}^2\tilde{T}(\tilde{W}^2)}{\hat{T}} \\ \hat{X}(0) = x, \qquad 0 \le t \le T, \end{cases}$$

x is a real constant. Let also, $\hat{Y}$ be a solution to the iso-stochastic differential equation

$$\begin{cases} d\hat{Y} = h(\hat{Y})dt + d\frac{\tilde{W}^2\tilde{T}(\tilde{W}^2)}{\hat{T}} \\ \hat{Y}(0) = y, \qquad 0 \le t \le T, \end{cases}$$

where $y \in \mathbb{R}$. Find a relation between h, f, g, x, y and u so that

$$\hat{X} = u(\hat{Y}).$$

Problem 2.3.2. *Let $\hat{X}$ be a solution to the iso-stochastic differential equation*

$$\begin{cases} d\hat{X} = f(\hat{X})dt + g(\hat{X})d\frac{\frac{\tilde{W}^3}{\tilde{T}(\tilde{W}^3)}}{\hat{T}} \\ \hat{X}(0) = x, \qquad 0 \le t \le T, \end{cases}$$

x is a real constant. Let also, $\hat{Y}$ be a solution to the iso-stochastic differential equation

$$\begin{cases} d\hat{Y} = h(\hat{Y})dt + d\frac{\frac{\tilde{W}^3}{\tilde{T}(\tilde{W}^3)}}{\hat{T}} \\ \hat{Y}(0) = y, \qquad 0 \le t \le T, \end{cases}$$

where $y \in \mathbb{R}$. Find a relation between h, f, g, x, y and u so that

$$\hat{X} = u(\hat{Y}).$$

Problem 2.3.3. *Let $\hat{X}$ be a solution to the problem*

$$\begin{cases} d\hat{X} = dt + \cos^2(\hat{X})d\frac{\tilde{W}^1}{\hat{T}} \\ \hat{X}(0) = x, \qquad 0 \le t \le T. \end{cases}$$

Let also, $\hat{Y}$ be a solution to the iso-stochastic differential equation

$$\begin{cases} d\hat{Y} = \left(\tan(\hat{Y})\frac{\hat{T}(\Omega)}{\hat{T}} + \frac{1}{\cos^2(\hat{Y})}\right)dt + d\frac{\tilde{W}^1}{\hat{T}} \\ \hat{Y}(0) = y, \qquad 0 \le t \le T, \end{cases}$$

$y \in \mathbb{R}$, such that

$$\hat{X} = u(\hat{Y}).$$

Prove that

$$\tan(u(\hat{Y})) = \hat{Y} + C, \qquad \tan(x) = y + C,$$

where C is a real constant.

Problem 2.3.4. *Let $\hat{X}$ be a solution to the iso-stochastic differential equation*

$$\begin{cases} d\hat{X} = f(\hat{X})dt + \hat{X}d\frac{\tilde{W}^2\tilde{T}(\tilde{W}^2)}{\hat{T}} \\ \hat{X}(0) = x, \qquad 0 \le t \le T, \end{cases}$$

where $x \in \mathbb{R}$. Let also, $\hat{Y}$ be a solution to the iso-stochastic differential equation

$$\begin{cases} d\hat{Y} = \hat{Y}dt + d\frac{\tilde{W}^2\tilde{T}(\tilde{W}^2)}{\hat{T}} \\ \hat{Y}(0) = y, \qquad 0 \le t \le T, \end{cases}$$

where $y \in \mathbb{R}$, such that

$$\hat{X} = e^{\hat{Y}}.$$

Prove that

$$f\left(e^{\hat{Y}}\right) = e^{\hat{Y}}\hat{Y} + \frac{1}{2}e^{\hat{Y}}\frac{\hat{T}(\Omega)}{\hat{T}}.$$

Problem 2.3.5. *Let $\hat{X}$ be a solution to the iso-stochastic differential equation*

$$\begin{cases} d\hat{X} = f(\hat{X})dt + 2\sqrt{\hat{X}}d\frac{\frac{\tilde{W}^3}{\tilde{T}(\tilde{W}^3)}}{\hat{T}} \\ \hat{X}(0) = x, \qquad 0 \le t \le T, \end{cases}$$

where $x \in \mathbb{R}$. Let also, $\hat{Y}$ be a solution to the iso-stochastic differential equation

$$\begin{cases} d\hat{Y} = \hat{Y}dt + d\frac{\frac{\tilde{W}^3}{\tilde{T}(\tilde{W}^3)}}{\hat{T}} \\ \hat{Y}(0) = y, \qquad 0 \le t \le T, \end{cases}$$

$y \in \mathbb{R}$, such that

$$\hat{X} = \hat{Y}^2.$$

Prove that

$$f(\hat{X}) = 2\hat{X} + \frac{\hat{T}(\Omega)}{\hat{T}}.$$

Problem 2.3.6. *Let $\hat{X}$ be a solution to the iso-stochastic differential equation*

$$\begin{cases} d\hat{X} = \sin(\hat{X})dt + \cos^2(\hat{X})d\frac{\frac{\tilde{W}^3}{\tilde{T}(\tilde{W}^3)}}{\hat{T}} \\ \hat{X}(0) = 1, \qquad 0 \le t \le T. \end{cases}$$

Let also, $\hat{Y}$ be a solution to the iso-stochastic differential equation

$$\begin{cases} d\hat{Y} = e^{\hat{Y}}dt + d\frac{\frac{\tilde{W}^3}{\tilde{T}(\tilde{W}^3)}}{\hat{T}} \\ \hat{Y}(0) = 2, \qquad 0 \le t \le T. \end{cases}$$

Find a relation between $\hat{X}$ and $\hat{Y}$.

Answer. There is no a relation between $\hat{X}$ and $\hat{Y}$.

Problem 2.3.7. *Let $\hat{X}$ be a solution to the iso-stochastic differential equation*

$$\begin{cases} d\hat{X} = \sin^2(\hat{X})dt - 3\tan(\hat{X})\cos^2(\hat{X})d\frac{\tilde{W}^2\tilde{T}(\tilde{W}^2)}{\hat{T}} \\ \hat{X}(0) = 1, \qquad 0 \le t \le T. \end{cases}$$

Let also, $\hat{Y}$ be a solution to the iso-stochastic differential equation

$$\begin{cases} d\hat{Y} = \hat{Y}dt + e^{\hat{Y}}d\frac{\tilde{W}^2\tilde{T}(\tilde{W}^2)}{\hat{T}} \\ \hat{Y}(0) = -2, \qquad 0 \le t \le T. \end{cases}$$

Find a relation between $\hat{X}$ and $\hat{Y}$.

Answer. There is no a relation between $\hat{X}$ and $\hat{Y}$.

Problem 2.3.8. *Let $\hat{X}$ be a solution to the iso-stochastic differential equation*

$$\begin{cases} d\hat{X} = e^{\hat{X}}dt + \tan(\hat{X})d\frac{\tilde{W}^1}{\hat{T}} \\ \hat{X}(0) = -1, \qquad 0 \le t \le T. \end{cases}$$

Let also, $\hat{Y}$ be a solution to the iso-stochastic differential equation

$$\begin{cases} d\hat{Y} = e^{2\hat{Y}}dt + e^{-\hat{Y}}d\frac{\tilde{W}^1}{\hat{T}} \\ \hat{Y}(0) = 1, \qquad 0 \le t \le T. \end{cases}$$

Find a relation between $\hat{X}$ and $\hat{Y}$.

Answer. There is no a relation between $\hat{X}$ and $\hat{Y}$.

2.4 Properties of Solutions

We will start with the following lemma.

Lemma 2.4.1. *For every $a \geq 0$ and for every $n \in \mathbb{N}$ we have*

$$(1+a)a^{n-1} \leq 1+2a^n.$$

Proof. The given inequality is equivalent to the inequality

$$1-a^{n-1}+a^n \geq 0.$$

1.case. Let $a \geq 1$. Then for every $n \in \mathbb{N}$ we have

$$a^n \geq a^{n-1}$$

or

$$-a^{n-1}+a^n \geq 0.$$

From here

$$1-a^{n-1}+a^n \geq 0$$

for every $n \in \mathbb{N}$.

2.case. Let $0 \leq a < 1$. Then for every $n \in \mathbb{N}$

$$1 \geq a^{n-1}$$

or

$$1-a^{n-1} \geq 0.$$

Hence,

$$1-a^{n-1}+a^n \geq 0$$

for every $n \in \mathbb{N}$.

□

Theorem 2.4.2. *We suppose that $\hat{f}$, $\hat{g} : \mathbb{R} \times [0,T] \longrightarrow \mathbb{R}$ are continuous and satisfy*

$$|\hat{f}(x,t)| \leq L(1+|x|), \qquad |\hat{g}(x,t)| \leq L(1+|x|)$$

for all $t \in [0,T]$, $x \in \mathbb{R}$, for some positive constant L. Let also, $\hat{T} \geq N$ on Ω and U. If, in addition,

$$\hat{E}(|\hat{X}_0|^{2p}) < \infty$$

for some integer p, then the solution $\hat{X}$ of the iso-stochastic differential equation

$$\begin{cases} d\hat{X} = \hat{f}(\hat{X},t)dt + \hat{g}(\hat{X},t)d\frac{\tilde{W}^1}{\hat{T}} \\ \hat{X}(0) = \hat{X}_0, \qquad 0 \leq t \leq T, \end{cases}$$

satisfies the estimates

$$\hat{E}\left(|\hat{X}(t)|^{2p}\right) \leq C_0\left(1+\hat{E}\left(|\hat{X}_0|^{2p}\right)\right)e^{C_1 t}$$

and

$$\hat{E}\left(|\hat{X}(t)-\hat{X}_0|^{2p}\right) \leq C_0\left(1+\hat{E}\left(|\hat{X}_0|^{2p}\right)\right)t^p e^{C_1 t},$$

where C_1 and C_2 are constants depending only on T and L.

Proof. First of all, we will note that

$$d\hat{X}^{2p} = 2p\hat{X}^{2p-1}d\hat{X} + \tfrac{1}{2}2p(2p-1)\hat{X}^{2p-2}\hat{g}^2(\hat{X},t)\tfrac{\hat{T}(\Omega)}{\hat{T}}dt$$

$$= 2p\hat{X}^{2p-1}\left(\hat{f}(\hat{X},t)dt + \hat{g}(\hat{X},t)d\tfrac{\tilde{W}^1}{\hat{T}}\right) + p(2p-1)\hat{g}^2(\hat{X},t)\tfrac{\hat{T}(\Omega)}{\hat{T}}dt$$

$$= \left(2p\hat{X}^{2p-1}\hat{f}(\hat{X},t) + p(2p-1)\hat{X}^{2p-2}\hat{g}^2(\hat{X},t)\tfrac{\hat{T}(\Omega)}{\hat{T}}\right)dt + 2p\hat{X}^{2p-1}\hat{g}(\hat{X},t)d\tfrac{\tilde{W}^1}{\hat{T}}.$$

Hence,

$$\hat{X}^{2p} = \hat{X}_0^{2p} + \textstyle\int_0^t\left(2p\hat{X}^{2p-1}\hat{f}(\hat{X},t) + p(2p-1)\hat{X}^{2p-2}\hat{g}^2(\hat{X},t)\tfrac{\hat{T}(\Omega)}{\hat{T}}\right)ds$$

$$+2p\textstyle\int_0^t\hat{X}^{2p-1}\hat{g}(\hat{X},t)d\tfrac{\tilde{W}^1}{\hat{T}},$$

whereupon

$$\left|\hat{X}^{2p}\right| = \Big|\hat{X}_0^{2p} + \textstyle\int_0^t\left(2p\hat{X}^{2p-1}\hat{f}(\hat{X},t) + p(2p-1)\hat{X}^{2p-2}\hat{g}^2(\hat{X},t)\tfrac{\hat{T}(\Omega)}{\hat{T}}\right)ds$$

$$+2p\textstyle\int_0^t\hat{X}^{2p-1}\hat{g}(\hat{X},t)d\tfrac{\tilde{W}^1}{\hat{T}}\Big|$$

$$\leq |X_0|^{2p} + \textstyle\int_0^t\left(2p|\hat{X}|^{2p-1}|\hat{f}(\hat{X},t)| + p(2p-1)|\hat{X}|^{2p-2}\hat{g}^2(\hat{X},t)\tfrac{\hat{T}(\Omega)}{\hat{T}}\right)ds$$

$$+2p\textstyle\int_0^t|\hat{X}|^{2p-1}|\hat{g}(\hat{X},t)|d\tfrac{\tilde{W}^1}{\hat{T}}.$$

From here,

$$\hat{E}\left(\left|\hat{X}^{2p}\right|\right) \leq \hat{E}\Big(|X_0|^{2p} + \textstyle\int_0^t\left(2p|\hat{X}|^{2p-1}|\hat{f}(\hat{X},t)| + p(2p-1)|\hat{X}|^{2p-2}\hat{g}^2(\hat{X},t)\tfrac{\hat{T}(\Omega)}{\hat{T}}\right)ds$$

$$+2p\textstyle\int_0^t|\hat{X}|^{2p-1}|\hat{g}(\hat{X},t)|d\tfrac{\tilde{W}^1}{\hat{T}}\Big)$$

$$= \hat{E}\left(|X_0|^{2p}\right) + \hat{E}\left(\textstyle\int_0^t\left(2p|\hat{X}|^{2p-1}|\hat{f}(\hat{X},t)| + p(2p-1)|\hat{X}|^{2p-2}\hat{g}^2(\hat{X},t)\tfrac{\hat{T}(\Omega)}{\hat{T}}\right)ds\right)$$

$$+2p\hat{E}\left(\textstyle\int_0^t|\hat{X}|^{2p-1}|\hat{g}(\hat{X},t)|d\tfrac{\tilde{W}^1}{\hat{T}}\right)$$

$$= \hat{E}\left(|X_0|^{2p}\right) + \hat{E}\left(\textstyle\int_0^t\left(2p|\hat{X}|^{2p-1}|\hat{f}(\hat{X},t)| + p(2p-1)|\hat{X}|^{2p-2}\hat{g}^2(\hat{X},t)\tfrac{\hat{T}(\Omega)}{\hat{T}}\right)ds\right),$$

i.e.,

$$\hat{E}\left(|X|^{2p}\right) \leq \hat{E}\left(|X_0|^{2p}\right)$$

(16)
$$+\hat{E}\left(\int_0^t \left(2p|\hat{X}|^{2p-1}|\hat{f}(\hat{X},t)| + p(2p-1)|\hat{X}|^{2p-2}\hat{g}^2(\hat{X},t)\frac{\hat{T}(\Omega)}{\hat{T}}\right) ds\right)$$

We observe that

$$|\hat{X}|^{2p-1}|\hat{f}(\hat{X},t)| \leq L|\hat{X}|^{2p-1}(1+|\hat{X}|)$$

$$\leq 2L(1+|\hat{X}|^{2p}),$$

$$|\hat{X}|^{2p-2}\hat{g}^2(\hat{X},t) \leq L^2|\hat{X}|^{2p-2}(1+|\hat{X})^2$$

$$\leq 2L^2|\hat{X}|^{2p-2}(1+|\hat{X}|^2)$$

$$\leq 4L^2(1+|\hat{X}|^{2p}).$$

Therefore, from the last inequality and from (16) we get that

$$\hat{E}\left(|\hat{X}|^{2p}\right) \leq \hat{E}\left(|\hat{X}_0|^{2p}\right)$$

$$+\hat{E}\left(\int_0^t \left(4pL(1+|\hat{X}|^{2p}) + 4p(2p-1)L^2(1+|\hat{X}|^{2p})\frac{\hat{T}(\Omega)}{\hat{T}}\right)\right) dt$$

$$\leq C_0\left(1+\hat{E}\left(|\hat{X}_0|^{2p}\right)\right) + C_1\hat{E}\left(\int_0^t |\hat{X}|^{2p} ds\right)$$

$$\leq C_0\left(1+\hat{E}\left(|\hat{X}_0|^{2p}\right)\right) + C_1\int_0^t \hat{E}\left(|\hat{X}|^{2p} ds\right).$$

In other words

$$\hat{E}\left(|\hat{X}|^{2p}\right) \leq C_0\left(1+\hat{E}\left(|\hat{X}_0|^{2p}\right)\right) + C_1\int_0^t \hat{E}\left(|\hat{X}|^{2p} ds\right).$$

From the last inequality and from Gronwall's inequality it follows that

$$\hat{E}\left(|\hat{X}|^{2p}\right) \leq C_0\left(1+\hat{E}\left(|\hat{X}_0|^{2p}\right)\right) e^{C_1 t}$$

for all $0 \leq t \leq T$.

To obtain the second inequality we use similar arguments as above and

$$d\left(\hat{X}(t) - \hat{X}_0\right) = \hat{f}(\hat{X},t)dt + \hat{g}(\hat{X},t)d\frac{\tilde{W}^1}{\hat{T}}.$$

□

Problem 2.4.3. *We suppose that* $\hat{f}$, $\hat{g} : \mathbb{R} \times [0,T] \longrightarrow \mathbb{R}$ *are continuous and satisfy*

$$|\hat{f}(x,t)| \leq L(1+|x|), \qquad |\hat{g}(x,t)| \leq L(1+|x|)$$

for all $t \in [0,T]$, $x \in \mathbb{R}$, *for some positive constant L. Let also,* $\hat{T} \geq N$ *on* Ω *and* U,

$$\hat{E}(|\hat{X}_0|^{2p}) < \infty$$

for some integer p. Prove that the solution $\hat{X}$ of the iso-stochastic differential equation

$$\begin{cases} d\hat{X} = \hat{f}(\hat{X},t)dt + \hat{g}(\hat{X},t)d\frac{\tilde{W}^2\tilde{T}(\tilde{W}^2)}{\hat{T}} \\ \hat{X}(0) = \hat{X}_0, \qquad 0 \le t \le T, \end{cases}$$

satisfies the estimates

$$\hat{E}\left(|\hat{X}(t)|^{2p}\right) \le C_0\left(1 + \hat{E}\left(|\hat{X}_0|^{2p}\right)\right)e^{C_1 t}$$

and

$$\hat{E}\left(|\hat{X}(t) - \hat{X}_0|^{2p}\right) \le C_0\left(1 + \hat{E}\left(|\hat{X}_0|^{2p}\right)\right)t^p e^{C_1 t},$$

where C_1 and C_2 are constants depending only on T and L.

Problem 2.4.4. *We suppose that $\hat{f}$, $\hat{g} : \mathbb{R} \times [0,T] \longrightarrow \mathbb{R}$ are continuous and satisfy*

$$|\hat{f}(x,t)| \le L(1+|x|), \qquad |\hat{g}(x,t)| \le L(1+|x|)$$

for all $t \in [0,T]$, $x \in \mathbb{R}$, for some positive constant L. Let also, $\hat{T} \ge N$ on Ω and U,

$$\hat{E}(|\hat{X}_0|^{2p}) < \infty$$

for some integer p. Prove that the solution $\hat{X}$ of the iso-stochastic differential equation

$$\begin{cases} d\hat{X} = \hat{f}(\hat{X},t)dt + \hat{g}(\hat{X},t)d\frac{\frac{\tilde{W}^3}{\tilde{T}(\tilde{W}^3)}}{\hat{T}} \\ \hat{X}(0) = \hat{X}_0, \qquad 0 \le t \le T, \end{cases}$$

satisfies the estimates

$$\hat{E}\left(|\hat{X}(t)|^{2p}\right) \le C_0\left(1 + \hat{E}\left(|\hat{X}_0|^{2p}\right)\right)e^{C_1 t}$$

and

$$\hat{E}\left(|\hat{X}(t) - \hat{X}_0|^{2p}\right) \le C_0\left(1 + \hat{E}\left(|\hat{X}_0|^{2p}\right)\right)t^p e^{C_1 t},$$

where C_1 and C_2 are constants depending only on T and L.

Problem 2.4.5. *We suppose that $\hat{f}$, $\hat{g} : \mathbb{R} \times [0,T] \longrightarrow \mathbb{R}$ are continuous and satisfy*

$$|\hat{f}(x,t)| \le L|x|, \qquad |\hat{g}(x,t)| \le L|x|$$

for all $t \in [0,T]$, $x \in \mathbb{R}$, for some positive constant L. Let also, $\hat{T} \ge N$ on Ω and U,

$$\hat{E}(|\hat{X}_0|^{2p}) < \infty$$

for some integer p. Prove that the solution $\hat{X}$ of the iso-stochastic differential equation

$$\begin{cases} d\hat{X} = \hat{f}(\hat{X},t)dt + \hat{g}(\hat{X},t)d\frac{\tilde{W}^1}{\hat{T}} \\ \hat{X}(0) = \hat{X}_0, \qquad 0 \le t \le T, \end{cases}$$

satisfies the estimates

$$\hat{E}\left(|\hat{X}(t)|^{2p}\right) \leq C_0\left(1+\hat{E}\left(|\hat{X}_0|^{2p}\right)\right)e^{C_1 t}$$

and

$$\hat{E}\left(|\hat{X}(t)-\hat{X}_0|^{2p}\right) \leq C_0\left(1+\hat{E}\left(|\hat{X}_0|^{2p}\right)\right)t^p e^{C_1 t},$$

where C_1 and C_2 are constants depending only on T and L.

Problem 2.4.6. *We suppose that $\hat{f}$, $\hat{g}:\mathbb{R}\times[0,T]\longrightarrow\mathbb{R}$ are continuous and satisfy*

$$|\hat{f}(x,t)| \leq L|x|, \qquad |\hat{g}(x,t)| \leq L|x|$$

for all $t\in[0,T]$, $x\in\mathbb{R}$, for some positive constant L. Let also, $\hat{T}\geq N$ on Ω and U,

$$\hat{E}(|\hat{X}_0|^{2p}) < \infty$$

for some integer p. Prove that the solution $\hat{X}$ of the iso-stochastic differential equation

$$\begin{cases} d\hat{X} = \hat{f}(\hat{X},t)dt + \hat{g}(\hat{X},t)d\frac{\tilde{W}^2\hat{T}(\tilde{W}^2)}{\hat{T}} \\ \hat{X}(0) = \hat{X}_0, \qquad 0\leq t\leq T, \end{cases}$$

satisfies the estimates

$$\hat{E}\left(|\hat{X}(t)|^{2p}\right) \leq C_0\left(1+\hat{E}\left(|\hat{X}_0|^{2p}\right)\right)e^{C_1 t}$$

and

$$\hat{E}\left(|\hat{X}(t)-\hat{X}_0|^{2p}\right) \leq C_0\left(1+\hat{E}\left(|\hat{X}_0|^{2p}\right)\right)t^p e^{C_1 t},$$

where C_1 and C_2 are constants depending only on T and L.

Problem 2.4.7. *We suppose that $\hat{f}$, $\hat{g}:\mathbb{R}\times[0,T]\longrightarrow\mathbb{R}$ are continuous and satisfy*

$$|\hat{f}(x,t)| \leq L|x|, \qquad |\hat{g}(x,t)| \leq L|x|$$

for all $t\in[0,T]$, $x\in\mathbb{R}$, for some positive constant L. Let also, $\hat{T}\geq N$ on Ω and U,

$$\hat{E}(|\hat{X}_0|^{2p}) < \infty$$

for some integer p. Prove that the solution $\hat{X}$ of the iso-stochastic differential equation

$$\begin{cases} d\hat{X} = \hat{f}(\hat{X},t)dt + \hat{g}(\hat{X},t)d\frac{\frac{\tilde{W}^3}{\hat{T}(\tilde{W}^3)}}{\hat{T}} \\ \hat{X}(0) = \hat{X}_0, \qquad 0\leq t\leq T, \end{cases}$$

satisfies the estimates

$$\hat{E}\left(|\hat{X}(t)|^{2p}\right) \leq C_0\left(1+\hat{E}\left(|\hat{X}_0|^{2p}\right)\right)e^{C_1 t}$$

and

$$\hat{E}\left(|\hat{X}(t)-\hat{X}_0|^{2p}\right) \leq C_0\left(1+\hat{E}\left(|\hat{X}_0|^{2p}\right)\right)t^p e^{C_1 t},$$

where C_1 and C_2 are constants depending only on T and L.

Chapter 3

Linear Iso-Stochastic Differential Equations

Definition 3.0.8. *The iso-stochastic differential equation*

$$(1) \qquad \begin{cases} d\hat{X} = \big(a_1(t)+a_2(t)\hat{X}(t)\big)\,dt + \big(b_1(t)+b_2(t)\hat{X}(t)\big)\,d\frac{\tilde{W}^1}{\hat{T}} \\ \hat{X}(0) = \hat{X}_0, \qquad 0 \le t \le T, \end{cases}$$

where a_1, a_2, b_1, $b_2 : [0,T] \longrightarrow \mathbb{R}$ are real-valued continuous functions, will be called linear iso-stochastic differential equation of the first kind.

We will search a solution to the equation (1) in the form

$$\hat{X}(t) = \hat{X}_1(t)\hat{X}_2(t),$$

where $\hat{X}_1$ satisfies

$$(2) \qquad \begin{cases} d\hat{X}_1 = a_2(t)\hat{X}_1(t)dt + b_2(t)\hat{X}_1(t)d\frac{\tilde{W}^1}{\hat{T}} \\ \hat{X}_1(0) = 1, \qquad 0 \le t \le T, \end{cases}$$

and $\hat{X}_2$ satisfies

$$(3) \qquad \begin{array}{l} d\hat{X}_2 = A(t)dt + B(t)d\frac{\tilde{W}^1}{\hat{T}} \\ \hat{X}_2(0) = \hat{X}_0, \qquad 0 \le t \le T. \end{array}$$

The functions A, $B : [0,T] \longrightarrow \mathbb{R}$ are continuous functions which will be selected below.

We have

$$d\hat{X} = d(\hat{X}_1\hat{X}_2)$$

$$= \hat{X}_2(t)d\hat{X}_1 + \hat{X}_1(t)d\hat{X}_2 + b_1(t)B(t)\hat{X}_1(t)dt$$

$$= \hat{X}_2(t)\left(a_1(t)\hat{X}_1(t)dt + b_2(t)\hat{X}_2(t)d\tfrac{\tilde{W}^1}{\hat{T}}\right)$$

$$+\hat{X}_1(t)\left(A(t)dt + B(t)d\tfrac{\tilde{W}^1}{\hat{T}}\right)$$

$$+b_2(t)B(t)\hat{X}_1(t)dt$$

$$= a_2(t)\hat{X}(t)dt + b_2(t)\hat{X}(t)d\tfrac{\tilde{W}^1}{\hat{T}} + \hat{X}_1(t)\left(A(t) + b_2(t)B(t)\right)dt$$

$$+\hat{X}_1(t)B(t)d\tfrac{\tilde{W}^1}{\hat{T}}.$$

From here and from (1) we get

$$\left(a_1(t) + a_2(t)\hat{X}(t)\right)dt + \left(b_1(t) + b_2(t)\hat{X}(t)\right)d\tfrac{\tilde{W}^1}{\hat{T}}$$

$$= a_2(t)\hat{X}(t)dt + b_2(t)\hat{X}(t)d\tfrac{\tilde{W}^1}{\hat{T}} + \hat{X}_1(t)\left(A(t) + b_2(t)B(t)\right)dt$$

$$+\hat{X}_1(t)B(t)d\tfrac{\tilde{W}^1}{\hat{T}}$$

or

$$a_1(t)dt + b_1(t)d\frac{\tilde{W}^1}{\hat{T}} = \hat{X}_1(t)\left(A(t) + b_2(t)B(t)\right)dt + \hat{X}_1(t)B(t)d\frac{\tilde{W}^1}{\hat{T}},$$

whereupon

$$\begin{cases} a_1(t) = \hat{X}_1(t)\left(A(t) + b_2(t)B(t)\right) \\ b_1(t) = \hat{X}_1(t)B(t). \end{cases}$$

Therefore

$$\begin{cases} B(t) = \hat{X}_1^{-1}(t)b_1(t) \\ A(t) + b_2(t)B(t) = \hat{X}_1^{-1}(t)a_1(t) \end{cases}$$

or

$$(4) \qquad \begin{cases} A(t) = \hat{X}_1^{-1}(t)\left(a_1(t) - b_1(t)b_2(t)\right) \\ B(t) = \hat{X}_1^{-1}(t)b_1(t), \qquad 0 \le t \le T. \end{cases}$$

From (2) we get

$$\hat{X}_1^{-1}(t) = e^{\int_0^t \left(a_2(s)ds + b_2(s)d\frac{\tilde{W}^1}{\hat{T}}\right)}, \qquad 0 \le t \le T,$$

and from (3)

$$\hat{X}_2(t) = \hat{X}_0 + \int_0^t \left(A(s)ds + B(s)d\frac{\tilde{W}^1}{\hat{T}}\right), \qquad 0 \le t \le T.$$

Definition 3.0.9. *The iso-stochastic differential equation*

$$(5)\qquad \begin{cases} d\hat{X} = \big(a_1(t)+a_2(t)\hat{X}(t)\big)\,dt + \big(b_1(t)+b_2(t)\hat{X}(t)\big)\,d\dfrac{\tilde{W}^2\tilde{T}(\tilde{W}^2)}{\hat{T}} \\ \hat{X}(0)=\hat{X}_0, \qquad 0\le t\le T, \end{cases}$$

where a_1, a_2, b_1, $b_2 : [0,T] \longrightarrow \mathbb{R}$ are real-valued continuous functions, will be called linear iso-stochastic differential equation of the second kind.

We will search a solution to the equation (5) in the form

$$\hat{X}(t)=\hat{X}_1(t)\hat{X}_2(t),$$

where $\hat{X}_1$ satisfies

$$(6)\qquad \begin{cases} d\hat{X}_1 = a_2(t)\hat{X}_1(t)dt + b_2(t)\hat{X}_1(t)d\dfrac{\tilde{W}^2\tilde{T}(\tilde{W}^2)}{\hat{T}} \\ \hat{X}_1(0)=1, \qquad 0\le t\le T, \end{cases}$$

and $\hat{X}_2$ satisfies

$$(7)\qquad \begin{cases} d\hat{X}_2 = A(t)dt + B(t)d\dfrac{\tilde{W}^2\tilde{T}(\tilde{W}^2)}{\hat{T}} \\ \hat{X}_2(0)=\hat{X}_0, \qquad 0\le t\le T. \end{cases}$$

The functions A, $B : [0,T] \longrightarrow \mathbb{R}$ are continuous functions which will be selected below.

Then

$$\begin{aligned} d\hat{X} &= d(\hat{X}_1\hat{X}_2) \\ &= \hat{X}_2(t)d\hat{X}_1 + \hat{X}_1(t)d\hat{X}_2 + b_1(t)B(t)\hat{X}_1(t)dt \\ &= \hat{X}_2(t)\left(a_1(t)\hat{X}_1(t)dt + b_2(t)\hat{X}_2(t)d\frac{\tilde{W}^2\tilde{T}(\tilde{W}^2)}{\hat{T}}\right) \\ &\quad +\hat{X}_1(t)\left(A(t)dt + B(t)d\frac{\tilde{W}^2\tilde{T}(\tilde{W}^2)}{\hat{T}}\right) \\ &\quad +b_2(t)B(t)\hat{X}_1(t)dt \\ &= a_2(t)\hat{X}(t)dt + b_2(t)\hat{X}(t)d\frac{\tilde{W}^2\tilde{T}(\tilde{W}^2)}{\hat{T}} + \hat{X}_1(t)\left(A(t)+b_2(t)B(t)\right)dt \\ &\quad +\hat{X}_1(t)B(t)d\frac{\tilde{W}^2\tilde{T}(\tilde{W}^2)}{\hat{T}}. \end{aligned}$$

Hence and (5) we find

$$\begin{aligned} &\big(a_1(t)+a_2(t)\hat{X}(t)\big)\,dt + \big(b_1(t)+b_2(t)\hat{X}(t)\big)\,d\frac{\tilde{W}^2\tilde{T}(\tilde{W}^2)}{\hat{T}} \\ &= a_2(t)\hat{X}(t)dt + b_2(t)\hat{X}(t)d\frac{\tilde{W}^2\tilde{T}(\tilde{W}^2)}{\hat{T}} + \hat{X}_1(t)\left(A(t)+b_2(t)B(t)\right)dt \\ &\quad +\hat{X}_1(t)B(t)d\frac{\tilde{W}^2\tilde{T}(\tilde{W}^2)}{\hat{T}} \end{aligned}$$

or

$$a_1(t)dt + b_1(t)d\frac{\tilde{W}^2\tilde{T}(\tilde{W}^2)}{\hat{T}} = \hat{X}_1(t)\left(A(t) + b_2(t)B(t)\right)dt + \hat{X}_1(t)B(t)d\frac{\tilde{W}^2\tilde{T}(\tilde{W}^2)}{\hat{T}},$$

from where we obtain the system (4).

From (6) we have

$$\hat{X}_1^{-1}(t) = e^{\int_0^t \left(a_2(s)ds + b_2(s)d\frac{\tilde{W}^2\tilde{T}(\tilde{W}^2)}{\hat{T}}\right)}, \qquad 0 \le t \le T,$$

and using (7)

$$\hat{X}_2(t) = \hat{X}_0 + \int_0^t \left(A(s)ds + B(s)d\frac{\tilde{W}^2\tilde{T}(\tilde{W}^2)}{\hat{T}}\right), \qquad 0 \le t \le T.$$

Definition 3.0.10. *The iso-stochastic differential equation*

$$(8) \qquad \begin{cases} d\hat{X} = \left(a_1(t) + a_2(t)\hat{X}(t)\right)dt + \left(b_1(t) + b_2(t)\hat{X}(t)\right)d\frac{\frac{\tilde{W}^3}{\tilde{T}(\tilde{W}^3)}}{\hat{T}} \\ \hat{X}(0) = \hat{X}_0, \qquad 0 \le t \le T, \end{cases}$$

where a_1, a_2, b_1, $b_2 : [0,T] \longrightarrow \mathbb{R}$ are real-valued continuous functions, will be called linear iso-stochastic differential equation of the third kind.

We will find a solution to the equation (8) in the form

$$\hat{X}(t) = \hat{X}_1(t)\hat{X}_2(t),$$

where $\hat{X}_1$ satisfies

$$(9) \qquad \begin{cases} d\hat{X}_1 = a_2(t)\hat{X}_1(t)dt + b_2(t)\hat{X}_1(t)d\frac{\frac{\tilde{W}^3}{\tilde{T}(\tilde{W}^3)}}{\hat{T}} \\ \hat{X}_1(0) = 1, \qquad 0 \le t \le T, \end{cases}$$

and $\hat{X}_2$ satisfies

$$(10) \qquad \begin{cases} d\hat{X}_2 = A(t)dt + B(t)d\frac{\frac{\tilde{W}^3}{\tilde{T}(\tilde{W}^3)}}{\hat{T}} \\ \hat{X}_2(0) = \hat{X}_0, \qquad 0 \le t \le T. \end{cases}$$

The functions A, $B : [0,T] \longrightarrow \mathbb{R}$ are continuous functions which will be selected below.

Then

$$d\hat{X} = d(\hat{X}_1\hat{X}_2)$$

$$= \hat{X}_2(t)d\hat{X}_1 + \hat{X}_1(t)d\hat{X}_2 + b_1(t)B(t)\hat{X}_1(t)dt$$

$$= \hat{X}_2(t)\left(a_1(t)\hat{X}_1(t)dt + b_2(t)\hat{X}_2(t)d\frac{\frac{\tilde{W}^3}{\tilde{T}(\tilde{W}^3)}}{\hat{T}}\right)$$

$$+\hat{X}_1(t)\left(A(t)dt + B(t)d\frac{\frac{\tilde{W}^3}{\tilde{T}(\tilde{W}^3)}}{\hat{T}}\right)$$

$$+b_2(t)B(t)\hat{X}_1(t)dt$$

$$= a_2(t)\hat{X}(t)dt + b_2(t)\hat{X}(t)d\frac{\frac{\tilde{W}^3}{\tilde{T}(\tilde{W}^3)}}{\hat{T}} + \hat{X}_1(t)\left(A(t) + b_2(t)B(t)\right)dt$$

$$+\hat{X}_1(t)B(t)d\frac{\frac{\tilde{W}^3}{\tilde{T}(\tilde{W}^3)}}{\hat{T}}.$$

From here and (8) we find

$$\left(a_1(t) + a_2(t)\hat{X}(t)\right)dt + \left(b_1(t) + b_2(t)\hat{X}(t)\right)d\frac{\frac{\tilde{W}^3}{\tilde{T}(\tilde{W}^3)}}{\hat{T}}$$

$$= a_2(t)\hat{X}(t)dt + b_2(t)\hat{X}(t)d\frac{\frac{\tilde{W}^3}{\tilde{T}(\tilde{W}^3)}}{\hat{T}} + \hat{X}_1(t)\left(A(t) + b_2(t)B(t)\right)dt$$

$$+\hat{X}_1(t)B(t)d\frac{\frac{\tilde{W}^3}{\tilde{T}(\tilde{W}^3)}}{\hat{T}}$$

or

$$a_1(t)dt + b_1(t)d\frac{\frac{\tilde{W}^3}{\tilde{T}(\tilde{W}^3)}}{\hat{T}} = \hat{X}_1(t)\left(A(t) + b_2(t)B(t)\right)dt + \hat{X}_1(t)B(t)d\frac{\frac{\tilde{W}^3}{\tilde{T}(\tilde{W}^3)}}{\hat{T}},$$

from where we obtain the system (4).

From (9) we have

$$\hat{X}_1^{-1}(t) = e^{\int_0^t\left(a_2(s)ds + b_2(s)d\frac{\frac{\tilde{W}^3}{\tilde{T}(\tilde{W}^3)}}{\hat{T}}\right)}, \qquad 0 \le t \le T,$$

and from (10)

$$\hat{X}_2(t) = \hat{X}_0 + \int_0^t\left(A(s)ds + B(s)d\frac{\frac{\tilde{W}^3}{\tilde{T}(\tilde{W}^3)}}{\hat{T}}\right), \qquad 0 \le t \le T.$$

Chapter 4

Dependence on Parameters

Theorem 4.0.11. *Suppose for $k = 1,2,\ldots$, that $\hat{f}$, $\hat{g}$, $\hat{X}_0$ and $\hat{X}_0^k$ satisfy the hypotheses of the existence and uniqueness theorem for the iso-stochastic differential equations of the first kind. Let also,*

$$\lim_{k\longrightarrow\infty} \hat{E}\left(|\hat{X}_0^k - \hat{X}_0|^{\hat{2}}\right) = 0.$$

Finally suppose that $\hat{X}^k$ solves the iso-stochastic differential equation of the first kind

$$\begin{cases} d\hat{X}^k = \hat{f}(\hat{X}^k,t)dt + \hat{g}(\hat{X}^k,t)d\frac{\tilde{W}^1}{\hat{T}} \\ \hat{X}^k(0) = \hat{X}_0^k, \qquad 0 \le t \le T, \qquad\qquad k = 1,2,\ldots. \end{cases}$$

Then

$$\lim_{k\longrightarrow\infty} \hat{E}\left(\max_{0\le t\le T} |\hat{X}^k(t) - \hat{X}(t)|^{\hat{2}}\right) = 0,$$

where

$$\begin{cases} d\hat{X} = \hat{f}(\hat{X},t)dt + \hat{g}(\hat{X},t)d\frac{\tilde{W}^1}{\hat{T}} \\ \hat{X}(0) = \hat{X}_0, \qquad 0 \le t \le T. \end{cases}$$

Proof. For the solutions $\hat{X}^k$, $k = 1,2,\ldots$, and $\hat{X}$ we have the following representations

$$\hat{X}^k(t) = \hat{X}_0^k + \int_0^t \hat{f}(\hat{X}^k(s),s)ds + \int_0^t \hat{g}(\hat{X}^k(s),s)d\frac{\tilde{W}^1}{\hat{T}}, \qquad k = 1,2,\ldots,$$

$$\hat{X}(t) = \hat{X}_0 + \int_0^t \hat{f}(\hat{X}(s),s)ds + \int_0^t \hat{g}(\hat{X}(s),s)d\frac{\tilde{W}^1}{\hat{T}}, \qquad 0 \le t \le T.$$

Then

$$\hat{X}^k(t) - \hat{X}(t) = \hat{X}_0^k + \int_0^t \hat{f}(\hat{X}^k(s),s)ds + \int_0^t \hat{g}(\hat{X}^k(s),s)d\frac{\tilde{W}^1}{\hat{T}}$$

$$-\hat{X}_0 - \int_0^t \hat{f}(\hat{X}(s),s)ds - \int_0^t \hat{g}(\hat{X}(s),s)d\frac{\tilde{W}^1}{\hat{T}}$$

$$= \hat{X}_0^k - \hat{X}_0 + \int_0^t \left(\hat{f}(\hat{X}^k(s),s) - \hat{f}(\hat{X}(s),s)\right)ds$$

$$+ \int_0^t \left(\hat{g}(\hat{X}^k(s),s) - \hat{g}(\hat{X}(s),s)\right)d\frac{\tilde{W}^1}{\hat{T}},$$

from here

$$|\hat{X}^k(t)-\hat{X}(t)| = \left|\hat{X}_0^k-\hat{X}_0+\int_0^t\left(\hat{f}(\hat{X}^k(s),s)-\hat{f}(\hat{X}(s),s)\right)ds\right.$$

$$\left.+\int_0^t\left(\hat{g}(\hat{X}^k(s),s)-\hat{g}(\hat{X}(s),s)\right)d\frac{\tilde{W}^1}{\hat{T}}\right|$$

$$\leq \left|\hat{X}_0^k-\hat{X}_0\right|+\int_0^t\left|\hat{f}(\hat{X}^k(s),s)-\hat{f}(\hat{X}(s),s)\right|ds$$

$$+\int_0^t\left|\hat{g}(\hat{X}^k(s),s)-\hat{g}(\hat{X}(s),s)\right|d\frac{\tilde{W}^1}{\hat{T}}$$

and

$$|\hat{X}^k(t)-\hat{X}(t)|^{\hat{2}} = |\hat{X}^k(t)-\hat{X}(t)|^2\hat{T}$$

$$=\hat{T}\left(\left|\hat{X}_0^k-\hat{X}_0\right|+\int_0^t\left|\hat{f}(\hat{X}^k(s),s)-\hat{f}(\hat{X}(s),s)\right|ds\right.$$

$$\left.+\int_0^t\left|\hat{g}(\hat{X}^k(s),s)-\hat{g}(\hat{X}(s),s)\right|d\frac{\tilde{W}^1}{\hat{T}}\right|\Bigg)^2$$

$$\leq 2\hat{T}|_0^k-\hat{X}_0|^2+2\hat{T}\left(\int_0^t|\hat{f}(\hat{X}^k(s),s)-\hat{f}(\hat{X}(s),s)|ds\right)^2$$

$$+2\hat{T}\left(\int_0^t\left|\hat{g}(\hat{X}^k(s),s)-\hat{g}(\hat{X}(s),s)\right|d\frac{\tilde{W}^1}{\hat{T}}\right)^2$$

$$\leq 2\hat{T}|_0^k-\hat{X}_0|^2+2t\hat{T}\int_0^t|\hat{f}(\hat{X}^k(s),s)-\hat{f}(\hat{X}(s),s)|^2ds$$

$$+2\hat{T}\left(\int_0^t\left|\hat{g}(\hat{X}^k(s),s)-\hat{g}(\hat{X}(s),s)\right|d\frac{\tilde{W}^1}{\hat{T}}\right)^2$$

$$=2|\hat{X}_0^k-\hat{X}_0|^{\hat{2}}+2t\int_0^t|\hat{f}(\hat{X}^k(s),s)-\hat{f}(\hat{X}(s),s)|^{\hat{2}}ds$$

$$+2\left(\int_0^t\frac{1}{\tilde{T}}\tilde{T}(\hat{g}(\hat{X}^k(s),s)-\hat{g}(\hat{X}(s),s))\frac{\tilde{T}^2}{\tilde{T}^2}d\frac{\tilde{W}^1}{\hat{T}}\right)^{\hat{2}}$$

$$\leq 2|\hat{X}_0^k-\hat{X}_0|^{\hat{2}}+2TL^2\int_0^t|\hat{X}^k(s)-\hat{X}(s)|^{\hat{2}}ds$$

$$+2\left(\hat{\int}_0^t\frac{\hat{g}(\hat{X}^k(s),s)-\hat{g}(\hat{X}(s),s)}{\tilde{T}}\hat{\times}\hat{d}\frac{\tilde{W}^1}{\hat{T}}\right)^{\hat{2}},$$

i.e.,

$$|\hat{X}^k(t)-\hat{X}(t)|^{\hat{2}}\leq 2|\hat{X}_0^k-\hat{X}_0|^{\hat{2}}+2TL^2\int_0^t|\hat{X}^k(s)-\hat{X}(s)|^{\hat{2}}ds$$

$$+2\left(\hat{\int}_0^t\frac{\hat{g}(\hat{X}^k(s),s)-\hat{g}(\hat{X}(s),s)}{\tilde{T}}\hat{\times}\hat{d}\frac{\tilde{W}^1}{\hat{T}}\right)^{\hat{2}}.$$

Therefore

$$\hat{E}(|\hat{X}^k(t)-\hat{X}(t)|^{\hat{2}}) \le 2\hat{E}(|\hat{X}_0^k-\hat{X}_0|^{\hat{2}}) + 2TL^2\int_0^t \hat{E}(|\hat{X}^k(s)-\hat{X}(s)|^{\hat{2}})ds$$

$$+2\hat{E}\left(\left(\hat{\int}_0^t \frac{\hat{g}(\hat{X}^k(s),s)-\hat{g}(\hat{X}(s),s)}{\tilde{T}} \hat{\times} \hat{d}\frac{\tilde{W}^1}{\hat{T}}\right)^{\hat{2}}\right)$$

$$= 2\hat{E}(|\hat{X}_0^k-\hat{X}_0|^{\hat{2}}) + 2TL^2\int_0^t \hat{E}(|\hat{X}^k(s)-\hat{X}(s)|^{\hat{2}})ds$$

$$+2\hat{E}\left(\int_0^t \frac{(\hat{g}(\hat{X}^k(s),s)-\hat{g}(\hat{X}(s),s))^2\tilde{T}^2}{\tilde{T}^2\hat{T}}dt\right)$$

$$= 2\hat{E}(|\hat{X}_0^k-\hat{X}_0|^{\hat{2}}) + 2TL^2\int_0^t \hat{E}(|\hat{X}^k(s)-\hat{X}(s)|^{\hat{2}})ds$$

$$+2\hat{E}\left(\int_0^t \frac{(\hat{g}(\hat{X}^k(s),s)-\hat{g}(\hat{X}(s),s))^2}{\hat{T}}dt\right)$$

$$\le 2\hat{E}(|\hat{X}_0^k-\hat{X}_0|^{\hat{2}}) + 2TL^2\int_0^t \hat{E}(|\hat{X}^k(s)-\hat{X}(s)|^{\hat{2}})ds$$

$$+2\frac{L^2}{N}\hat{E}\left(\int_0^t |\hat{X}^k(s)-\hat{X}(s)|^2 ds\right)$$

$$= 2\hat{E}(|\hat{X}_0^k-\hat{X}_0|^{\hat{2}}) + 2TL^2\int_0^t \hat{E}(|\hat{X}^k(s)-\hat{X}(s)|^{\hat{2}})ds$$

$$+2\frac{L^2}{N}\hat{E}\left(\int_0^t \frac{\hat{T}|\hat{X}^k(s)\ \ \hat{X}(s)|^2}{\hat{T}}ds\right)$$

$$\le \hat{E}(|\hat{X}_0^k-\hat{X}_0|^{\hat{2}}) + 2TL^2\int_0^t \hat{E}(|\hat{X}^k(s)-\hat{X}(s)|^{\hat{2}})ds$$

$$+2\frac{L^2}{N^2}\hat{E}\left(\int_0^t |\hat{X}^k(s)-\hat{X}(s)|^{\hat{2}} ds\right)$$

$$= 2\hat{E}(|\hat{X}_0^k-\hat{X}_0|^{\hat{2}}) + \left(2TL^2+2\frac{L^2}{N^2}\right)\int_0^t \hat{E}(|\hat{X}^k(s)-\hat{X}(s)|^{\hat{2}})ds,$$

i.e.,

$$\hat{E}\left(|\hat{X}^k(t)-\hat{X}(t)|^{\hat{2}}\right) \le 2\hat{E}(|\hat{X}_0^k-\hat{X}_0|^{\hat{2}}) + \left(2TL^2+2\frac{L^2}{N^2}\right)\int_0^t \hat{E}(|\hat{X}^k(s)-\hat{X}(s)|^{\hat{2}})ds.$$

From here and from Gronwall's lemma it follows that

$$\hat{E}\left(|\hat{X}^k(t)-\hat{X}(t)|^{\hat{2}}\right) \le 2\hat{E}\left(|\hat{X}_0^k-\hat{X}_0|^{\hat{2}}\right)e^{\left(2TL^2+2\frac{L^2}{N^2}\right)t}.$$

Because

$$\lim_{k\longrightarrow\infty}\hat{E}\left(|\hat{X}_0^k-\hat{X}_0|^{\hat{2}}\right)=0$$

we find that

$$\lim_{k\longrightarrow\infty}\hat{E}\left(\max_{0\le t\le T}|\hat{X}^k(t)-\hat{X}(t)|^{\hat{2}}\right)=0.$$

□

Theorem 4.0.12. *Suppose for $k=1,2,\ldots$, that $\hat{f}$, $\hat{g}$, $\hat{X}_0$ and $\hat{X}_0^k$ satisfy the hypotheses of the existence and uniqueness theorem for the iso-stochastic differential equations of the second kind. Let also,*

$$\lim_{k\longrightarrow\infty}\hat{E}\left(|\hat{X}_0^k-\hat{X}_0|^{\hat{2}}\right)=0.$$

Finally suppose that $\hat{X}^k$ solves the iso-stochastic differential equation of the second kind

$$\begin{cases} d\hat{X}^k=\hat{f}(\hat{X}^k,t)dt+\hat{g}(\hat{X}^k,t)d\frac{\tilde{W}^2\tilde{T}(\tilde{W}^2)}{\hat{T}} \\ \hat{X}^k(0)=\hat{X}_0^k, \qquad 0\le t\le T, \qquad\qquad k=1,2,\ldots. \end{cases}$$

Then

$$\lim_{k\longrightarrow\infty}\hat{E}\left(\max_{0\le t\le T}|\hat{X}^k(t)-\hat{X}(t)|^{\hat{2}}\right)=0,$$

where

$$\begin{cases} d\hat{X}=\hat{f}(\hat{X},t)dt+\hat{g}(\hat{X},t)d\frac{\tilde{W}^2\tilde{T}(\tilde{W}^2)}{\hat{T}} \\ \hat{X}(0)=\hat{X}_0, \qquad 0\le t\le T. \end{cases}$$

Proof. For the solutions $\hat{X}^k$, $k=1,2,\ldots$, and $\hat{X}$ we will use the following representations

$$\hat{X}^k(t)=\hat{X}_0^k+\int_0^t\hat{f}(\hat{X}^k(s),s)ds+\int_0^t\hat{g}(\hat{X}^k(s),s)d\frac{\tilde{W}^2\tilde{T}(\tilde{W}^2)}{\hat{T}}, \qquad k=1,2,\ldots,$$

$$\hat{X}(t)=\hat{X}_0+\int_0^t\hat{f}(\hat{X}(s),s)ds+\int_0^t\hat{g}(\hat{X}(s),s)d\frac{\tilde{W}^2\tilde{T}(\tilde{W}^2)}{\hat{T}}, \qquad 0\le t\le T.$$

From here,

$$\hat{X}^k(t)-\hat{X}(t)=\hat{X}_0^k+\int_0^t\hat{f}(\hat{X}^k(s),s)ds+\int_0^t\hat{g}(\hat{X}^k(s),s)d\frac{\tilde{W}^2\tilde{T}(\tilde{W}^2)}{\hat{T}}$$

$$-\hat{X}_0-\int_0^t\hat{f}(\hat{X}(s),s)ds-\int_0^t\hat{g}(\hat{X}(s),s)d\frac{\tilde{W}^2\tilde{T}(\tilde{W}^2)}{\hat{T}}$$

$$=\hat{X}_0^k-\hat{X}_0+\int_0^t\left(\hat{f}(\hat{X}^k(s),s)-\hat{f}(\hat{X}(s),s)\right)ds$$

$$+\int_0^t\left(\hat{g}(\hat{X}^k(s),s)-\hat{g}(\hat{X}(s),s)\right)d\frac{\tilde{W}^2\tilde{T}(\tilde{W}^2)}{\hat{T}},$$

whereupon

$$|\hat{X}^k(t)-\hat{X}(t)|=\Big|\hat{X}_0^k-\hat{X}_0+\int_0^t\left(\hat{f}(\hat{X}^k(s),s)-\hat{f}(\hat{X}(s),s)\right)ds$$

$$+\int_0^t\left(\hat{g}(\hat{X}^k(s),s)-\hat{g}(\hat{X}(s),s)\right)d\frac{\tilde{W}^2\tilde{T}(\tilde{W}^2)}{\hat{T}}\Big|$$

$$\le\Big|\hat{X}_0^k-\hat{X}_0\Big|+\int_0^t\Big|\hat{f}(\hat{X}^k(s),s)-\hat{f}(\hat{X}(s),s)\Big|ds$$

$$+\int_0^t\Big|\hat{g}(\hat{X}^k(s),s)-\hat{g}(\hat{X}(s),s)\Big|d\frac{\tilde{W}^2\tilde{T}(\tilde{W}^2)}{\hat{T}},$$

also,

$$|\hat{X}^k(t)-\hat{X}(t)|^{\hat{2}} = |\hat{X}^k(t)-\hat{X}(t)|^2\hat{T}$$

$$= \hat{T}\left(\left|\hat{X}_0^k-\hat{X}_0\right| + \int_0^t\left|\hat{f}(\hat{X}^k(s),s)-\hat{f}(\hat{X}(s),s)\right|ds\right.$$

$$\left.+\int_0^t\left|\hat{g}(\hat{X}^k(s),s)-\hat{g}(\hat{X}(s),s)\right|d\frac{\tilde{W}^2\tilde{T}(\tilde{W}^2)}{\hat{T}}\right|\Big)^2$$

$$\leq 2\hat{T}|_0^k-\hat{X}_0|^2+2\hat{T}\left(\int_0^t|\hat{f}(\hat{X}^k(s),s)-\hat{f}(\hat{X}(s),s)|ds\right)^2$$

$$+2\hat{T}\left(\int_0^t\left|\hat{g}(\hat{X}^k(s),s)-\hat{g}(\hat{X}(s),s)\right|d\frac{\tilde{W}^2\tilde{T}(\tilde{W}^2)}{\hat{T}}\right)^2$$

$$\leq 2\hat{T}|_0^k-\hat{X}_0|^2+2t\hat{T}\int_0^t|\hat{f}(\hat{X}^k(s),s)-\hat{f}(\hat{X}(s),s)|^2ds$$

$$+2\hat{T}\left(\int_0^t\left|\hat{g}(\hat{X}^k(s),s)-\hat{g}(\hat{X}(s),s)\right|d\frac{\tilde{W}^2\tilde{T}(\tilde{W}^2)}{\hat{T}}\right)^2$$

$$= 2|\hat{X}_0^k-\hat{X}_0|^{\hat{2}}+2t\int_0^t|\hat{f}(\hat{X}^k(s),s)-\hat{f}(\hat{X}(s),s)|^{\hat{2}}ds$$

$$+2\left(\int_0^t\frac{1}{\tilde{T}}\tilde{T}(\hat{g}(\hat{X}^k(s),s)-\hat{g}(\hat{X}(s),s))\frac{\tilde{T}^2}{\tilde{T}^2}d\frac{\tilde{W}^2\tilde{T}(\tilde{W}^2)}{\hat{T}}\right)^{\hat{2}}$$

$$\leq 2|\hat{X}_0^k-\hat{X}_0|^{\hat{2}}+2TL^2\int_0^t|\hat{X}^k(s)-\hat{X}(s)|^{\hat{2}}ds$$

$$+2\left(\hat{\int}_0^t\frac{\hat{g}(\hat{X}^k(s),s)-\hat{g}(\hat{X}(s),s)}{\tilde{T}}\hat{\times}\hat{d}\frac{\tilde{W}^2\tilde{T}(\tilde{W}^2)}{\hat{T}}\right)^{\hat{2}},$$

or,

$$|\hat{X}^k(t)-\hat{X}(t)|^{\hat{2}} \leq 2|\hat{X}_0^k-\hat{X}_0|^{\hat{2}}+2TL^2\int_0^t|\hat{X}^k(s)-\hat{X}(s)|^{\hat{2}}ds$$

$$+2\left(\hat{\int}_0^t\frac{\hat{g}(\hat{X}^k(s),s)-\hat{g}(\hat{X}(s),s)}{\tilde{T}}\hat{\times}\hat{d}\frac{\tilde{W}^2\tilde{T}(\tilde{W}^2)}{\hat{T}}\right)^{\hat{2}}.$$

Therefore

$$\hat{E}(|\hat{X}^k(t)-\hat{X}(t)|^{\hat{2}}) \leq 2\hat{E}(|\hat{X}_0^k-\hat{X}_0|^{\hat{2}}) + 2TL^2\int_0^t \hat{E}(|\hat{X}^k(s)-\hat{X}(s)|^{\hat{2}})ds$$

$$+2\hat{E}\left(\left(\hat{\int}_0^t \frac{\hat{g}(\hat{X}^k(s),s)-\hat{g}(\hat{X}(s),s)}{\tilde{T}} \hat{\times} \hat{d}\frac{\tilde{W}^2\tilde{T}(\tilde{W}^2)}{\hat{T}}\right)^{\hat{2}}\right)$$

$$= 2\hat{E}(|\hat{X}_0^k-\hat{X}_0|^{\hat{2}}) + 2TL^2\int_0^t \hat{E}(|\hat{X}^k(s)-\hat{X}(s)|^{\hat{2}})ds$$

$$+2\hat{E}\left(\int_0^t \frac{(\hat{g}(\hat{X}^k(s),s)-\hat{g}(\hat{X}(s),s))^2\tilde{T}^2}{\tilde{T}^2\hat{T}}dt\right)$$

$$= 2\hat{E}(|\hat{X}_0^k-\hat{X}_0|^{\hat{2}}) + 2TL^2\int_0^t \hat{E}(|\hat{X}^k(s)-\hat{X}(s)|^{\hat{2}})ds$$

$$+2\hat{E}\left(\int_0^t \frac{(\hat{g}(\hat{X}^k(s),s)-\hat{g}(\hat{X}(s),s))^2}{\hat{T}}dt\right)$$

$$\leq 2\hat{E}(|\hat{X}_0^k-\hat{X}_0|^{\hat{2}}) + 2TL^2\int_0^t \hat{E}(|\hat{X}^k(s)-\hat{X}(s)|^{\hat{2}})ds$$

$$+2\frac{L^2}{N}\hat{E}\left(\int_0^t |\hat{X}^k(s)-\hat{X}(s)|^2 ds\right)$$

$$= 2\hat{E}(|\hat{X}_0^k-\hat{X}_0|^{\hat{2}}) + 2TL^2\int_0^t \hat{E}(|\hat{X}^k(s)-\hat{X}(s)|^{\hat{2}})ds$$

$$+2\frac{L^2}{N}\hat{E}\left(\int_0^t \frac{\hat{T}|\hat{X}^k(s)-\hat{X}(s)|^2}{\hat{T}}ds\right)$$

$$\leq \hat{E}(|\hat{X}_0^k-\hat{X}_0|^{\hat{2}}) + 2TL^2\int_0^t \hat{E}(|\hat{X}^k(s)-\hat{X}(s)|^{\hat{2}})ds$$

$$+2\frac{L^2}{N^2}\hat{E}\left(\int_0^t |\hat{X}^k(s)-\hat{X}(s)|^{\hat{2}} ds\right)$$

$$= 2\hat{E}(|\hat{X}_0^k-\hat{X}_0|^{\hat{2}}) + \left(2TL^2+2\frac{L^2}{N^2}\right)\int_0^t \hat{E}(|\hat{X}^k(s)-\hat{X}(s)|^{\hat{2}})ds,$$

or,

$$\hat{E}\left(|\hat{X}^k(t)-\hat{X}(t)|^{\hat{2}}\right) \leq 2\hat{E}(|\hat{X}_0^k-\hat{X}_0|^{\hat{2}}) + \left(2TL^2+2\frac{L^2}{N^2}\right)\int_0^t \hat{E}(|\hat{X}^k(s)-\hat{X}(s)|^{\hat{2}})ds.$$

From the last inequality and from Gronwall's we get

$$\hat{E}\left(|\hat{X}^k(t)-\hat{X}(t)|^{\hat{2}}\right) \leq 2\hat{E}\left(|\hat{X}_0^k-\hat{X}_0|^{\hat{2}}\right)e^{\left(2TL^2+2\frac{L^2}{N^2}\right)t}.$$

Since

$$\lim_{k\longrightarrow\infty}\hat{E}\left(|\hat{X}_0^k-\hat{X}_0|^{\hat{2}}\right)=0$$

we obtain

$$\lim_{k\longrightarrow\infty}\hat{E}\left(\max_{0\leq t\leq T}|\hat{X}^k(t)-\hat{X}(t)|^{\hat{2}}\right)=0.$$

□

Theorem 4.0.13. *Suppose for* $k = 1,2,\ldots$, *that* $\hat{f}$, $\hat{g}$, $\hat{X}_0$ *and* $\hat{X}_0^k$ *satisfy the hypotheses of the existence and uniqueness theorem for the iso-stochastic differential equations of the third kind. Let also,*

$$\lim_{k\longrightarrow\infty} \hat{E}\left(|\hat{X}_0^k - \hat{X}_0|^{\hat{2}}\right) = 0.$$

Finally suppose that $\hat{X}^k$ *solves the iso-stochastic differential equation of the third kind*

$$\begin{cases} d\hat{X}^k = \hat{f}(\hat{X}^k,t)dt + \hat{g}(\hat{X}^k,t)d\frac{\frac{\tilde{W}^3}{\tilde{T}(\tilde{W}^3)}}{\hat{T}} \\ \hat{X}^k(0) = \hat{X}_0^k, \qquad 0 \le t \le T, \qquad\qquad k = 1,2,\ldots. \end{cases}$$

Then

$$\lim_{k\longrightarrow\infty} \hat{E}\left(\max_{0\le t\le T} |\hat{X}^k(t) - \hat{X}(t)|^{\hat{2}}\right) = 0,$$

where

$$\begin{cases} d\hat{X} = \hat{f}(\hat{X},t)dt + \hat{g}(\hat{X},t)d\frac{\frac{\tilde{W}^3}{\tilde{T}(\tilde{W}^3)}}{\hat{T}} \\ \hat{X}(0) = \hat{X}_0, \qquad 0 \le t \le T. \end{cases}$$

Proof. For the solutions $\hat{X}^k$, $k = 1,2,\ldots$, and $\hat{X}$ we have the following representations

$$\hat{X}^k(t) = \hat{X}_0^k + \int_0^t \hat{f}(\hat{X}^k(s),s)ds + \int_0^t \hat{g}(\hat{X}^k(s),s)d\frac{\frac{\tilde{W}^3}{\tilde{T}(\tilde{W}^3)}}{\hat{T}}, \qquad k = 1,2,\ldots,$$

$$\hat{X}(t) = \hat{X}_0 + \int_0^t \hat{f}(\hat{X}(s),s)ds + \int_0^t \hat{g}(\hat{X}(s),s)d\frac{\frac{\tilde{W}^3}{\tilde{T}(\tilde{W}^3)}}{\hat{T}}, \qquad 0 \le t \le T.$$

Then

$$\hat{X}^k(t) - \hat{X}(t) = \hat{X}_0^k + \int_0^t \hat{f}(\hat{X}^k(s),s)ds + \int_0^t \hat{g}(\hat{X}^k(s),s)d\frac{\frac{\tilde{W}^3}{\tilde{T}(\tilde{W}^3)}}{\hat{T}}$$

$$-\hat{X}_0 - \int_0^t \hat{f}(\hat{X}(s),s)ds - \int_0^t \hat{g}(\hat{X}(s),s)d\frac{\frac{\tilde{W}^3}{\tilde{T}(\tilde{W}^3)}}{\hat{T}}$$

$$= \hat{X}_0^k - \hat{X}_0 + \int_0^t \left(\hat{f}(\hat{X}^k(s),s) - \hat{f}(\hat{X}(s),s)\right)ds$$

$$+ \int_0^t \left(\hat{g}(\hat{X}^k(s),s) - \hat{g}(\hat{X}(s),s)\right)d\frac{\frac{\tilde{W}^3}{\tilde{T}(\tilde{W}^3)}}{\hat{T}},$$

from here

$$|\hat{X}^k(t) - \hat{X}(t)| = \Big|\hat{X}_0^k - \hat{X}_0 + \int_0^t \left(\hat{f}(\hat{X}^k(s),s) - \hat{f}(\hat{X}(s),s)\right)ds$$

$$+ \int_0^t \left(\hat{g}(\hat{X}^k(s),s) - \hat{g}(\hat{X}(s),s)\right)d\frac{\frac{\tilde{W}^3}{\tilde{T}(\tilde{W}^3)}}{\hat{T}}\Big|$$

$$\le \left|\hat{X}_0^k - \hat{X}_0\right| + \int_0^t \left|\hat{f}(\hat{X}^k(s),s) - \hat{f}(\hat{X}(s),s)\right|ds$$

$$+ \int_0^t \left|\hat{g}(\hat{X}^k(s),s) - \hat{g}(\hat{X}(s),s)\right|d\frac{\frac{\tilde{W}^3}{\tilde{T}(\tilde{W}^3)}}{\hat{T}}$$

and

$$
\begin{aligned}
&|\hat{X}^k(t)-\hat{X}(t)|^{\hat{2}} = |\hat{X}^k(t)-\hat{X}(t)|^2\hat{T}\\
&= \hat{T}\Big(\Big|\hat{X}_0^k-\hat{X}_0\Big| + \int_0^t \Big|\hat{f}(\hat{X}^k(s),s)-\hat{f}(\hat{X}(s),s)\Big| ds\\
&+ \int_0^t \big|\hat{g}(\hat{X}^k(s),s)-\hat{g}(\hat{X}(s),s)\big| d\frac{\frac{\tilde{W}^3}{\tilde{T}(\tilde{W}^3)}}{\hat{T}}\Big|\Big)^2\\
&\le 2\hat{T}|_0^k-\hat{X}_0|^2 + 2\hat{T}\left(\int_0^t |\hat{f}(\hat{X}^k(s),s)-\hat{f}(\hat{X}(s),s)|ds\right)^2\\
&+2\hat{T}\left(\int_0^t \big|\hat{g}(\hat{X}^k(s),s)-\hat{g}(\hat{X}(s),s)\big| d\frac{\frac{\tilde{W}^3}{\tilde{T}(\tilde{W}^3)}}{\hat{T}}\right)^2\\
&\le 2\hat{T}|_0^k-\hat{X}_0|^2 + 2t\hat{T}\int_0^t |\hat{f}(\hat{X}^k(s),s)-\hat{f}(\hat{X}(s),s)|^2 ds\\
&+2\hat{T}\left(\int_0^t \big|\hat{g}(\hat{X}^k(s),s)-\hat{g}(\hat{X}(s),s)\big| d\frac{\frac{\tilde{W}^3}{\tilde{T}(\tilde{W}^3)}}{\hat{T}}\right)^2\\
&= 2|\hat{X}_0^k-\hat{X}_0|^{\hat{2}} + 2t\int_0^t |\hat{f}(\hat{X}^k(s),s)-\hat{f}(\hat{X}(s),s)|^{\hat{2}} ds\\
&+2\left(\int_0^t \frac{1}{\tilde{T}}\tilde{T}(\hat{g}(\hat{X}^k(s),s)-\hat{g}(\hat{X}(s),s))\frac{\tilde{T}^2}{\tilde{T}^2} d\frac{\frac{\tilde{W}^3}{\tilde{T}(\tilde{W}^3)}}{\hat{T}}\right)^{\hat{2}}\\
&\le 2|\hat{X}_0^k-\hat{X}_0|^{\hat{2}} + 2TL^2\int_0^t |\hat{X}^k(s)-\hat{X}(s)|^{\hat{2}} ds\\
&+2\left(\hat{\int}_0^t \frac{\hat{g}(\hat{X}^k(s),s)-\hat{g}(\hat{X}(s),s)}{\tilde{T}}\,\hat{\times}\hat{d}\frac{\frac{\tilde{W}^3}{\tilde{T}(\tilde{W}^3)}}{\hat{T}}\right)^{\hat{2}},
\end{aligned}
$$

i.e.,

$$
\begin{aligned}
&|\hat{X}^k(t)-\hat{X}(t)|^{\hat{2}} \le 2|\hat{X}_0^k-\hat{X}_0|^{\hat{2}} + 2TL^2\int_0^t |\hat{X}^k(s)-\hat{X}(s)|^{\hat{2}} ds\\
&+2\left(\hat{\int}_0^t \frac{\hat{g}(\hat{X}^k(s),s)-\hat{g}(\hat{X}(s),s)}{\tilde{T}}\,\hat{\times}\hat{d}\frac{\frac{\tilde{W}^3}{\tilde{T}(\tilde{W}^3)}}{\hat{T}}\right)^{\hat{2}}.
\end{aligned}
$$

Therefore

$$\hat{E}(|\hat{X}^k(t)-\hat{X}(t)|^{\hat{2}}) \leq 2\hat{E}(|\hat{X}_0^k-\hat{X}_0|^{\hat{2}}) + 2TL^2\int_0^t \hat{E}(|\hat{X}^k(s)-\hat{X}(s)|^{\hat{2}})ds$$

$$+2\hat{E}\left(\left(\hat{\int}_0^t \frac{\hat{g}(\hat{X}^k(s),s)-\hat{g}(\hat{X}(s),s)}{\tilde{T}} \hat{\times} \hat{d}\frac{\frac{\tilde{W}^3}{\tilde{T}(\tilde{W}^3)}}{\hat{T}}\right)^{\hat{2}}\right)$$

$$= 2\hat{E}(|\hat{X}_0^k-\hat{X}_0|^{\hat{2}}) + 2TL^2\int_0^t \hat{E}(|\hat{X}^k(s)-\hat{X}(s)|^{\hat{2}})ds$$

$$+2\hat{E}\left(\int_0^t \frac{(\hat{g}(\hat{X}^k(s),s)-\hat{g}(\hat{X}(s),s))^2\tilde{T}^2}{\tilde{T}^2\hat{T}}dt\right)$$

$$= 2\hat{E}(|\hat{X}_0^k-\hat{X}_0|^{\hat{2}}) + 2TL^2\int_0^t \hat{E}(|\hat{X}^k(s)-\hat{X}(s)|^{\hat{2}})ds$$

$$+2\hat{E}\left(\int_0^t \frac{(\hat{g}(\hat{X}^k(s),s)-\hat{g}(\hat{X}(s),s))^2}{\hat{T}}dt\right)$$

$$\leq 2\hat{E}(|\hat{X}_0^k-\hat{X}_0|^{\hat{2}}) + 2TL^2\int_0^t \hat{E}(|\hat{X}^k(s)-\hat{X}(s)|^{\hat{2}})ds$$

$$+2\frac{L^2}{N}\hat{E}\left(\int_0^t |\hat{X}^k(s)-\hat{X}(s)|^2 ds\right)$$

$$= 2\hat{E}(|\hat{X}_0^k-\hat{X}_0|^{\hat{2}}) + 2TL^2\int_0^t \hat{E}(|\hat{X}^k(s)-\hat{X}(s)|^{\hat{2}})ds$$

$$+2\frac{L^2}{N}\hat{E}\left(\int_0^t \frac{\hat{T}|\hat{X}^k(s)-\hat{X}(s)|^2}{\hat{T}}ds\right)$$

$$\leq \hat{E}(|\hat{X}_0^k-\hat{X}_0|^{\hat{2}}) + 2TL^2\int_0^t \hat{E}(|\hat{X}^k(s)-\hat{X}(s)|^{\hat{2}})ds$$

$$+2\frac{L^2}{N^2}\hat{E}\left(\int_0^t |\hat{X}^k(s)-\hat{X}(s)|^{\hat{2}} ds\right)$$

$$= 2\hat{E}(|\hat{X}_0^k-\hat{X}_0|^{\hat{2}}) + \left(2TL^2+2\frac{L^2}{N^2}\right)\int_0^t \hat{E}(|\hat{X}^k(s)-\hat{X}(s)|^{\hat{2}})ds,$$

i.e.,

$$\hat{E}\left(|\hat{X}^k(t)-\hat{X}(t)|^{\hat{2}}\right) \leq 2\hat{E}(|\hat{X}_0^k-\hat{X}_0|^{\hat{2}}) + \left(2TL^2+2\frac{L^2}{N^2}\right)\int_0^t \hat{E}(|\hat{X}^k(s)-\hat{X}(s)|^{\hat{2}})ds.$$

From here and from Gronwall's lemma it follows that

$$\hat{E}\left(|\hat{X}^k(t)-\hat{X}(t)|^{\hat{2}}\right) \leq 2\hat{E}\left(|\hat{X}_0^k-\hat{X}_0|^{\hat{2}}\right) e^{\left(2TL^2+2\frac{L^2}{N^2}\right)t}.$$

Because

$$\lim_{k\longrightarrow\infty}\hat{E}\left(|\hat{X}_0^k-\hat{X}_0|^{\hat{2}}\right) = 0$$

we find that

$$\lim_{k\longrightarrow\infty}\hat{E}\left(\max_{0\leq t\leq T}|\hat{X}^k(t)-\hat{X}(t)|^{\hat{2}}\right) = 0.$$

□

Theorem 4.0.14. *Suppose for* $k = 1, 2, \ldots$, *that* $\hat{f}^k$, $\hat{g}^k$, $\hat{f}$, $\hat{g}$, $\hat{X}_0$ *and* $\hat{X}_0^k$ *satisfy the hypotheses of the existence and uniqueness theorem for the iso-stochastic differential equations of the first kind. Let also,*

$$\lim_{k\longrightarrow\infty} \hat{E}\left(|\hat{X}_0^k - \hat{X}_0|^{\hat{2}}\right) = 0$$

and for each $M > 0$

$$\lim_{k\longrightarrow\infty} \max_{0\le t\le T, |x|\le M} \left(|\hat{f}^k(x,t) - \hat{f}(x,t)| + |\hat{g}^k(x,t) - \hat{g}(x,t)|\right) = 0.$$

Finally suppose that $\hat{X}^k$ *solves the iso-stochastic differential equation of the first kind*

$$\begin{cases} d\hat{X}^k = \hat{f}^k(\hat{X}^k, t)dt + \hat{g}^k(\hat{X}^k, t)d\frac{\tilde{W}^1}{\hat{T}} \\ \hat{X}^k(0) = \hat{X}_0^k, \qquad 0 \le t \le T, \qquad\qquad k = 1, 2, \ldots. \end{cases}$$

Then

$$\lim_{k\longrightarrow\infty} \hat{E}\left(\max_{0\le t\le T} |\hat{X}^k(t) - \hat{X}(t)|^{\hat{2}}\right) = 0,$$

where

$$\begin{cases} d\hat{X} = \hat{f}(\hat{X}, t)dt + \hat{g}(\hat{X}, t)d\frac{\tilde{W}^1}{\hat{T}} \\ \hat{X}(0) = \hat{X}_0, \qquad 0 \le t \le T. \end{cases}$$

Proof. For the solutions $\hat{X}^k$, $k = 1, 2, \ldots$, and $\hat{X}$ we have the following representations

$$\hat{X}^k(t) = \hat{X}_0^k + \int_0^t \hat{f}^k(\hat{X}^k(s), s)ds + \int_0^t \hat{g}^k(\hat{X}^k(s), s)d\frac{\tilde{W}^1}{\hat{T}}, \qquad k = 1, 2, \ldots,$$

$$\hat{X}(t) = \hat{X}_0 + \int_0^t \hat{f}(\hat{X}(s), s)ds + \int_0^t \hat{g}(\hat{X}(s), s)d\frac{\tilde{W}^1}{\hat{T}}, \qquad 0 \le t \le T.$$

Then

$$\hat{X}^k(t) - \hat{X}(t) = \hat{X}_0^k + \int_0^t \hat{f}^k(\hat{X}^k(s), s)ds + \int_0^t \hat{g}^k(\hat{X}^k(s), s)d\frac{\tilde{W}^1}{\hat{T}}$$

$$-\hat{X}_0 - \int_0^t \hat{f}(\hat{X}(s), s)ds - \int_0^t \hat{g}(\hat{X}(s), s)d\frac{\tilde{W}^1}{\hat{T}}$$

$$= \hat{X}_0^k - \hat{X}_0 + \int_0^t \left(\hat{f}^k(\hat{X}^k(s), s) - \hat{f}(\hat{X}(s), s)\right)ds$$

$$+ \int_0^t \left(\hat{g}^k(\hat{X}^k(s), s) - \hat{g}(\hat{X}(s), s)\right)d\frac{\tilde{W}^1}{\hat{T}},$$

$$= \hat{X}_0^k - \hat{X}_0 + \int_0^t \left(\hat{f}^k(\hat{X}^k(s), s) - \hat{f}(\hat{X}^k(s), s) + \hat{f}(\hat{X}^k(s), s) - \hat{f}(\hat{X}(s), s)\right)ds$$

$$+ \int_0^t \left(\hat{g}^k(\hat{X}^k(s), s) - \hat{g}(\hat{X}^k(s), s) + \hat{g}(\hat{X}^k(s), s) - \hat{g}(\hat{X}(s), s)\right)d\frac{\tilde{W}^1}{\hat{T}}$$

$$= \hat{X}_0^k - \hat{X}_0 + \int_0^t \left(\hat{f}^k(\hat{X}^k(s), s) - \hat{f}(\hat{X}^k(s), s)\right)ds + \int_0^t \left(\hat{f}(\hat{X}^k(s), s) - \hat{f}(\hat{X}(s), s)\right)ds$$

$$+ \int_0^t \left(\hat{g}^k(\hat{X}^k(s), s) - \hat{g}(\hat{X}^k(s), s)\right)d\frac{\tilde{W}^1}{\hat{T}} + \int_0^t \left(\hat{g}(\hat{X}^k(s), s) - \hat{g}(\hat{X}(s), s)\right)d\frac{\tilde{W}^1}{\hat{T}}$$

from here

$$|\hat{X}^k(t)-\hat{X}(t)| = \Big|\hat{X}_0^k-\hat{X}_0+\int_0^t\big(\hat{f}(\hat{X}^k(s),s)-\hat{f}(\hat{X}(s),s)\big)ds$$

$$+\int_0^t\big(\hat{f}^k(\hat{X}^k(s),s)-\hat{f}(\hat{X}^k(s),s)\big)ds+\int_0^t\big(\hat{g}^k(\hat{X}^k(s),s)-\hat{g}(\hat{X}^k(s),s)\big)d\tfrac{\tilde{W}^1}{\hat{T}}$$

$$+\int_0^t\big(\hat{g}(\hat{X}^k(s),s)-\hat{g}(\hat{X}(s),s)\big)d\tfrac{\tilde{W}^1}{\hat{T}}\Big|$$

$$\leq\Big|\hat{X}_0^k-\hat{X}_0\Big|+\int_0^t\Big|\hat{f}(\hat{X}^k(s),s)-\hat{f}(\hat{X}(s),s)\Big|ds$$

$$+\int_0^t\big|\hat{f}^k(\hat{X}^k(s),s)-\hat{f}(\hat{X}^k(s),s)\big|ds+\int_0^t\big|\hat{g}^k(\hat{X}^k(s),s)-\hat{g}(\hat{X}^k(s),s)\big|d\tfrac{\tilde{W}^1}{\hat{T}}$$

$$+\int_0^t\big|\hat{g}(\hat{X}^k(s),s)-\hat{g}(\hat{X}(s),s)\big|d\tfrac{\tilde{W}^1}{\hat{T}}.$$

Let $\varepsilon>0$ be arbitrarily chosen. Since

$$\lim_{k\longrightarrow\infty}\max_{0\leq t\leq T,|x|\leq M}\Big(|\hat{f}^k(x,t)-\hat{f}(x,t)|+|\hat{g}^k(x,t)-\hat{g}(x,t)|\Big)=0,$$

then there exists enough large $k\in\mathbb{N}$ such that

$$\int_0^t\Big(\hat{f}^k(\hat{X}^k(s),s)-\hat{f}(\hat{X}^k(s),s)\Big)ds+\int_0^t\Big(\hat{g}^k(\hat{X}^k(s),s)-\hat{g}(\hat{X}^k(s),s)\Big)d\frac{\tilde{W}^1}{\hat{T}}\leq\varepsilon T.$$

Thereforc

$$|\hat{X}^k(t)-\hat{X}(t)|\leq\Big|\hat{X}_0^k-\hat{X}_0\Big|+\int_0^t\Big|\hat{f}(\hat{X}^k(s),s)-\hat{f}(\hat{X}(s),s)\Big|ds+\varepsilon T$$

$$+\int_0^t\big|\hat{g}(\hat{X}^k(s),s)-\hat{g}(\hat{X}(s),s)\big|d\tfrac{\tilde{W}^1}{\hat{T}}.$$

From here

$$|\hat{X}^k(t)-\hat{X}(t)|^{\hat{2}}=|\hat{X}^k(t)-\hat{X}(t)|^2\hat{T}$$

$$=\hat{T}\Big(\Big|\hat{X}_0^k-\hat{X}_0\Big|+\int_0^t\Big|\hat{f}(\hat{X}^k(s),s)-\hat{f}(\hat{X}(s),s)\Big|ds+\varepsilon T$$

$$+\int_0^t\big|\hat{g}(\hat{X}^k(s),s)-\hat{g}(\hat{X}(s),s)\big|d\tfrac{\tilde{W}^1}{\hat{T}}\Big|\Big)^2$$

$$\leq 2\hat{T}|_0^k-\hat{X}_0|^2+2\hat{T}\left(\int_0^t|\hat{f}(\hat{X}^k(s),s)-\hat{f}(\hat{X}(s),s)|ds\right)^2+2\varepsilon^2T^2\hat{T}$$

$$+2\hat{T}\left(\int_0^t\big|\hat{g}(\hat{X}^k(s),s)-\hat{g}(\hat{X}(s),s)\big|d\tfrac{\tilde{W}^1}{\hat{T}}\right)^2$$

$$\leq 2\hat{T}|\hat{X}_0^k - \hat{X}_0|^2 + 2t\hat{T}\int_0^t |\hat{f}(\hat{X}^k(s),s) - \hat{f}(\hat{X}(s),s)|^2 ds + 2\varepsilon^2 T^{\hat{2}}$$

$$+2\hat{T}\left(\int_0^t \left|\hat{g}(\hat{X}^k(s),s) - \hat{g}(\hat{X}(s),s)\right| d\frac{\tilde{W}^1}{\hat{T}}\right)^2$$

$$= 2|\hat{X}_0^k - \hat{X}_0|^{\hat{2}} + 2t\int_0^t |\hat{f}(\hat{X}^k(s),s) - \hat{f}(\hat{X}(s),s)|^{\hat{2}} ds + 2\varepsilon^2 T^{\hat{2}}$$

$$+2\left(\int_0^t \frac{1}{\tilde{T}}\tilde{T}(\hat{g}(\hat{X}^k(s),s) - \hat{g}(\hat{X}(s),s))\frac{\tilde{T}^2}{\tilde{T}^2} d\frac{\tilde{W}^1}{\hat{T}}\right)^{\hat{2}}$$

$$\leq 2|\hat{X}_0^k - \hat{X}_0|^{\hat{2}} + 2TL^2\int_0^t |\hat{X}^k(s) - \hat{X}(s)|^{\hat{2}} ds + 2\varepsilon^2 T^{\hat{2}}$$

$$+2\left(\hat{\int}_0^t \frac{\hat{g}(\hat{X}^k(s),s) - \hat{g}(\hat{X}(s),s)}{\tilde{T}} \hat{\times} d\hat{} \frac{\tilde{W}^1}{\hat{T}}\right)^{\hat{2}},$$

i.e.,

$$|\hat{X}^k(t) - \hat{X}(t)|^{\hat{2}} \leq 2|\hat{X}_0^k - \hat{X}_0|^{\hat{2}} + 2TL^2\int_0^t |\hat{X}^k(s) - \hat{X}(s)|^{\hat{2}} ds + 2\varepsilon^2 T^{\hat{2}}$$

$$+2\left(\hat{\int}_0^t \frac{\hat{g}(\hat{X}^k(s),s) - \hat{g}(\hat{X}(s),s)}{\tilde{T}} \hat{\times} d\hat{} \frac{\tilde{W}^1}{\hat{T}}\right)^{\hat{2}}.$$

Therefore

$$\hat{E}(|\hat{X}^k(t) - \hat{X}(t)|^{\hat{2}}) \leq 2\hat{E}(|\hat{X}_0^k - \hat{X}_0|^{\hat{2}}) + 2TL^2\int_0^t \hat{E}(|\hat{X}^k(s) - \hat{X}(s)|^{\hat{2}}) ds$$

$$+2\varepsilon^2\hat{E}(T^{\hat{2}}) + 2\hat{E}\left(\left(\hat{\int}_0^t \frac{\hat{g}(\hat{X}^k(s),s) - \hat{g}(\hat{X}(s),s)}{\tilde{T}} \hat{\times} d\frac{\tilde{W}^1}{\hat{T}}\right)^{\hat{2}}\right)$$

$$= 2\hat{E}(|\hat{X}_0^k - \hat{X}_0|^{\hat{2}}) + 2TL^2\int_0^t \hat{E}(|\hat{X}^k(s) - \hat{X}(s)|^{\hat{2}}) ds$$

$$+2\varepsilon^2\hat{E}(T^{\hat{2}}) + 2\hat{E}\left(\int_0^t \frac{(\hat{g}(\hat{X}^k(s),s) - \hat{g}(\hat{X}(s),s))^2\tilde{T}^2}{\tilde{T}^2\hat{T}} dt\right)$$

$$= 2\hat{E}(|\hat{X}_0^k - \hat{X}_0|^{\hat{2}}) + 2TL^2\int_0^t \hat{E}(|\hat{X}^k(s) - \hat{X}(s)|^{\hat{2}}) ds$$

$$+2\varepsilon^2\hat{E}(T^{\hat{2}}) + 2\hat{E}\left(\int_0^t \frac{(\hat{g}(\hat{X}^k(s),s) - \hat{g}(\hat{X}(s),s))^2}{\hat{T}} dt\right)$$

$$\leq 2\hat{E}(|\hat{X}_0^k - \hat{X}_0|^{\hat{2}}) + 2TL^2\int_0^t \hat{E}(|\hat{X}^k(s) - \hat{X}(s)|^{\hat{2}}) ds$$

$$+2\varepsilon^2\hat{E}(T^{\hat{2}}) + 2\frac{L^2}{N}\hat{E}\left(\int_0^t |\hat{X}^k(s) - \hat{X}(s)|^2 ds\right)$$

$$= 2\hat{E}(|\hat{X}_0^k - \hat{X}_0|^{\hat{2}}) + 2TL^2\int_0^t \hat{E}(|\hat{X}^k(s) - \hat{X}(s)|^{\hat{2}}) ds$$

$$+2\varepsilon^2\hat{E}(T^{\hat{2}}) + 2\frac{L^2}{N}\hat{E}\left(\int_0^t \frac{\hat{T}|\hat{X}^k(s) - \hat{X}(s)|^2}{\hat{T}} ds\right)$$

$$\leq \hat{E}(|\hat{X}_0^k - \hat{X}_0|^{\hat{2}}) + 2TL^2 \int_0^t \hat{E}(|\hat{X}^k(s) - \hat{X}(s)|^{\hat{2}})ds$$

$$+2\varepsilon^2 \hat{E}(T^{\hat{2}}) + 2\frac{L^2}{N^2}\hat{E}\left(\int_0^t |\hat{X}^k(s) - \hat{X}(s)|^{\hat{2}}ds\right)$$

$$= 2\hat{E}(|\hat{X}_0^k - \hat{X}_0|^{\hat{2}}) + 2\varepsilon^2 \hat{E}(T^{\hat{2}})$$

$$+\left(2TL^2 + 2\frac{L^2}{N^2}\right)\int_0^t \hat{E}(|\hat{X}^k(s) - \hat{X}(s)|^{\hat{2}})ds,$$

i.e.,

$$\hat{E}\left(|\hat{X}^k(t) - \hat{X}(t)|^{\hat{2}}\right) \leq 2\hat{E}(|\hat{X}_0^k - \hat{X}_0|^{\hat{2}}) + 2\varepsilon^2 \hat{E}(T^{\hat{2}})$$

$$+\left(2TL^2 + 2\frac{L^2}{N^2}\right)\int_0^t \hat{E}(|\hat{X}^k(s) - \hat{X}(s)|^{\hat{2}})ds.$$

From here and from Gronwall's lemma it follows that

$$\hat{E}\left(|\hat{X}^k(t) - \hat{X}(t)|^{\hat{2}}\right) \leq 2\left(\hat{E}\left(|\hat{X}_0^k - \hat{X}_0|^{\hat{2}}\right) + \varepsilon^2 \hat{E}(T^{\hat{2}})\right)e^{\left(2TL^2 + 2\frac{L^2}{N^2}\right)t}.$$

Because

$$\lim_{k\longrightarrow\infty} \hat{E}\left(|\hat{X}_0^k - \hat{X}_0|^{\hat{2}}\right) = 0,$$

$$\lim_{k\longrightarrow\infty} \varepsilon^2 \hat{E}(T^{\hat{2}}) = 0,$$

we find that

$$\lim_{k\longrightarrow\infty} \hat{E}\left(\max_{0\leq t\leq T} |\hat{X}^k(t) - \hat{X}(t)|^{\hat{2}}\right) = 0.$$

□

Theorem 4.0.15. *Suppose for $k = 1, 2, \ldots$, that $\hat{f}^k$, $\hat{g}^k$, $\hat{f}$, $\hat{g}$, $\hat{X}_0$ and $\hat{X}_0^k$ satisfy the hypotheses of the existence and uniqueness theorem for the iso-stochastic differential equations of the second kind. Let also,*

$$\lim_{k\longrightarrow\infty} \hat{E}\left(|\hat{X}_0^k - \hat{X}_0|^{\hat{2}}\right) = 0$$

and for each $M > 0$

$$\lim_{k\longrightarrow\infty} \max_{0\leq t\leq T, |x|\leq M} \left(|\hat{f}^k(x,t) - \hat{f}(x,t)| + |\hat{g}^k(x,t) - \hat{g}(x,t)|\right) = 0.$$

Finally suppose that $\hat{X}^k$ solves the iso-stochastic differential equation of the second kind

$$\begin{cases} d\hat{X}^k = \hat{f}^k(\hat{X}^k, t)dt + \hat{g}^k(\hat{X}^k, t)d\frac{\tilde{W}^2\tilde{T}(\tilde{W}^2)}{\hat{T}} \\ \hat{X}^k(0) = \hat{X}_0^k, \qquad 0 \leq t \leq T, \qquad\qquad k = 1, 2, \ldots. \end{cases}$$

Then

$$\lim_{k\longrightarrow\infty} \hat{E}\left(\max_{0\leq t\leq T} |\hat{X}^k(t) - \hat{X}(t)|^{\hat{2}}\right) = 0,$$

where

$$\begin{cases} d\hat{X} = \hat{f}(\hat{X},t)dt + \hat{g}(\hat{X},t)d\frac{\tilde{W}^2\tilde{T}(\tilde{W}^2)}{\hat{T}} \\ \hat{X}(0) = \hat{X}_0, \qquad 0 \le t \le T. \end{cases}$$

Proof. For the solutions $\hat{X}^k$, $k = 1,2,\ldots$, and $\hat{X}$ we have the following representations

$$\hat{X}^k(t) = \hat{X}_0^k + \int_0^t \hat{f}^k(\hat{X}^k(s),s)ds + \int_0^t \hat{g}^k(\hat{X}^k(s),s)d\frac{\tilde{W}^2\tilde{T}(\tilde{W}^2)}{\hat{T}}, \qquad k = 1,2,\ldots,$$

$$\hat{X}(t) = \hat{X}_0 + \int_0^t \hat{f}(\hat{X}(s),s)ds + \int_0^t \hat{g}(\hat{X}(s),s)d\frac{\tilde{W}^2\tilde{T}(\tilde{W}^2)}{\hat{T}}, \qquad 0 \le t \le T.$$

Therefore

$$\hat{X}^k(t) - \hat{X}(t) = \hat{X}_0^k + \int_0^t \hat{f}^k(\hat{X}^k(s),s)ds + \int_0^t \hat{g}^k(\hat{X}^k(s),s)d\frac{\tilde{W}^2\tilde{T}(\tilde{W}^2)}{\hat{T}}$$

$$-\hat{X}_0 - \int_0^t \hat{f}(\hat{X}(s),s)ds - \int_0^t \hat{g}(\hat{X}(s),s)d\frac{\tilde{W}^2\tilde{T}(\tilde{W}^2)}{\hat{T}}$$

$$= \hat{X}_0^k - \hat{X}_0 + \int_0^t \left(\hat{f}^k(\hat{X}^k(s),s) - \hat{f}(\hat{X}(s),s)\right)ds$$

$$+ \int_0^t \left(\hat{g}^k(\hat{X}^k(s),s) - \hat{g}(\hat{X}(s),s)\right)d\frac{\tilde{W}^2\tilde{T}(\tilde{W}^2)}{\hat{T}},$$

$$= \hat{X}_0^k - \hat{X}_0 + \int_0^t \left(\hat{f}^k(\hat{X}^k(s),s) - \hat{f}(\hat{X}^k(s),s) + \hat{f}(\hat{X}^k(s),s) - \hat{f}(\hat{X}(s),s)\right)ds$$

$$+ \int_0^t \left(\hat{g}^k(\hat{X}^k(s),s) - \hat{g}(\hat{X}^k(s),s) + \hat{g}(\hat{X}^k(s),s) - \hat{g}(\hat{X}(s),s)\right)d\frac{\tilde{W}^2\tilde{T}(\tilde{W}^2)}{\hat{T}}$$

$$= \hat{X}_0^k - \hat{X}_0 + \int_0^t \left(\hat{f}^k(\hat{X}^k(s),s) - \hat{f}(\hat{X}^k(s),s)\right)ds + \int_0^t \left(\hat{f}(\hat{X}^k(s),s) - \hat{f}(\hat{X}(s),s)\right)ds$$

$$+ \int_0^t \left(\hat{g}^k(\hat{X}^k(s),s) - \hat{g}(\hat{X}^k(s),s)\right)d\frac{\tilde{W}^2\tilde{T}(\tilde{W}^2)}{\hat{T}} + \int_0^t \left(\hat{g}(\hat{X}^k(s),s) - \hat{g}(\hat{X}(s),s)\right)d\frac{\tilde{W}^2\tilde{T}(\tilde{W}^2)}{\hat{T}}$$

from where

$$|\hat{X}^k(t) - \hat{X}(t)| = \Big|\hat{X}_0^k - \hat{X}_0 + \int_0^t \left(\hat{f}(\hat{X}^k(s),s) - \hat{f}(\hat{X}(s),s)\right)ds$$

$$+ \int_0^t \left(\hat{f}^k(\hat{X}^k(s),s) - \hat{f}(\hat{X}^k(s),s)\right)ds + \int_0^t \left(\hat{g}^k(\hat{X}^k(s),s) - \hat{g}(\hat{X}^k(s),s)\right)d\frac{\tilde{W}^2\tilde{T}(\tilde{W}^2)}{\hat{T}}$$

$$+ \int_0^t \left(\hat{g}(\hat{X}^k(s),s) - \hat{g}(\hat{X}(s),s)\right)d\frac{\tilde{W}^2\tilde{T}(\tilde{W}^2)}{\hat{T}}\Big|$$

$$\le \left|\hat{X}_0^k - \hat{X}_0\right| + \int_0^t \left|\hat{f}(\hat{X}^k(s),s) - \hat{f}(\hat{X}(s),s)\right|ds$$

$$+ \int_0^t \left|\hat{f}^k(\hat{X}^k(s),s) - \hat{f}(\hat{X}^k(s),s)\right|ds + \int_0^t \left|\hat{g}^k(\hat{X}^k(s),s) - \hat{g}(\hat{X}^k(s),s)\right|d\frac{\tilde{W}^2\tilde{T}(\tilde{W}^2)}{\hat{T}}$$

$$+ \int_0^t \left|\hat{g}(\hat{X}^k(s),s) - \hat{g}(\hat{X}(s),s)\right|d\frac{\tilde{W}^2\tilde{T}(\tilde{W}^2)}{\hat{T}}.$$

Let $\varepsilon > 0$ be arbitrarily chosen. Since

$$\lim_{k\longrightarrow\infty} \max_{0\le t\le T, |x|\le M} \left(|\hat{f}^k(x,t) - \hat{f}(x,t)| + |\hat{g}^k(x,t) - \hat{g}(x,t)| \right) = 0,$$

then there exists enough large $k \in \mathbb{N}$ such that

$$\int_0^t \left(\hat{f}^k(\hat{X}^k(s),s) - \hat{f}(\hat{X}^k(s),s) \right) ds + \int_0^t \left(\hat{g}^k(\hat{X}^k(s),s) - \hat{g}(\hat{X}^k(s),s) \right) d\frac{\tilde{W}^2\tilde{T}(\tilde{W}^2)}{\hat{T}} \le \varepsilon T.$$

Then

$$|\hat{X}^k(t) - \hat{X}(t)| \le \left|\hat{X}_0^k - \hat{X}_0\right| + \int_0^t \left|\hat{f}(\hat{X}^k(s),s) - \hat{f}(\hat{X}(s),s)\right| ds + \varepsilon T$$

$$+ \int_0^t \left|\hat{g}(\hat{X}^k(s),s) - \hat{g}(\hat{X}(s),s)\right| d\frac{\tilde{W}^2\tilde{T}(\tilde{W}^2)}{\hat{T}}.$$

From here

$$|\hat{X}^k(t) - \hat{X}(t)|^{\hat{2}} = |\hat{X}^k(t) - \hat{X}(t)|^2\hat{T}$$

$$= \hat{T}\left(\left|\hat{X}_0^k - \hat{X}_0\right| + \int_0^t \left|\hat{f}(\hat{X}^k(s),s) - \hat{f}(\hat{X}(s),s)\right| ds + \varepsilon T \right.$$

$$\left. + \int_0^t \left|\hat{g}(\hat{X}^k(s),s) - \hat{g}(\hat{X}(s),s)\right| d\frac{\tilde{W}^2\tilde{T}(\tilde{W}^2)}{\hat{T}} \right| \Big)^2$$

$$\le 2\hat{T}|_0^k - \hat{X}_0|^2 + 2\hat{T}\left(\int_0^t |\hat{f}(\hat{X}^k(s),s) - \hat{f}(\hat{X}(s),s)| ds \right)^2 + 2\varepsilon^2 T^2 \hat{T}$$

$$+ 2\hat{T}\left(\int_0^t \left|\hat{g}(\hat{X}^k(s),s) - \hat{g}(\hat{X}(s),s)\right| d\frac{\tilde{W}^2\tilde{T}(\tilde{W}^2)}{\hat{T}} \right)^2$$

$$\le 2\hat{T}|_0^k - \hat{X}_0|^2 + 2t\hat{T}\int_0^t |\hat{f}(\hat{X}^k(s),s) - \hat{f}(\hat{X}(s),s)|^2 ds + 2\varepsilon^{\hat{2}} T^{\hat{2}}$$

$$+ 2\hat{T}\left(\int_0^t \left|\hat{g}(\hat{X}^k(s),s) - \hat{g}(\hat{X}(s),s)\right| d\frac{\tilde{W}^2\tilde{T}(\tilde{W}^2)}{\hat{T}} \right)^2$$

$$= 2|\hat{X}_0^k - \hat{X}_0|^{\hat{2}} + 2t\int_0^t |\hat{f}(\hat{X}^k(s),s) - \hat{f}(\hat{X}(s),s)|^{\hat{2}} ds + 2\varepsilon^2 T^{\hat{2}}$$

$$+ 2\left(\int_0^t \frac{1}{\tilde{T}}\tilde{T}(\hat{g}(\hat{X}^k(s),s) - \hat{g}(\hat{X}(s),s))\frac{\tilde{T}^2}{\tilde{T}^2} d\frac{\tilde{W}^2\tilde{T}(\tilde{W}^2)}{\hat{T}} \right)^{\hat{2}}$$

$$\le 2|\hat{X}_0^k - \hat{X}_0|^{\hat{2}} + 2TL^2\int_0^t |\hat{X}^k(s) - \hat{X}(s)|^{\hat{2}} ds + 2\varepsilon^2 T^{\hat{2}}$$

$$+ 2\left(\hat{\int}_0^t \frac{\hat{g}(\hat{X}^k(s),s) - \hat{g}(\hat{X}(s),s)}{\tilde{T}} \hat{\times} \hat{d}\frac{\tilde{W}^2\tilde{T}(\tilde{W}^2)}{\hat{T}} \right)^{\hat{2}},$$

i.e.,

$$|\hat{X}^k(t) - \hat{X}(t)|^{\hat{2}} \le 2|\hat{X}_0^k - \hat{X}_0|^{\hat{2}} + 2TL^2\int_0^t |\hat{X}^k(s) - \hat{X}(s)|^{\hat{2}} ds + 2\varepsilon^2 T^{\hat{2}}$$

$$+ 2\left(\hat{\int}_0^t \frac{\hat{g}(\hat{X}^k(s),s) - \hat{g}(\hat{X}(s),s)}{\tilde{T}} \hat{\times} \hat{d}\frac{\tilde{W}^2\tilde{T}(\tilde{W}^2)}{\hat{T}} \right)^{\hat{2}}.$$

Consequently

$$\hat{E}(|\hat{X}^k(t)-\hat{X}(t)|^{\hat{2}}) \leq 2\hat{E}(|\hat{X}^k_0-\hat{X}_0|^{\hat{2}})+2TL^2\int_0^t \hat{E}(|\hat{X}^k(s)-\hat{X}(s)|^{\hat{2}})ds$$

$$+2\varepsilon^2\hat{E}(T^{\hat{2}})+2\hat{E}\left(\left(\hat{\int}_0^t \frac{\hat{g}(\hat{X}^k(s),s)-\hat{g}(\hat{X}(s),s)}{\tilde{T}}\hat{\times}\hat{d}\frac{\tilde{W}^2\tilde{T}(\tilde{W}^2)}{\hat{T}}\right)^{\hat{2}}\right)$$

$$=2\hat{E}(|\hat{X}^k_0-\hat{X}_0|^{\hat{2}})+2TL^2\int_0^t \hat{E}(|\hat{X}^k(s)-\hat{X}(s)|^{\hat{2}})ds$$

$$+2\varepsilon^2\hat{E}(T^{\hat{2}})+2\hat{E}\left(\int_0^t \frac{(\hat{g}(\hat{X}^k(s),s)-\hat{g}(\hat{X}(s),s))^2\tilde{T}^2}{\tilde{T}^2\hat{T}}dt\right)$$

$$=2\hat{E}(|\hat{X}^k_0-\hat{X}_0|^{\hat{2}})+2TL^2\int_0^t \hat{E}(|\hat{X}^k(s)-\hat{X}(s)|^{\hat{2}})ds$$

$$+2\varepsilon^2\hat{E}(T^{\hat{2}})+2\hat{E}\left(\int_0^t \frac{(\hat{g}(\hat{X}^k(s),s)-\hat{g}(\hat{X}(s),s))^2}{\hat{T}}dt\right)$$

$$\leq 2\hat{E}(|\hat{X}^k_0-\hat{X}_0|^{\hat{2}})+2TL^2\int_0^t \hat{E}(|\hat{X}^k(s)-\hat{X}(s)|^{\hat{2}})ds$$

$$+2\varepsilon^2\hat{E}(T^{\hat{2}})+2\frac{L^2}{N}\hat{E}\left(\int_0^t |\hat{X}^k(s)-\hat{X}(s)|^2ds\right)$$

$$=2\hat{E}(|\hat{X}^k_0-\hat{X}_0|^{\hat{2}})+2TL^2\int_0^t \hat{E}(|\hat{X}^k(s)-\hat{X}(s)|^{\hat{2}})ds$$

$$+2\varepsilon^2\hat{E}(T^{\hat{2}})+2\frac{L^2}{N}\hat{E}\left(\int_0^t \frac{\hat{T}|\hat{X}^k(s)-\hat{X}(s)|^2}{\hat{T}}ds\right)$$

$$\leq \hat{E}(|\hat{X}^k_0-\hat{X}_0|^{\hat{2}})+2TL^2\int_0^t \hat{E}(|\hat{X}^k(s)-\hat{X}(s)|^{\hat{2}})ds$$

$$+2\varepsilon^2\hat{E}(T^{\hat{2}})+2\frac{L^2}{N^2}\hat{E}\left(\int_0^t |\hat{X}^k(s)-\hat{X}(s)|^{\hat{2}}ds\right)$$

$$=2\hat{E}(|\hat{X}^k_0-\hat{X}_0|^{\hat{2}})+2\varepsilon^2\hat{E}(T^{\hat{2}})$$

$$+\left(2TL^2+2\frac{L^2}{N^2}\right)\int_0^t \hat{E}(|\hat{X}^k(s)-\hat{X}(s)|^{\hat{2}})ds,$$

or,

$$\hat{E}\left(|\hat{X}^k(t)-\hat{X}(t)|^{\hat{2}}\right) \leq 2\hat{E}(|\hat{X}^k_0-\hat{X}_0|^{\hat{2}})+2\varepsilon^2\hat{E}(T^{\hat{2}})$$

$$+\left(2TL^2+2\frac{L^2}{N^2}\right)\int_0^t \hat{E}(|\hat{X}^k(s)-\hat{X}(s)|^{\hat{2}})ds.$$

From the last inequality and from Gronwall's lemma we get

$$\hat{E}\left(|\hat{X}^k(t)-\hat{X}(t)|^{\hat{2}}\right) \leq 2\left(\hat{E}\left(|\hat{X}^k_0-\hat{X}_0|^{\hat{2}}\right)+\varepsilon^2\hat{E}(T^{\hat{2}})\right)e^{\left(2TL^2+2\frac{L^2}{N^2}\right)t}.$$

Since

$$\lim_{k\longrightarrow\infty}\hat{E}\left(|\hat{X}^k_0-\hat{X}_0|^{\hat{2}}\right)=0,$$

$$\lim_{k\longrightarrow\infty}\varepsilon^2\hat{E}(T^{\hat{2}})=0,$$

we obtain that

$$\lim_{k\longrightarrow\infty} \hat{E}\left(\max_{0\leq t\leq T}|\hat{X}^k(t)-\hat{X}(t)|^{\hat{2}}\right)=0.$$

□

Theorem 4.0.16. *Suppose for $k=1,2,\ldots$, that $\hat{f}^k$, $\hat{g}^k$, $\hat{f}$, $\hat{g}$, $\hat{X}_0$ and $\hat{X}_0^k$ satisfy the hypotheses of the existence and uniqueness theorem for the iso-stochastic differential equations of the third kind. Let also,*

$$\lim_{k\longrightarrow\infty} \hat{E}\left(|\hat{X}_0^k-\hat{X}_0|^{\hat{2}}\right)=0$$

and for each $M>0$

$$\lim_{k\longrightarrow\infty}\max_{0\leq t\leq T,|x|\leq M}\left(|\hat{f}^k(x,t)-\hat{f}(x,t)|+|\hat{g}^k(x,t)-\hat{g}(x,t)|\right)=0.$$

Finally suppose that $\hat{X}^k$ solves the iso-stochastic differential equation of the third kind

$$\begin{cases} d\hat{X}^k=\hat{f}^k(\hat{X}^k,t)dt+\hat{g}^k(\hat{X}^k,t)d\frac{\frac{\tilde{W}^3}{\tilde{T}(\tilde{W}^3)}}{\hat{T}} \\ \hat{X}^k(0)=\hat{X}_0^k, \qquad 0\leq t\leq T, \qquad\qquad k=1,2,\ldots. \end{cases}$$

Then

$$\lim_{k\longrightarrow\infty} \hat{E}\left(\max_{0\leq t\leq T}|\hat{X}^k(t)-\hat{X}(t)|^{\hat{2}}\right)=0,$$

where

$$\begin{cases} d\hat{X}=\hat{f}(\hat{X},t)dt+\hat{g}(\hat{X},t)d\frac{\frac{\tilde{W}^3}{\tilde{T}(\tilde{W}^3)}}{\hat{T}} \\ \hat{X}(0)=\hat{X}_0, \qquad 0\leq t\leq T. \end{cases}$$

Proof. For the solutions $\hat{X}^k$, $k=1,2,\ldots$, and $\hat{X}$ we have the following representations

$$\hat{X}^k(t)=\hat{X}_0^k+\int_0^t\hat{f}^k(\hat{X}^k(s),s)ds+\int_0^t\hat{g}^k(\hat{X}^k(s),s)d\frac{\frac{\tilde{W}^3}{\tilde{T}(\tilde{W}^3)}}{\hat{T}}, \qquad k=1,2,\ldots,$$

$$\hat{X}(t)=\hat{X}_0+\int_0^t\hat{f}(\hat{X}(s),s)ds+\int_0^t\hat{g}(\hat{X}(s),s)d\frac{\frac{\tilde{W}^3}{\tilde{T}(\tilde{W}^3)}}{\hat{T}}, \qquad 0\leq t\leq T.$$

Then

$$\hat{X}^k(t)-\hat{X}(t)=\hat{X}_0^k+\int_0^t \hat{f}^k(\hat{X}^k(s),s)ds+\int_0^t \hat{g}^k(\hat{X}^k(s),s)d\frac{\frac{\tilde{W}^3}{\tilde{T}(\tilde{W}^3)}}{\hat{T}}$$

$$-\hat{X}_0-\int_0^t \hat{f}(\hat{X}(s),s)ds-\int_0^t \hat{g}(\hat{X}(s),s)d\frac{\frac{\tilde{W}^3}{\tilde{T}(\tilde{W}^3)}}{\hat{T}}$$

$$=\hat{X}_0^k-\hat{X}_0+\int_0^t \left(\hat{f}^k(\hat{X}^k(s),s)-\hat{f}(\hat{X}(s),s)\right)ds$$

$$+\int_0^t \left(\hat{g}^k(\hat{X}^k(s),s)-\hat{g}(\hat{X}(s),s)\right)d\frac{\frac{\tilde{W}^3}{\tilde{T}(\tilde{W}^3)}}{\hat{T}},$$

$$=\hat{X}_0^k-\hat{X}_0+\int_0^t \left(\hat{f}^k(\hat{X}^k(s),s)-\hat{f}(\hat{X}^k(s),s)+\hat{f}(\hat{X}^k(s),s)-\hat{f}(\hat{X}(s),s)\right)ds$$

$$+\int_0^t \left(\hat{g}^k(\hat{X}^k(s),s)-\hat{g}(\hat{X}^k(s),s)+\hat{g}(\hat{X}^k(s),s)-\hat{g}(\hat{X}(s),s)\right)d\frac{\frac{\tilde{W}^3}{\tilde{T}(\tilde{W}^3)}}{\hat{T}}$$

$$=\hat{X}_0^k-\hat{X}_0+\int_0^t \left(\hat{f}^k(\hat{X}^k(s),s)-\hat{f}(\hat{X}^k(s),s)\right)ds+\int_0^t \left(\hat{f}(\hat{X}^k(s),s)-\hat{f}(\hat{X}(s),s)\right)ds$$

$$+\int_0^t \left(\hat{g}^k(\hat{X}^k(s),s)-\hat{g}(\hat{X}^k(s),s)\right)d\frac{\frac{\tilde{W}^3}{\tilde{T}(\tilde{W}^3)}}{\hat{T}}+\int_0^t \left(\hat{g}(\hat{X}^k(s),s)-\hat{g}(\hat{X}(s),s)\right)d\frac{\frac{\tilde{W}^3}{\tilde{T}(\tilde{W}^3)}}{\hat{T}}$$

from here

$$|\hat{X}^k(t)-\hat{X}(t)|=\Bigg|\hat{X}_0^k-\hat{X}_0+\int_0^t \left(\hat{f}(\hat{X}^k(s),s)-\hat{f}(\hat{X}(s),s)\right)ds$$

$$+\int_0^t \left(\hat{f}^k(\hat{X}^k(s),s)-\hat{f}(\hat{X}^k(s),s)\right)ds+\int_0^t \left(\hat{g}^k(\hat{X}^k(s),s)-\hat{g}(\hat{X}^k(s),s)\right)d\frac{\frac{\tilde{W}^3}{\tilde{T}(\tilde{W}^3)}}{\hat{T}}$$

$$+\int_0^t \left(\hat{g}(\hat{X}^k(s),s)-\hat{g}(\hat{X}(s),s)\right)d\frac{\frac{\tilde{W}^3}{\tilde{T}(\tilde{W}^3)}}{\hat{T}}\Bigg|$$

$$\leq \left|\hat{X}_0^k-\hat{X}_0\right|+\int_0^t \left|\hat{f}(\hat{X}^k(s),s)-\hat{f}(\hat{X}(s),s)\right|ds$$

$$+\int_0^t \left|\hat{f}^k(\hat{X}^k(s),s)-\hat{f}(\hat{X}^k(s),s)\right|ds+\int_0^t \left|\hat{g}^k(\hat{X}^k(s),s)-\hat{g}(\hat{X}^k(s),s)\right|d\frac{\frac{\tilde{W}^3}{\tilde{T}(\tilde{W}^3)}}{\hat{T}}$$

$$+\int_0^t \left|\hat{g}(\hat{X}^k(s),s)-\hat{g}(\hat{X}(s),s)\right|d\frac{\frac{\tilde{W}^3}{\tilde{T}(\tilde{W}^3)}}{\hat{T}}.$$

Let $\varepsilon>0$ be arbitrarily chosen. Since

$$\lim_{k\longrightarrow\infty}\max_{0\leq t\leq T,|x|\leq M}\left(|\hat{f}^k(x,t)-\hat{f}(x,t)|+|\hat{g}^k(x,t)-\hat{g}(x,t)|\right)=0,$$

then there exists enough large $k\in\mathbb{N}$ such that

$$\int_0^t \left(\hat{f}^k(\hat{X}^k(s),s)-\hat{f}(\hat{X}^k(s),s)\right)ds+\int_0^t \left(\hat{g}^k(\hat{X}^k(s),s)-\hat{g}(\hat{X}^k(s),s)\right)d\frac{\frac{\tilde{W}^3}{\tilde{T}(\tilde{W}^3)}}{\hat{T}}\leq\varepsilon T.$$

Therefore

$$|\hat{X}^k(t)-\hat{X}(t)| \le \left|\hat{X}_0^k - \hat{X}_0\right| + \int_0^t \left|\hat{f}(\hat{X}^k(s),s)-\hat{f}(\hat{X}(s),s)\right| ds + \varepsilon T$$

$$+ \int_0^t \left|\hat{g}(\hat{X}^k(s),s)-\hat{g}(\hat{X}(s),s)\right| d\frac{\frac{\tilde{W}^3}{\tilde{T}(\tilde{W}^3)}}{\hat{T}}.$$

From here

$$|\hat{X}^k(t)-\hat{X}(t)|^{\hat{2}} = |\hat{X}^k(t)-\hat{X}(t)|^2\hat{T}$$

$$= \hat{T}\Bigg(\left|\hat{X}_0^k - \hat{X}_0\right| + \int_0^t \left|\hat{f}(\hat{X}^k(s),s)-\hat{f}(\hat{X}(s),s)\right| ds + \varepsilon T$$

$$+ \int_0^t \left|\hat{g}(\hat{X}^k(s),s)-\hat{g}(\hat{X}(s),s)\right| d\frac{\frac{\tilde{W}^3}{\tilde{T}(\tilde{W}^3)}}{\hat{T}}\Bigg|\Bigg)^2$$

$$\le 2\hat{T}|_0^k - \hat{X}_0|^2 + 2\hat{T}\left(\int_0^t |\hat{f}(\hat{X}^k(s),s)-\hat{f}(\hat{X}(s),s)| ds\right)^2 + 2\varepsilon^2 T^2 \hat{T}$$

$$+2\hat{T}\left(\int_0^t \left|\hat{g}(\hat{X}^k(s),s)-\hat{g}(\hat{X}(s),s)\right| d\frac{\frac{\tilde{W}^3}{\tilde{T}(\tilde{W}^3)}}{\hat{T}}\right)^2$$

$$\le 2\hat{T}|_0^k - \hat{X}_0|^2 + 2t\hat{T}\int_0^t |\hat{f}(\hat{X}^k(s),s)-\hat{f}(\hat{X}(s),s)|^2 ds + 2\varepsilon^2 T^{\hat{2}}$$

$$+2\hat{T}\left(\int_0^t \left|\hat{g}(\hat{X}^k(s),s)-\hat{g}(\hat{X}(s),s)\right| d\frac{\frac{\tilde{W}^3}{\tilde{T}(\tilde{W}^3)}}{\hat{T}}\right)^2$$

$$= 2|\hat{X}_0^k - \hat{X}_0|^{\hat{2}} + 2t\int_0^t |\hat{f}(\hat{X}^k(s),s)-\hat{f}(\hat{X}(s),s)|^{\hat{2}} ds + 2\varepsilon^{\hat{2}} T^{\hat{2}}$$

$$+2\left(\int_0^t \frac{1}{\tilde{T}}\tilde{T}(\hat{g}(\hat{X}^k(s),s)-\hat{g}(\hat{X}(s),s))\frac{\tilde{T}^2}{\tilde{T}^2} d\frac{\frac{\tilde{W}^3}{\tilde{T}(\tilde{W}^3)}}{\hat{T}}\right)^{\hat{2}}$$

$$\le 2|\hat{X}_0^k - \hat{X}_0|^{\hat{2}} + 2TL^2\int_0^t |\hat{X}^k(s)-\hat{X}(s)|^{\hat{2}} ds + 2\varepsilon^2 T^{\hat{2}}$$

$$+2\left(\hat{\int}_0^t \frac{\hat{g}(\hat{X}^k(s),s)-\hat{g}(\hat{X}(s),s)}{\tilde{T}} \hat{\times} \hat{d}\frac{\frac{\tilde{W}^3}{\tilde{T}(\tilde{W}^3)}}{\hat{T}}\right)^{\hat{2}},$$

i.e.,

$$|\hat{X}^k(t)-\hat{X}(t)|^{\hat{2}} \le 2|\hat{X}_0^k - \hat{X}_0|^{\hat{2}} + 2TL^2\int_0^t |\hat{X}^k(s)-\hat{X}(s)|^{\hat{2}} ds + 2\varepsilon^2 T^{\hat{2}}$$

$$+2\left(\hat{\int}_0^t \frac{\hat{g}(\hat{X}^k(s),s)-\hat{g}(\hat{X}(s),s)}{\tilde{T}} \hat{\times} \hat{d}\frac{\frac{\tilde{W}^3}{\tilde{T}(\tilde{W}^3)}}{\hat{T}}\right)^{\hat{2}}.$$

Therefore

$$\hat{E}(|\hat{X}^k(t)-\hat{X}(t)|^{\hat{2}}) \leq 2\hat{E}(|\hat{X}_0^k-\hat{X}_0|^{\hat{2}})+2TL^2\int_0^t \hat{E}(|\hat{X}^k(s)-\hat{X}(s)|^{\hat{2}})ds$$

$$+2\varepsilon^2\hat{E}(T^{\hat{2}})+2\hat{E}\left(\left(\hat{\int}_0^t \frac{\hat{g}(\hat{X}^k(s),s)-\hat{g}(\hat{X}(s),s)}{\tilde{T}}\hat{\times}\hat{d}\frac{\frac{\tilde{W}^3}{\tilde{T}(\tilde{W}^3)}}{\hat{T}}\right)^{\hat{2}}\right)$$

$$=2\hat{E}(|\hat{X}_0^k-\hat{X}_0|^{\hat{2}})+2TL^2\int_0^t \hat{E}(|\hat{X}^k(s)-\hat{X}(s)|^{\hat{2}})ds$$

$$+2\varepsilon^2\hat{E}(T^{\hat{2}})+2\hat{E}\left(\int_0^t \frac{(\hat{g}(\hat{X}^k(s),s)-\hat{g}(\hat{X}(s),s))^2\tilde{T}^2}{\tilde{T}^2\hat{T}}dt\right)$$

$$=2\hat{E}(|\hat{X}_0^k-\hat{X}_0|^{\hat{2}})+2TL^2\int_0^t \hat{E}(|\hat{X}^k(s)-\hat{X}(s)|^{\hat{2}})ds$$

$$+2\varepsilon^2\hat{E}(T^{\hat{2}})+2\hat{E}\left(\int_0^t \frac{(\hat{g}(\hat{X}^k(s),s)-\hat{g}(\hat{X}(s),s))^2}{\hat{T}}dt\right)$$

$$\leq 2\hat{E}(|\hat{X}_0^k-\hat{X}_0|^{\hat{2}})+2TL^2\int_0^t \hat{E}(|\hat{X}^k(s)-\hat{X}(s)|^{\hat{2}})ds$$

$$+2\varepsilon^2\hat{E}(T^{\hat{2}})+2\frac{L^2}{N}\hat{E}\left(\int_0^t |\hat{X}^k(s)-\hat{X}(s)|^2 ds\right)$$

$$=2\hat{E}(|\hat{X}_0^k-\hat{X}_0|^{\hat{2}})+2TL^2\int_0^t \hat{E}(|\hat{X}^k(s)-\hat{X}(s)|^{\hat{2}})ds$$

$$+2\varepsilon^2\hat{E}(T^{\hat{2}})+2\frac{L^2}{N}\hat{E}\left(\int_0^t \frac{\hat{T}|\hat{X}^k(s)-\hat{X}(s)|^2}{\hat{T}}ds\right)$$

$$\leq \hat{E}(|\hat{X}_0^k-\hat{X}_0|^{\hat{2}})+2TL^2\int_0^t \hat{E}(|\hat{X}^k(s)-\hat{X}(s)|^{\hat{2}})ds$$

$$+2\varepsilon^2\hat{E}(T^{\hat{2}})+2\frac{L^2}{N^2}\hat{E}\left(\int_0^t |\hat{X}^k(s)-\hat{X}(s)|^{\hat{2}} ds\right)$$

$$=2\hat{E}(|\hat{X}_0^k-\hat{X}_0|^{\hat{2}})+2\varepsilon^2\hat{E}(T^{\hat{2}})$$

$$+\left(2TL^2+2\frac{L^2}{N^2}\right)\int_0^t \hat{E}(|\hat{X}^k(s)-\hat{X}(s)|^{\hat{2}})ds,$$

i.e.,

$$\hat{E}\left(|\hat{X}^k(t)-\hat{X}(t)|^{\hat{2}}\right) \leq 2\hat{E}(|\hat{X}_0^k-\hat{X}_0|^{\hat{2}})+2\varepsilon^2\hat{E}(T^{\hat{2}})$$

$$+\left(2TL^2+2\frac{L^2}{N^2}\right)\int_0^t \hat{E}(|\hat{X}^k(s)-\hat{X}(s)|^{\hat{2}})ds.$$

From here and from Gronwall's lemma it follows that

$$\hat{E}\left(|\hat{X}^k(t)-\hat{X}(t)|^{\hat{2}}\right) \leq 2\left(\hat{E}\left(|\hat{X}_0^k-\hat{X}_0|^{\hat{2}}\right)+\varepsilon^2\hat{E}(T^{\hat{2}})\right)e^{\left(2TL^2+2\frac{L^2}{N^2}\right)t}.$$

Because

$$\lim_{k\longrightarrow\infty} \hat{E}\left(|\hat{X}_0^k - \hat{X}_0|^{\hat{2}}\right) = 0,$$

$$\lim_{k\longrightarrow\infty} \varepsilon^2 \hat{E}(T^{\hat{2}}) = 0,$$

we find that

$$\lim_{k\longrightarrow\infty} \hat{E}\left(\max_{0\leq t\leq T} |\hat{X}^k(t) - \hat{X}(t)|^{\hat{2}}\right) = 0.$$

□

Chapter 5

Stability of Iso-Stochastic Differential Equations

5.1 Stability of Deterministic Equations

Suppose that

$$(1) \qquad \begin{cases} \dot{x}(t) = f(x,t) \\ x(t_0) = c, \qquad t \geq t_0, \end{cases}$$

where $c \in \mathbb{R}$, is the ordinary differential equation for an 1 - dimensional state function $x(t)$. We assume that for every initial value $c \in \mathbb{R}$, there exists a global solution defined on $[t_0, \infty)$, $x(t,c)$ and $f(x,\cdot)$ is continuous. Furthermore, we suppose that

$$f(0,t) = 0$$

for all $t \geq t_0$, so that (1) has the solution $x(t) = 0$ corresponding to the initial condition $c = 0$. We shall refer to this solution as the equilibrium position.

Definition 5.1.1. *The equilibrium position is said to be stable if, for every* $\varepsilon > 0$, *there exists a* $\delta = \delta(\varepsilon, t_0) > 0$ *such that*

$$\sup_{t_0 \leq t < \infty} |x(t,c)| < \varepsilon$$

whenever $|c| \leq \delta$. *Otherwise, it is said to be unstable. The equilibrium position is said to be asymptotically stable if it is stable and if*

$$\lim_{t \longrightarrow \infty} x(t,c) = 0$$

for all c in some neighbourhood of $x = 0$.

Definition 5.1.2. *A continuous scalar function* $v(x)$ *defined on* $U_h = [-h,h]$ *is said to be positive-definite if*

$$v(0) = 0, \qquad v(x) > 0$$

for all $x \neq 0$.

Definition 5.1.3. *A continuous function $v(x,t)$ defined on $U_h \times [t_0,\infty)$ is said to be positive-definite if $v(0,t)=0$ and there exists a positive-definite function $w(x)$ such that*

$$v(x,t) \geq w(x)$$

for all $t \geq t_0$.

Definition 5.1.4. *A continuous function $v(x,t)$ defined on $U_h \times [t_0,\infty)$ is said to be negative-definite if $-v$ is positive-definite.*

Definition 5.1.5. *A continuous nonnegative function $v(x,t)$ is said to be decrescent if there exists a positive-definite function $u(x)$ such that*

$$v(x,t) \leq u(x)$$

for all $t \geq t_0$.

Definition 5.1.6. *A continuous nonnegative function $v(x,t)$ is said to be radially unbounded if*

$$\inf_{t\geq t_0} v(x,t) \longrightarrow_{x\longrightarrow\infty} \infty.$$

Every positive-definite function $v(x)$ that is independent of t is also decrescent.

If $x(t)$ is a solution to the equation (1) and $v(x,t)$ is a positive-definite function with continuous first partial derivatives with respect to t and x. Let

$$V(t) = v(x(t),t).$$

Then

$$\begin{aligned} \dot{V}(t) &= v_x(x(t),t)\dot{x}(t) + v_t(x(t),t) \\ &= v_x(x(t),t)f(x(t),t) + v_t(x(t),t). \end{aligned}$$

If $\dot{V}(t) \leq 0$, then $x(t)$ varies in such a way that the values of $V(t)$ do not increase, that is, the distance of $x(t)$ from the equilibrium point measured by $v(x(t),t)$ does not increase.

This observation leads to the following sufficient criteria discovered by Lyapunov.

Theorem 5.1.7. **1.** *If there exists a positive-definite function $v(x,t)$ with continuous first partial derivatives such that the derivative formed along the trajectories of*

$$\dot{x}(t) = f(x(t),t), \qquad t \geq t_0, \qquad f(0,t)=0,$$

satisfies the inequality

$$\dot{v}(x(t),t) = v_t(x(t),t) + v_x(x(t),t)f(x(t),t) \leq 0$$

in

$$\{(x,t) : |x| \leq h, \qquad t \geq t_0\},$$

then the equilibrium position of the differential equation is stable.

2. *If there exists a positive-definite decrescent function $v(x,t)$ such that $\dot{v}(x(t),t)$ is negative-definite, then the equilibrium position is asymptotically stable.*

Definition 5.1.8. *A function $v(x,t)$ that satisfies the stability condition of the last theorem is said to be a Lyapunov function corresponding to the differential equation* (1).

Example 5.1.9. *We consider the linear equation*

$$\dot{x}(t) = \lambda x(t), \qquad t \geq t_0,$$

$$x(t_0) = c,$$

where $c \in \mathbb{R}$. It has the solution

$$x(t) = ce^{\lambda t}.$$

Then the equilibrium position is asymptotically stable if and only if $\lambda < 0$.

5.2 Basic Ideas of Iso-Stochastic Stability Theory

We suppose that

$$(2) \qquad \begin{cases} d\hat{X}(t) = \hat{f}(\hat{X}(t),t)dt + \hat{g}(\hat{X}(t),t)d\frac{\tilde{W}^1}{\hat{T}} \\ \hat{X}(t_0) = c, \qquad t \geq t_0, \end{cases}$$

where c is a real constant, is an iso-stochastic differential equation of the first kind that satisfies all assumptions of the existence and uniqueness theorem for the iso-stochastic differential equations of the first kind, and has continuous coefficients with respect to t. Finally, we assume that

$$\hat{f}(0,t) = \hat{g}(0,t) = 0 \qquad \text{for} \qquad t \geq t_0.$$

With $\hat{X}(t)$ we will denote the solution of the equation (2).

Let $v(x,t)$ denotes a positive definite function defined everywhere on $[t_0,\infty) \times \mathbb{R}$ that has continuous partial derivatives v_t, v_x and v_{xx}.

We consider the iso-stochastic process

$$V(t) = v(\hat{X}(t),t).$$

We have

$$dV(t) = \frac{\partial v(\hat{X}(t),t)}{\partial t}dt + \frac{\partial v(\hat{X}(t),t)}{\partial x}d\hat{X}(t) + \frac{1}{2}\frac{\partial^2 v(\hat{X}(t),t)}{\partial x^2}\hat{g}^2(\hat{X}(t),t)\frac{\hat{T}(\Omega)}{\hat{T}}dt$$

$$= \frac{\partial v(\hat{X}(t),t)}{\partial t}dt + \frac{\partial v(\hat{X}(t),t)}{\partial x}\left(\hat{f}(\hat{X}(t),t)dt + \hat{g}(\hat{X}(t),t)d\frac{\tilde{W}^1}{\hat{T}}\right) + \frac{1}{2}\frac{\partial^2 v(\hat{X}(t),t)}{\partial x^2}\hat{g}^2(\hat{X}(t),t)\frac{\hat{T}(\Omega)}{\hat{T}}dt$$

$$= \left(\frac{\partial v(\hat{X}(t),t)}{\partial t} + \frac{\partial v(\hat{X}(t),t)}{\partial x}\hat{f}(\hat{X}(t),t) + \frac{1}{2}\frac{\partial^2 v(\hat{X}(t),t)}{\partial x^2}\hat{g}^2(\hat{X}(t),t)\frac{\hat{T}(\Omega)}{\hat{T}}\right)dt$$

$$+ \frac{\partial v(\hat{X}(t),t)}{\partial x}\hat{g}^2(\hat{X}(t),t)d\frac{\tilde{W}^1}{\hat{T}},$$

whereupon

$$v(\hat{X}(t),t) = v(c,t_0) + \int_{t_0}^{t} \left(\frac{\partial v(\hat{X}(s),s)}{\partial t} + \frac{\partial v(\hat{X}(s),s)}{\partial x} \hat{f}(\hat{X}(s),s) + \frac{1}{2} \frac{\partial^2 v(\hat{X}(s),s)}{\partial x^2} \hat{g}^2(\hat{X}(s),s) \frac{\hat{T}(\Omega)}{\hat{T}} \right) ds$$

$$+ \int_{t_0}^{t} \frac{\partial v(\hat{X}(s),s)}{\partial x} \hat{g}^2(\hat{X}(s),s) d\frac{\tilde{W}^1}{\hat{T}}.$$

Let

$$Lv(\hat{X}(t),t) := \frac{\partial v(\hat{X}(t),t)}{\partial t} + \frac{\partial v(\hat{X}(t),t)}{\partial x} \hat{f}(\hat{X}(t),t) + \frac{1}{2} \frac{\partial^2 v(\hat{X}(t),t)}{\partial x^2} \hat{g}^2(\hat{X}(t),t) \frac{\hat{T}(\Omega)}{\hat{T}}.$$

Then

$$dV(t) = Lv(\hat{X}(t),t)dt + \frac{\partial v(\hat{X}(t),t)}{\partial x} \hat{g}(\hat{X}(t),t) d\frac{\tilde{W}^1}{\hat{T}}.$$

Hence,

$$V(t) = v(c,t_0) + \int_{t_0}^{t} Lv(\hat{X}(s),s)ds + \int_{t_0}^{t} \frac{\partial v(\hat{X}(s),s)}{\partial x} \hat{g}^2(\hat{X}(s),s) d\frac{\tilde{W}^1}{\hat{T}}.$$

We note that

$$\hat{E}(V(t)) = \hat{E}(v(c,t_0)) + \hat{E}\left(\int_{t_0}^{t} Lv(\hat{X}(s),s)ds + \int_{t_0}^{t} \frac{\partial v(\hat{X}(s),s)}{\partial x} \hat{g}^2(\hat{X}(s),s) d\frac{\tilde{W}^1}{\hat{T}} \right)$$

$$= \hat{E}(v(c,t_0)) + \int_{t_0}^{t} \hat{E}(Lv(\hat{X}(s),s))ds + \hat{E}\left(\int_{t_0}^{t} \frac{\partial v(\hat{X}(s),s)}{\partial x} \hat{g}^2(\hat{X}(s),s) d\frac{\tilde{W}^1}{\hat{T}} \right)$$

$$= \hat{E}(v(c,t_0)) + \int_{t_0}^{t} \hat{E}(Lv(\hat{X}(s),s))ds.$$

We can rewrite this in the following manner

$$\hat{E}(dV(t)) = \hat{E}(Lv(\hat{X}(t),t)dt).$$

Then

$$\hat{E}(dV(t)) \leq 0$$

if

$$Lv(\hat{X}(t),t) \leq 0$$

for all $t \geq t_0$, $x \in \mathbb{R}$.

This is the stochastic analogue of the requirement that $\dot{V} \leq 0$ in the deterministic case and it reduces to that case if $\hat{g}$ vanishes. We shall refer to the function $v(x,t)$ used here as a Lyapunov function corresponding to the iso-stochastic differential equation (2).

Definition 5.2.1. *The equilibrium point is said to be iso-stochastically stable if, for every* $\varepsilon > 0$,

$$\lim_{\varepsilon \longrightarrow 0} \hat{P}\left(\sup_{t_0 \leq t < \infty} |\hat{X}(t)| \geq \varepsilon \right) = 0.$$

Otherwise, it is said to be iso-stochastically unstable.

Definition 5.2.2. *The equilibrium position is said to be iso-stochastically asymptotically stable if it is iso-asymptotically stable and*

$$\lim_{c\longrightarrow\infty} \hat{P}\left(\lim_{t\longrightarrow\infty} \hat{X}(t,c)=0\right)=1.$$

Definition 5.2.3. *The equilibrium position is said to be iso-stochastically asymptotically stable in the large if it is iso-stochastically stable and*

$$\hat{P}\left(\lim_{t\longrightarrow\infty} \hat{X}(t,c)=0\right)=1$$

for all $c\in\mathbb{R}$.

We will note that, in the case $\hat{g}(x,t)=0$, those definitions reduce to the corresponding deterministic definitions. Therefore we have the following criteria.

Theorem 5.2.4. *Suppose that the iso-stochastic differential equation of the second kind* (2) *satisfies the existence and uniqueness theorem for the iso-stochastic differential equation of the second kind.*

(i) *Suppose that there exists a positive definite function* $v(x,t)$ *defined on* $U_h\times[t_0,\infty)$, *where* $h>0$, *that is everywhere, with the possible exception of the point* $x=0$, *continuously differentiable with respect to* t *and twice continuously differentiable with respect to* x. *Furthermore,*

$$Lv(x,t)\leq 0, \qquad t\geq t_0, \qquad 0<x\leq h,$$

where

$$L=\frac{\partial}{\partial t}+\hat{f}(x,t)\frac{\partial}{\partial x}+\frac{1}{2}\hat{g}^2(x,t)\frac{\partial^2}{\partial x^2}. \tag{3}$$

Then, the equilibrium position of (2) *is iso-stochastically stable.*

(ii) *If, in addition,* $v(x,t)$ *is decrescent and* $Lv(x,t)$ *is negative definite, then the equilibrium position of* (2) *is iso-stochastically asymptotically stable. Also, if* $v(x,t)$ *is radially unbounded function, defined everywhere on* $\mathbb{R}\times[t_0,\infty)$, *then the equilibrium position is iso-stochastically asymptotically stable in the large.*

Theorem 5.2.5. *Suppose that the iso-stochastic differential equation* (2) *satisfies all conditions of the existence and uniqueness theorem for the iso-stochastic differential equations of the second kind. Suppose that there exists a function* $v(x,t)$ *defined on*

$$\{0<x\leq h\}\times[t_0,\infty),$$

that has continuous partial derivatives with respect to t *and continuous second partial derivatives with respect to* x. *Then, if*

$$\lim_{x\longrightarrow 0}\inf_{t\geq t_0} v(x,t)=\infty$$

and

$$\sup_{t\geq t_0,\varepsilon<x\leq h} Lv(x,t)<0$$

for all $0 < \varepsilon < h$, *or only* $Lv(x,t) \leq 0$ *for* $x \neq 0$, *and in addition,*

$$y^2 \hat{g}^2(x,t) \geq m(x)|y|^2$$

for all $y \in \mathbb{R}$, $|x| \leq h$, $t \geq t_0$, *m is positive-definite. Then the equilibrium position of* (2) *is iso-stochastically unstable.*

Theorem 5.2.6. *Suppose that the iso-stochastic differential equation of the second kind*

$$(4) \qquad \begin{cases} d\hat{X}(t) = \hat{f}(\hat{X}(t),t)dt + \hat{g}(\hat{X}(t),t)d\frac{\tilde{W}^2\hat{T}(\tilde{W}^2)}{\hat{T}} \\ \hat{X}(t_0) = c, \qquad t \geq t_0 \end{cases}$$

satisfies the existence and uniqueness theorem for the iso-stochastic differential equation of the second kind.

(i) *Suppose that there exists a positive definite function* $v(x,t)$ *defined on* $U_h \times [t_0,\infty)$, *where* $h > 0$, *that is everywhere, with the possible exception of the point* $x = 0$, *continuously differentiable with respect to t and twice continuously differentiable with respect to x. Furthermore,*

$$Lv(x,t) \leq 0, \qquad t \geq t_0, \qquad 0 < x \leq h,$$

where L is defined by (3). *Then, the equilibrium position of* (4) *is iso-stochastically stable.*

(ii) *If, in addition,* $v(x,t)$ *is decrescent and* $Lv(x,t)$ *is negative definite, then the equilibrium position of* (4) *is iso-stochastically asymptotically stable. Also, if* $v(x,t)$ *is radially unbounded function, defined everywhere on* $\mathbb{R} \times [t_0,\infty)$, *then the equilibrium position is iso-stochastically asymptotically stable in the large.*

Theorem 5.2.7. *Suppose that the iso-stochastic differential equation* (4) *satisfies all conditions of the existence and uniqueness theorem for the iso-stochastic differential equations of the second kind. Suppose that there exists a function* $v(x,t)$ *defined on*

$$\{0 < x \leq h\} \times [t_0,\infty),$$

that has continuous partial derivatives with respect to t and continuous second partial derivatives with respect to x. Then, if

$$\lim_{x \longrightarrow 0} \inf_{t \geq t_0} v(x,t) = \infty$$

and

$$\sup_{t \geq t_0, \varepsilon < x \leq h} Lv(x,t) < 0$$

for all $0 < \varepsilon < h$, *or only* $Lv(x,t) \leq 0$ *for* $x \neq 0$, *and in addition,*

$$y^2 \hat{g}^2(x,t) \geq m(x)|y|^2$$

for all $y \in \mathbb{R}$, $|x| \leq h$, $t \geq t_0$, *m is positive-definite. Then the equilibrium position of* (4) *is iso-stochastically unstable.*

Theorem 5.2.8. *Suppose that the iso-stochastic differential equation of the third kind*

$$(5) \qquad \begin{cases} d\hat{X}(t) = \hat{f}(\hat{X}(t),t)dt + \hat{g}(\hat{X}(t),t)d\frac{\frac{\tilde{W}^3}{\tilde{T}(\tilde{W}^3)}}{\hat{T}} \\ \hat{X}(t_0) = c, \qquad t \geq t_0 \end{cases}$$

satisfies the existence and uniqueness theorem for the iso-stochastic differential equation of the third kind.

(i) *Suppose that there exists a positive definite function $v(x,t)$ defined on $U_h \times [t_0, \infty)$, where $h > 0$, that is everywhere, with the possible exception of the point $x = 0$, continuously differentiable with respect to t and twice continuously differentiable with respect to x. Furthermore,*

$$Lv(x,t) \leq 0, \qquad t \geq t_0, \qquad 0 < x \leq h,$$

where L is defined by (3)*. Then, the equilibrium position of* (5) *is iso-stochastically stable.*

(ii) *If, in addition, $v(x,t)$ is decrescent and $Lv(x,t)$ is negative definite, then the equilibrium position of* (5) *is iso-stochastically asymptotically stable. Also, if $v(x,t)$ is radially unbounded function, defined everywhere on $\mathbb{R} \times [t_0, \infty)$, then the equilibrium position is iso-stochastically asymptotically stable in the large.*

Theorem 5.2.9. *Suppose that the iso-stochastic differential equation* (5) *satisfies all conditions of the existence and uniqueness theorem for the iso-stochastic differential equations of the third kind. Suppose that there exists a function $v(x,t)$ defined on*

$$\{0 < x \leq h\} \times [t_0, \infty),$$

that has continuous partial derivatives with respect to t and continuous third partial derivatives with respect to x. Then, if

$$\lim_{x \longrightarrow 0} \inf_{t \geq t_0} v(x,t) = \infty$$

and

$$\sup_{t \geq t_0, \varepsilon < x \leq h} Lv(x,t) < 0$$

for all $0 < \varepsilon < h$, or only $Lv(x,t) \leq 0$ for $x \neq 0$, and in addition,

$$y^2 \hat{g}^2(x,t) \geq m(x)|y|^2$$

for all $y \in \mathbb{R}$, $|x| \leq h$, $t \geq t_0$, m is positive-definite. Then the equilibrium position of (5) *is iso-stochastically unstable.*

Problem 5.2.10. *Find the equilibrium positions to the iso-stochastic differential equations of the first kind*

$$\begin{cases} d\hat{X}(t) = \hat{X}^2(t)dt + \hat{X}(t)d\frac{\tilde{W}^1}{\hat{T}} \\ \hat{X}(0) = 0, \qquad t \geq 0, \end{cases}$$

Answer.

$$\hat{X}(t) = 0.$$

Problem 5.2.11. *Find the equilibrium positions to the iso-stochastic differential equations of the second kind*

$$\begin{cases} d\hat{X}(t) = \left(\hat{X}^2(t) - 1\right)\hat{X}(t)dt + \left(\hat{X}(t) - 1\right)d\frac{\tilde{W}^2\tilde{T}(\tilde{W}^2)}{\hat{T}} \\ \hat{X}(0) = 1, \qquad t \geq 0, \end{cases}$$

Answer.

$$\hat{X}(t) = 1.$$

Problem 5.2.12. *Find the equilibrium positions to the iso-stochastic differential equations of the third kind*

$$\begin{cases} d\hat{X}(t) = \left(\hat{X}^2(t) - 4\right)dt + \hat{X}(t)\left(\hat{X}(t) - 2\right)d\frac{\frac{\tilde{W}^3}{\tilde{T}(\tilde{W}^3)}}{\hat{T}} \\ \hat{X}(0) = 2, \qquad t \geq 0, \end{cases}$$

Answer.

$$\hat{X}(t) = 2.$$

Problem 5.2.13. *Prove that the equilibrium position to the iso-stochastic differential equation of the first kind*

$$\begin{cases} d\hat{X} = \hat{X}(t)d\frac{\tilde{W}^1}{\hat{T}} \\ \hat{X}(0) = 0, \qquad t \geq 0. \end{cases}$$

is iso-stochastically stable in the large.

Problem 5.2.14. *Prove that the equilibrium position to the iso-stochastic differential equation of the second kind*

$$\begin{cases} d\hat{X} = 2\hat{X}(t)dt + 3\hat{X}(t)d\frac{\tilde{W}^2\tilde{T}(\tilde{W}^2)}{\hat{T}} \\ \hat{X}(0) = 0, \qquad t \geq 0. \end{cases}$$

is iso-stochastically stable in the large.

Problem 5.2.15. *Prove that the equilibrium position to the iso-stochastic differential equation of the third kind*

$$\begin{cases} d\hat{X} = -\hat{X}(t)dt + 7\hat{X}(t)d\frac{\frac{\tilde{W}^3}{\tilde{T}(\tilde{W}^3)}}{\hat{T}} \\ \hat{X}(0) = 0, \qquad t \geq 0. \end{cases}$$

is iso-stochastically stable in the large.

Problem 5.2.16. *Prove that the equilibrium position to the iso-stochastic differential equation of the first kind*

$$\begin{cases} d\hat{X} = 10\hat{X}(t)dt + \hat{X}(t)d\frac{\tilde{W}^1}{\hat{T}} \\ \hat{X}(0) = 0, \qquad t \geq 0. \end{cases}$$

is iso-stochastically unstable.

Problem 5.2.17. *Prove that the equilibrium position to the iso-stochastic differential equation of the second kind*

$$\begin{cases} d\hat{X} = \hat{X}^2(t)dt - \frac{1}{10}\hat{X}(t)d\frac{\tilde{W}^2\tilde{T}(\tilde{W}^2)}{\hat{T}} \\ \hat{X}(0) = 0, \qquad t \geq 0. \end{cases}$$

is iso-stochastically unstable.

Problem 5.2.18. *Prove that the equilibrium position to the iso-stochastic differential equation of the third kind*

$$\begin{cases} d\hat{X} = 2\hat{X}^2(t)dt + \hat{X}(t)(\hat{X}(t)+1)d\frac{\frac{\tilde{W}^3}{\tilde{T}(\tilde{W}^3)}}{\hat{T}} \\ \hat{X}(0) = 0, \qquad t \geq 0. \end{cases}$$

is iso-stochastically unstable.

Problem 5.2.19. *Let us consider the iso-stochastic differential equation of the first kind*

$$\begin{cases} d\hat{X}(t) - a\hat{X}(t)dt + b\hat{X}(t)d\frac{\tilde{W}^1}{\hat{T}} \\ \hat{X}(t_0) = c, \qquad t \geq t_0, \end{cases}$$

where a, b and c are real constants.

(i) *Prove that*

$$\hat{X}(t) = ce^{\left(a-\frac{b^2}{2}\right)(t-t_0)+b\frac{\tilde{W}^1(t)-\tilde{W}^1(t_0)}{\hat{T}}}$$

is its solution.

(ii) *If*

$$a < \frac{b^2}{2},$$

prove that the equilibrium position is iso-stochastically stable in the large.

(iii) *If*

$$a \geq \frac{b^2}{2} > 0,$$

prove that the equilibrium position is iso-stochastically unstable.

Hint. Use the function $v(x,t) = |x|^{2r}$, $r > 0$.

Problem 5.2.20. *Let us consider the iso-stochastic differential equation of the second kind*

$$\begin{cases} d\hat{X}(t) = a\hat{X}(t)dt + b\hat{X}(t)d\frac{\tilde{W}^2\tilde{T}(\tilde{W}^2)}{\hat{T}} \\ \hat{X}(t_0) = c, \qquad t \geq t_0, \end{cases}$$

where a, b and c are real constants.

(i) *Prove that*

$$\hat{X}(t) = ce^{\left(a-\frac{b^2}{2}\right)(t-t_0)+b\frac{\tilde{W}^2(t)\tilde{T}(\tilde{W}^2(t))-\tilde{W}^2(t_0)\tilde{T}(\tilde{W}^2(t_0))}{\hat{T}}}$$

is its solution.

(ii) *If*

$$a < \frac{b^2}{2},$$

prove that the equilibrium position is iso-stochastically stable in the large.

(iii) *If*

$$a \geq \frac{b^2}{2} > 0,$$

prove that the equilibrium position is iso-stochastically unstable.

Problem 5.2.21. *Let us consider the iso-stochastic differential equation of the third kind*

$$\begin{cases} d\hat{X}(t) = a\hat{X}(t)dt + b\hat{X}(t)d\frac{\frac{\tilde{W}^3}{\tilde{T}(\tilde{W}^3)}}{\hat{T}} \\ \hat{X}(t_0) = c, \qquad t \geq t_0, \end{cases}$$

where a, b and c are real constants.

(i) *Prove that*

$$\hat{X}(t) = ce^{\left(a-\frac{b^2}{2}\right)(t-t_0)+b\frac{\frac{\tilde{W}^3(t)}{\tilde{T}(\tilde{W}^3(t))}-\frac{\tilde{W}^3(t_0)}{\tilde{T}(\tilde{W}^3(t_0))}}{\hat{T}}}$$

is its solution.

(ii) *If*

$$a < \frac{b^2}{2},$$

prove that the equilibrium position is iso-stochastically stable in the large.

(iii) *If*

$$a \geq \frac{b^2}{2} > 0,$$

prove that the equilibrium position is iso-stochastically unstable.

Definition 5.2.22. *Suppose that the iso-stochastic differential equation of the first kind* (2) *satisfies all conditions of the existence and uniqueness theorem for iso-stochastic differential equations of the first kind.*

The equilibrium position is said to be iso-stochastically stable in pth mean, where $p > 0$, *if, for every* $\varepsilon > 0$, *there exists a* $\delta > 0$ *such that*

$$\sup_{t_0 \leq t < \infty} \hat{E}(|\hat{X}(t,c)|^{\hat{p}}) \leq \varepsilon \qquad \text{for} \qquad |c| \leq \delta.$$

The equilibrium position is said to be iso-stochastically asymptotically stable in pth mean if it is iso-stochastically stable in pth mean and if

$$\lim_{c \longrightarrow \infty} \hat{E}(|\hat{X}(t,c)|^{\hat{p}}) = 0$$

for all c in a neighborhood of $x = 0$.

It is said to be iso-stochastically exponentially stable in pth mean if there exist positive constants c_1 *and* c_2 *such that, for sufficiently small c,*

$$\hat{E}(|\hat{X}(t,c)|^{\hat{p}}) \leq c_1 |c|^p e^{-c_2(t-t_0)}.$$

In the case $p = 1$ *or* $p = 2$, *we speak of iso-stochastically(asymptotic or exponential) stability in mean or in mean square, respectively.*

The equilibrium position is said to possess an iso-stochastically stable expectation value

$$\hat{m}(t) = \hat{E}(\hat{X}(t,c))$$

or an iso-stochastically stable second moment

$$\hat{P}(t) = \hat{E}(|\hat{X}(t,c)|^{\hat{2}})$$

if, for every $\varepsilon > 0$, *there exists a* $\delta > 0$ *such that*

$$\sup_{t_0 \leq t < \infty} |\hat{E}(\hat{X}(t,c))| \leq \varepsilon \qquad \text{for} \qquad \text{all} \qquad |c| \leq \delta,$$

or

$$\sup_{t_0 \leq t < \infty} |\hat{E}(\hat{X}^{\hat{2}}(t,c)| \leq \varepsilon \qquad \text{for} \qquad \text{all} \qquad |c| \leq \delta,$$

respectively.

Remark 5.2.23. *We note that an iso-stochastically(asymptotic, exponential) stability in pth mean implies the same stability in qth mean for all* $0 < q < p$. *In particular, stability in mean square implies stability in mean.*

Remark 5.2.24. *It follows from the iso-Chebyshev inequality, that*

$$\sup_{t_0 \leq t < \infty} \hat{P}(|\hat{X}(t,c)|^{\hat{p}}) \leq C \frac{1}{\varepsilon^p} \sup_{t_0 \leq t < \infty} \hat{E}(|\hat{X}(t,c)|^{\hat{p}}).$$

From here, the stability in pth mean implies

$$\lim_{c \longrightarrow \infty} \sup_{t_0 \leq t < \infty} \hat{P}(|\hat{X}(t,c)|^{\hat{p}} \geq \varepsilon) = 0$$

for all $\varepsilon > 0$.

This is called weak iso-stochastic stability.

Theorem 5.2.25. *A sufficient condition for the iso-stochastic exponential stability in pth mean for the iso-stochastic differential equations of the first kind* (2) *is the existence of a function* $v(x,t)$, *that is defined and continuous on* $\mathbb{R}\times[t_0,\infty)$, *that for all* $x\neq 0$ *is once continuously differentiable with respect to t and twice continuously differentiable with respect to x, and that satisfies the inequalities*

$$c_1|x|^p \le v(x,t) \le c_2|x|^p$$

and

$$Lv(x,t) \le -c_3|x|^p,$$

where L is defined by (3), *for certain positive constants* c_1, c_2 *and* c_3. *Then there exists a positive constant* c_4 *and an almost certainly finite random iso-variable* $\hat{K}(c)$ *dependent on* $c\in\mathbb{R}$ *such that*

$$|\hat{X}(t,c)| \le \hat{K}(c)e^{-c_4(t-t_0)} \qquad \text{for} \qquad \text{all} \qquad t\ge t_0,$$

for almost all sample iso-functions starting at c.

Theorem 5.2.26. *Suppose that the iso-stochastic differential equation of the first kind* (2) *satisfies all conditions of the existence and uniqueness theorem for iso-stochastic differential equations of the first kind. Let us suppose that*

$$|\hat{f}(x,t)-a(t)x| = o(|x|),$$

and

$$|\hat{g}(x,t)-b(t)x| = o(|x|),$$

uniformly in $t\ge t_0$ *as* $|x|\longrightarrow 0$. *Consider the linear iso-stochastic differential equation of the first kind*

$$(6) \qquad \begin{cases} d\hat{X}(t) = a(t)\hat{X}(t)dt + b(t)\hat{X}(t)d\frac{\tilde{W}^1}{\hat{T}} \\ \hat{X}(t_0) = c, \qquad t\ge t_0. \end{cases}$$

Suppose that, for the equilibrium points of (6)

$$\lim_{c\longrightarrow 0}\hat{P}(\sup_{s\le t<\infty}|\hat{X}(t,s)|>\varepsilon) = 0$$

for all $\varepsilon>0$ *uniformly in s and*

$$\hat{P}(\lim_{t\longrightarrow\infty}\hat{X}(t,s)=0) = 1.$$

Then the equilibrium position of (2) *is iso-stochastically asymptotically stable.*

For completeness, below we will formulate the corresponding definitions and results for the iso-stochastic differential equations of the second and third kind.

Definition 5.2.27. *Suppose that the iso-stochastic differential equation of the second kind (4) satisfies all conditions of the existence and uniqueness theorem for iso-stochastic differential equations of the second kind.*

The equilibrium position is said to be iso-stochastically stable in pth mean, where $p > 0$, if, for every $\varepsilon > 0$, there exists a $\delta > 0$ such that

$$\sup_{t_0 \le t < \infty} \hat{E}(|\hat{X}(t,c)|^{\hat{p}}) \le \varepsilon \qquad \text{for} \qquad |c| \le \delta.$$

The equilibrium position is said to be iso-stochastically asymptotically stable in pth mean if it is iso-stochastically stable in pth mean and if

$$\lim_{c \longrightarrow \infty} \hat{E}(|\hat{X}(t,c)|^{\hat{p}}) = 0$$

for all c in a neighborhood of $x = 0$.

It is said to be iso-stochastically exponentially stable in pth mean if there exist positive constants c_1 and c_2 such that, for sufficiently small c,

$$\hat{E}(|\hat{X}(t,c)|^{\hat{p}}) \le c_1 |c|^p e^{-c_2(t-t_0)}.$$

In the case $p = 1$ or $p = 2$, we speak of iso-stochastically(asymptotic or exponential) stability in mean or in mean square, respectively.

The equilibrium position is said to possess an iso-stochastically stable expectation value

$$\hat{m}(t) = \hat{E}(\hat{X}(t,c))$$

or an iso-stochastically stable second moment

$$\hat{P}(t) = \hat{E}(|\hat{X}(t,c)|^{\hat{2}})$$

if, for every $\varepsilon > 0$, there exists a $\delta > 0$ such that

$$\sup_{t_0 \le t < \infty} |\hat{E}(\hat{X}(t,c))| \le \varepsilon \qquad \text{for} \qquad \text{all} \qquad |c| \le \delta,$$

or

$$\sup_{t_0 \le t < \infty} |\hat{E}(\hat{X}^{\hat{2}}(t,c)| \le \varepsilon \qquad \text{for} \qquad \text{all} \qquad |c| \le \delta,$$

respectively.

Remark 5.2.28. *We note that an iso-stochastically(asymptotic, exponential) stability in pth mean implies the same stability in qth mean for all $0 < q < p$. In particular, stability in mean square implies stability in mean.*

Remark 5.2.29. *It follows from the iso-Chebyshev inequality, that*

$$\sup_{t_0 \le t < \infty} \hat{P}(|\hat{X}(t,c)|^{\hat{p}}) \le C \frac{1}{\varepsilon^p} \sup_{t_0 \le t < \infty} \hat{E}(|\hat{X}(t,c)|^{\hat{p}}).$$

From here, the stability in pth mean implies

$$\lim_{c \longrightarrow \infty} \sup_{t_0 \le t < \infty} \hat{P}(|\hat{X}(t,c)|^{\hat{p}} \ge \varepsilon) = 0$$

for all $\varepsilon > 0$.

This is called weak iso-stochastic stability.

Theorem 5.2.30. *A sufficient condition for the iso-stochastic exponential stability in pth mean for the iso-stochastic differential equations of the second kind* (4) *is the existence of a function* $v(x,t)$*, that is defined and continuous on* $\mathbb{R}\times[t_0,\infty)$*, that for all* $x\neq 0$ *is once continuously differentiable with respect to t and twice continuously differentiable with respect to x, and that satisfies the inequalities*

$$c_1|x|^p \leq v(x,t) \leq c_2|x|^p$$

and

$$Lv(x,t) \leq -c_3|x|^p,$$

where L is defined by (3)*, for certain positive constants* c_1*,* c_2 *and* c_3*. Then there exists a positive constant* c_4 *and an almost certainly finite random iso-variable* $\hat{K}(c)$ *dependent on* $c\in\mathbb{R}$ *such that*

$$|\hat{X}(t,c)| \leq \hat{K}(c)e^{-c_4(t-t_0)} \qquad \text{for} \qquad \text{all} \qquad t\geq t_0,$$

for almost all sample iso-functions starting at c.

Theorem 5.2.31. *Suppose that the iso-stochastic differential equation of the second kind* (4) *satisfies all conditions of the existence and uniqueness theorem for iso-stochastic differential equations of the second kind. Let us suppose that*

$$|\hat{f}(x,t)-a(t)x| = o(|x|),$$

and

$$|\hat{g}(x,t)-b(t)x| = o(|x|),$$

uniformly in $t\geq t_0$ *as* $|x|\longrightarrow 0$*. Consider the linear iso-stochastic differential equation of the second kind*

$$(7) \qquad \begin{cases} d\hat{X}(t) = a(t)\hat{X}(t)dt + b(t)\hat{X}(t)d\frac{\tilde{W}^2\tilde{T}(\tilde{W}^2)}{\hat{T}} \\ \hat{X}(t_0) = c, \qquad t\geq t_0. \end{cases}$$

Suppose that, for the equilibrium points of (7)

$$\lim_{c\longrightarrow 0}\hat{P}(\sup_{s\leq t<\infty}|\hat{X}(t,s)|>\varepsilon) = 0$$

for all $\varepsilon>0$ *uniformly in s and*

$$\hat{P}(\lim_{t\longrightarrow\infty}\hat{X}(t,s)=0) = 1.$$

Then the equilibrium position of (4) *is iso-stochastically asymptotically stable.*

Definition 5.2.32. *Suppose that the iso-stochastic differential equation of the third kind* (5) *satisfies all conditions of the existence and uniqueness theorem for iso-stochastic differential equations of the third kind.*

The equilibrium position is said to be iso-stochastically stable in pth mean, where $p > 0$, if, for every $\varepsilon > 0$, there exists a $\delta > 0$ such that

$$\sup_{t_0 \le t < \infty} \hat{E}(|\hat{X}(t,c)|^{\hat{p}}) \le \varepsilon \qquad \text{for} \qquad |c| \le \delta.$$

The equilibrium position is said to be iso-stochastically asymptotically stable in pth mean if it is iso-stochastically stable in pth mean and if

$$\lim_{c \longrightarrow \infty} \hat{E}(|\hat{X}(t,c)|^{\hat{p}}) = 0$$

for all c in a neighborhood of $x = 0$.

It is said to be iso-stochastically exponentially stable in pth mean if there exist positive constants c_1 and c_2 such that, for sufficiently small c,

$$\hat{E}(|\hat{X}(t,c)|^{\hat{p}}) \le c_1 |c|^p e^{-c_2(t-t_0)}.$$

In the case $p = 1$ or $p = 2$, we speak of iso-stochastically(asymptotic or exponential) stability in mean or in mean square, respectively.

The equilibrium position is said to possess an iso-stochastically stable expectation value

$$\hat{m}(t) = \hat{E}(\hat{X}(t,c))$$

or an iso-stochastically stable second moment

$$\hat{P}(t) - \hat{E}(|\hat{X}(t,c)|^{\hat{2}})$$

if, for every $\varepsilon > 0$, there exists a $\delta > 0$ such that

$$\sup_{t_0 \le t < \infty} |\hat{E}(\hat{X}(t,c))| \le \varepsilon \qquad \text{for} \qquad \text{all} \qquad |c| \le \delta,$$

or

$$\sup_{t_0 \le t < \infty} |\hat{E}(\hat{X}^{\hat{2}}(t,c)| \le \varepsilon \qquad \text{for} \qquad \text{all} \qquad |c| \le \delta,$$

respectively.

Remark 5.2.33. *We note that an iso-stochastically(asymptotic, exponential) stability in pth mean implies the same stability in qth mean for all $0 < q < p$. In particular, stability in mean square implies stability in mean.*

Remark 5.2.34. *It follows from the iso-Chebyshev inequality, that*

$$\sup_{t_0 \le t < \infty} \hat{P}(|\hat{X}(t,c)|^{\hat{p}}) \le C\frac{1}{\varepsilon^p} \sup_{t_0 \le t < \infty} \hat{E}(|\hat{X}(t,c)|^{\hat{p}}).$$

From here, the stability in pth mean implies

$$\lim_{c \longrightarrow \infty} \sup_{t_0 \le t < \infty} \hat{P}(|\hat{X}(t,c)|^{\hat{p}} \ge \varepsilon) = 0$$

for all $\varepsilon > 0$.

This is called weak iso-stochastic stability.

Theorem 5.2.35. *A sufficient condition for the iso-stochastic exponential stability in pth mean for the iso-stochastic differential equations of the third kind* (5) *is the existence of a function* $v(x,t)$*, that is defined and continuous on* $\mathbb{R}\times[t_0,\infty)$*, that for all* $x\neq 0$ *is once continuously differentiable with respect to t and twice continuously differentiable with respect to x, and that satisfies the inequalities*

$$c_1|x|^p \leq v(x,t) \leq c_2|x|^p$$

and

$$Lv(x,t) \leq -c_3|x|^p,$$

where L is defined by (3)*, for certain positive constants* c_1*,* c_2 *and* c_3*. Then there exists a positive constant* c_4 *and an almost certainly finite random iso-variable* $\hat{K}(c)$ *dependent on* $c\in\mathbb{R}$ *such that*

$$|\hat{X}(t,c)| \leq \hat{K}(c)e^{-c_4(t-t_0)} \qquad \text{for} \qquad \text{all} \qquad t \geq t_0,$$

for almost all sample iso-functions starting at c.

Theorem 5.2.36. *Suppose that the iso-stochastic differential equation of the third kind* (5) *satisfies all conditions of the existence and uniqueness theorem for iso-stochastic differential equations of the third kind. Let us suppose that*

$$|\hat{f}(x,t)-a(t)x| = o(|x|),$$

and

$$|\hat{g}(x,t)-b(t)x| = o(|x|),$$

uniformly in $t\geq t_0$ *as* $|x|\longrightarrow 0$*. Consider the linear iso-stochastic differential equation of the third kind*

$$(8) \qquad \begin{cases} d\hat{X}(t) = a(t)\hat{X}(t)dt + b(t)\hat{X}(t)d\dfrac{\frac{\tilde{W}^3}{\hat{T}(\tilde{W}^3)}}{\hat{T}} \\ \hat{X}(t_0) = c, \qquad t\geq t_0. \end{cases}$$

Suppose that, for the equilibrium points of (8)

$$\lim_{c\longrightarrow 0}\hat{P}(\sup_{s\leq t<\infty}|\hat{X}(t,s)| > \varepsilon) = 0$$

for all $\varepsilon > 0$ *uniformly in s and*

$$\hat{P}(\lim_{t\longrightarrow\infty}\hat{X}(t,s) = 0) = 1.$$

Then the equilibrium position of (5) *is iso-stochastically asymptotically stable.*

Chapter 6

Iso-Stratonovich Iso-Integral

We have defined for 1-dimensional iso-Brownian iso-motion of the first kind

$$\hat{\int}_0^T \frac{\tilde{W}^1}{\hat{T}} \hat{\times} \hat{d}\frac{\tilde{W}^1}{\hat{T}} := \lim_{|P^n| \longrightarrow 0} \sum_{k=1}^{m_n-1} \frac{1}{\hat{T}} \tilde{W}^1(\tau_k^n)\left(\tilde{W}^1(t_{k+1}^n) - \tilde{W}^1(t_k^n)\right),$$

where

$$\tau_k^n = (1-\lambda)t_k^n + \lambda t_{k+1}^n, \qquad \lambda \in [0,1],$$

$$P^n := \left\{0 = t_0^n < t_1^n < \cdots < t_{m_n}^n = T\right\}$$

is a partition of the interval $[0,T]$.

We have deducted that

$$(1) \qquad \hat{\int}_0^T \frac{\tilde{W}^1}{\hat{T}} \hat{\times} \hat{d}\frac{\tilde{W}^1}{\hat{T}} = \frac{\left(\tilde{W}^1(T)\right)^{\hat{2}}}{2\hat{T}} + \left(\lambda - \frac{1}{2}\right)\frac{T}{\hat{T}}\hat{T}(\Omega).$$

For $\lambda = 0$ we have that

$$(2) \qquad \hat{\int}_0^T \frac{\tilde{W}^1}{\hat{T}} \hat{\times} \hat{d}\frac{\tilde{W}^1}{\hat{T}} = \frac{\left(\tilde{W}^1(T)\right)^{\hat{2}}}{2\hat{T}} - \frac{1}{2}\frac{T}{\hat{T}}\hat{T}(\Omega).$$

When $\lambda = 1$ we get

$$(3) \qquad \hat{\int}_0^T \frac{\tilde{W}^1}{\hat{T}} \hat{\times} \hat{d}\frac{\tilde{W}^1}{\hat{T}} = \frac{\left(\tilde{W}^1(T)\right)^{\hat{2}}}{2\hat{T}} + \frac{1}{2}\frac{T}{\hat{T}}\hat{T}(\Omega).$$

Definition 6.0.37. *Iso-Stratonovich iso-integral of the first kind for $\tilde{W}^1$ we define as follows*

$$\int_0^T \tilde{W}^1 \circ d\frac{\tilde{W}^1}{\hat{T}} := \lim_{|P^n| \longrightarrow 0} \sum_{k=0}^{m_n-1} \frac{1}{\hat{T}} \frac{\tilde{W}^1(t_{k+1}^n) + \tilde{W}^1(t_k^n)}{2}\left(\tilde{W}^1(t_{k+1}^n) - \tilde{W}^1(t_k^n)\right).$$

From the definition, using (2) and (3), we get

$$\int_0^T \tilde{W}^1 \circ d\frac{\tilde{W}^1}{\hat{T}} = \frac{1}{2}\lim_{|P^n| \longrightarrow 0} \sum_{k=0}^{m_n-1} \frac{\tilde{W}^1(t_{k+1}^n)}{\hat{T}}\left(\tilde{W}^1(t_{k+1}^n) - \tilde{W}^1(t_k^n)\right)$$

$$+\frac{1}{2}\lim_{|P^n| \longrightarrow 0} \sum_{k=0}^{m_n-1} \frac{\tilde{W}^1(t_k^n)}{\hat{T}}\left(\tilde{W}^1(t_{k+1}^n) - \tilde{W}^1(t_k^n)\right)$$

$$= \frac{1}{2}\left(\frac{(\tilde{W}^1(T))^2}{2} - \frac{1}{2}\frac{T}{\hat{T}}\hat{T}(\Omega)\right) + \frac{1}{2}\left(\frac{(\tilde{W}^1(T))^2}{2} + \frac{1}{2}\frac{T}{P\hat{T}}\hat{T}(\Omega)\right)$$

$$= \frac{\left(\tilde{W}^1(T)\right)^2}{2}.$$

Definition 6.0.38. *Let $\hat{X}$ be an iso-stochastic process with values $\mathbb{R}$. We define*

$$\int_0^T \hat{f}(\hat{X}(t),t)\circ d\frac{\tilde{W}^1}{\hat{T}} := \lim_{|P^n|\longrightarrow 0}\sum_{k=0}^{m_n-1}\hat{f}\left(\frac{\hat{X}(t_{k+1}^n)+\hat{X}(t_k^n)}{2},t_k^n\right)\frac{\tilde{W}^1(t_{k+1}^n)-\tilde{W}^1(t_k^n)}{\hat{T}},$$

where $f:\mathbb{R}\times[0,T]\longrightarrow\mathbb{R}$ is given function.

From the definition it follows

$$\begin{aligned}
&\textstyle\int_0^T \hat{f}(\tilde{W}^1(t),t)\circ d\frac{\tilde{W}^1}{\hat{T}} - \int_0^T f(\tilde{W}^1(t),t)d\frac{\tilde{W}^1}{\hat{T}}\\
&= \textstyle\lim_{|P^n|\longrightarrow 0}\sum_{k=0}^{m_n-1}\hat{f}\left(\frac{\tilde{W}^1(t_{k+1}^n)+\tilde{W}^1(t_k^n)}{2},t_k^n\right)\frac{\tilde{W}^1(t_{k+1}^n)-\tilde{W}^1(t_k^n)}{\hat{T}}\\
&\textstyle -\lim_{|P^n|\longrightarrow 0}\sum_{k=0}^{m_n-1} f\left(\tilde{W}^1(\tau_k^n),t_k^n\right)\frac{\tilde{W}^1(t_{k+1}^n)-\tilde{W}^1(t_k^n)}{\hat{T}}\\
&= \textstyle\lim_{|P^n|\longrightarrow 0}\sum_{k=0}^{m_n-1}\left(\hat{f}\left(\frac{\tilde{W}^1(t_{k+1}^n)+\tilde{W}^1(t_k^n)}{2},t_k^n\right)-f\left(\tilde{W}^1(\tau_k^n),t_k^n\right)\right)\frac{\tilde{W}^1(t_{k+1}^n)-\tilde{W}^1(t_k^n)}{\hat{T}}\\
&= \textstyle\frac{1}{2}\int_0^T\frac{\partial f}{\partial x}(\tilde{W}^1(t),t)dt,
\end{aligned}$$

i.e.,

$$\int_0^T \hat{f}(\tilde{W}^1(t),t)\circ d\frac{\tilde{W}^1}{\hat{T}} = \int_0^T f(\tilde{W}^1(t),t)d\frac{\tilde{W}^1}{\hat{T}} + \frac{1}{2}\int_0^T\frac{\partial f}{\partial x}(\tilde{W}^1(t),t)dt.$$

Definition 6.0.39. *We suppose that the iso-process $\hat{X}(\cdot)$ solves the iso-stochastic iso-integral of the first kind*

$$\hat{X}(t) = \hat{X}(0) + \int_0^t \hat{f}(\hat{X}(s),s)ds + \int_0^t \hat{g}(\hat{X}(s),s)\circ d\frac{\tilde{W}^1}{\hat{T}}, \qquad 0\le t\le T,$$

for $\hat{f}$, $\hat{g}:\mathbb{R}\times[0,T]\longrightarrow\mathbb{R}$. We den write

$$d\hat{X}(t) = \hat{f}(\hat{X}(t),t)dt + \hat{g}(\hat{X}(t),t)\circ d\frac{\tilde{W}^1}{\hat{T}},$$

which is called the iso-Stratonovich iso-stochastic differential of the first kind.

Now we assume that

$$d\hat{X}(t) = \hat{f}(\hat{X}(t),t)dt + \hat{g}(\hat{X}(t),t)\circ d\frac{\tilde{W}^1}{\hat{T}}$$

and $u:\mathbb{R}\times[0,T]\longrightarrow\mathbb{R}$ is smooth. Let also,

$$\hat{Y}(t) = \hat{u}(\hat{X}(t),t).$$

Then

$$\begin{aligned}
d\hat{Y} &= \textstyle\frac{\partial u}{\partial t}dt + \frac{\partial u}{\partial x}\circ d\hat{X}\\
&= \textstyle\left(\frac{\partial u}{\partial t}+\hat{f}(\hat{X}(t),t)\right)dt + \frac{\partial u}{\partial x}\hat{g}(\hat{X}(t),t)\circ d\frac{\tilde{W}^1}{\hat{T}}.
\end{aligned}$$

For 1-dimensional iso-Brownian iso-motion of the second kind

$\hat{\int}_0^T \frac{\tilde{W}^2\tilde{T}(\tilde{W}^2)}{\tilde{T}} \hat{\times} \hat{d}\frac{\tilde{W}^2\tilde{T}(\tilde{W}^2)}{\hat{T}}$

$:= \lim_{|P^n|\longrightarrow 0} \sum_{k=1}^{m_n-1} \frac{1}{\hat{T}} \tilde{W}^2\tilde{T}(\tilde{W}^2)(\tau_k^n) \left(\tilde{W}^2\tilde{T}(\tilde{W}^2)(t_{k+1}^n) - \tilde{W}^2\tilde{T}(\tilde{W}^2)(t_k^n)\right).$

We have deducted that

$$\hat{\int}_0^T \frac{\tilde{W}^2\tilde{T}(\tilde{W}^2)}{\tilde{T}} \hat{\times} \hat{d}\frac{\tilde{W}^2\tilde{T}(\tilde{W}^2)}{\hat{T}} = \frac{\left(\tilde{W}^2\tilde{T}(\tilde{W}^2)(T)\right)^{\hat{2}}}{2\hat{T}} + \left(\lambda - \frac{1}{2}\right)\frac{T}{\hat{T}}\hat{T}(\Omega).$$

For $\lambda = 0$ we have that

(4) $\quad \hat{\int}_0^T \frac{\tilde{W}^2\tilde{T}(\tilde{W}^2)}{\tilde{T}} \hat{\times} \hat{d}\frac{\tilde{W}^2\tilde{T}(\tilde{W}^2)}{\hat{T}} = \frac{\left(\tilde{W}^2\tilde{T}(\tilde{W}^2)(T)\right)^{\hat{2}}}{2\hat{T}} - \frac{1}{2}\frac{T}{\hat{T}}\hat{T}(\Omega).$

When $\lambda = 1$ we get

(5) $\quad \hat{\int}_0^T \frac{\tilde{W}^2\tilde{T}(\tilde{W}^2)}{\tilde{T}} \hat{\times} \hat{d}\frac{\tilde{W}^2\tilde{T}(\tilde{W}^2)}{\hat{T}} = \frac{\left(\tilde{W}^2\tilde{T}(\tilde{W}^2)(T)\right)^{\hat{2}}}{2\hat{T}} + \frac{1}{2}\frac{T}{\hat{T}}\hat{T}(\Omega).$

Definition 6.0.40. *Iso-Stratonovich iso-integral of the second kind for $\tilde{W}^2\tilde{T}(\tilde{W}^2)$ we define as follows*

$\int_0^T \tilde{W}^2\tilde{T}(\tilde{W}^2) \circ d\frac{\tilde{W}^2\tilde{T}(\tilde{W}^2)}{\hat{T}}$

$:= \lim_{|P^n|\longrightarrow 0} \sum_{k=0}^{m_n-1} \frac{1}{\hat{T}} \frac{\tilde{W}^2\tilde{T}(\tilde{W}^2)(t_{k+1}^n) + \tilde{W}^2\tilde{T}(\tilde{W}^2)(t_k^n)}{2} \left(\tilde{W}^2\tilde{T}(\tilde{W}^2)(t_{k+1}^n) - \tilde{W}^2\tilde{T}(\tilde{W}^2)(t_k^n)\right).$

From the definition, using (4) and (5), we get

$\int_0^T \tilde{W}^2\tilde{T}(\tilde{W}^2) \circ d\frac{W^2\tilde{T}(\tilde{W}^2)}{\hat{T}}$

$= \frac{1}{2} \lim_{|P^n|\longrightarrow 0} \sum_{k=0}^{m_n-1} \frac{\tilde{W}^2\tilde{T}(\tilde{W}^2)(t_{k+1}^n)}{\hat{T}} \left(\tilde{W}^2\tilde{T}(\tilde{W}^2)(t_{k+1}^n) - \tilde{W}^2\tilde{T}(\tilde{W}^2)(t_k^n)\right)$

$+ \frac{1}{2} \lim_{|P^n|\longrightarrow 0} \sum_{k=0}^{m_n-1} \frac{\tilde{W}^2\tilde{T}(\tilde{W}^2)(t_k^n)}{\hat{T}} \left(\tilde{W}^2\tilde{T}(\tilde{W}^2)(t_{k+1}^n) - \tilde{W}^2\tilde{T}(\tilde{W}^2)(t_k^n)\right)$

$= \frac{1}{2}\left(\frac{(\tilde{W}^2\tilde{T}(\tilde{W}^2)(T))^2}{2} - \frac{1}{2}\frac{T}{\hat{T}}\hat{T}(\Omega)\right) + \frac{1}{2}\left(\frac{(\tilde{W}^2\tilde{T}(\tilde{W}^2)(T))^2}{2} + \frac{1}{2}\frac{T}{P\hat{T}}\hat{T}(\Omega)\right)$

$= \frac{\left(\tilde{W}^2\tilde{T}(\tilde{W}^2)(T)\right)^2}{2}.$

Definition 6.0.41. *Let $\hat{X}$ be an iso-stochastic process with values $\mathbb{R}$. We define*

$\int_0^T \hat{f}(\hat{X}(t),t) \circ d\frac{\tilde{W}^2\tilde{T}(\tilde{W}^2)}{\hat{T}}$

$:= \lim_{|P^n|\longrightarrow 0} \sum_{k=0}^{m_n-1} \hat{f}\left(\frac{\hat{X}(t_{k+1}^n) + \hat{X}(t_k^n)}{2}, t_k^n\right) \frac{\tilde{W}^2\tilde{T}(\tilde{W}^2)(t_{k+1}^n) - \tilde{W}^2\tilde{T}(\tilde{W}^2)(t_k^n)}{\hat{T}},$

where $f : \mathbb{R} \times [0,T] \longrightarrow \mathbb{R}$ is given function.

From the definition it follows

$$\int_0^T \hat{f}(\tilde{W}^2\tilde{T}(\tilde{W}^2)(t),t)\circ d\frac{\tilde{W}^2\tilde{T}(\tilde{W}^2)}{\hat{T}} - \int_0^T f(\tilde{W}^2\tilde{T}(\tilde{W}^2)(t),t)d\frac{\tilde{W}^2\tilde{T}(\tilde{W}^2)}{\hat{T}}$$

$$= \lim_{|P^n|\longrightarrow 0}\sum_{k=0}^{m_n-1}\hat{f}\left(\frac{\tilde{W}^2\tilde{T}(\tilde{W}^2)(t_{k+1}^n)+\tilde{W}^2\tilde{T}(\tilde{W}^2)(t_k^n)}{2},t_k^n\right)\frac{\tilde{W}^2\tilde{T}(\tilde{W}^2)(t_{k+1}^n)-\tilde{W}^2\tilde{T}(\tilde{W}^2)(t_k^n)}{\hat{T}}$$

$$-\lim_{|P^n|\longrightarrow 0}\sum_{k=0}^{m_n-1} f\left(\tilde{W}^2\tilde{T}(\tilde{W}^2)(\tau_k^n),t_k^n\right)\frac{\tilde{W}^2\tilde{T}(\tilde{W}^2)(t_{k+1}^n)-\tilde{W}^2\tilde{T}(\tilde{W}^2)(t_k^n)}{\hat{T}}$$

$$= \lim_{|P^n|\longrightarrow 0}\sum_{k=0}^{m_n-1}\left(\hat{f}\left(\frac{\tilde{W}^2\tilde{T}(\tilde{W}^2)(t_{k+1}^n)+\tilde{W}^2\tilde{T}(\tilde{W}^2)(t_k^n)}{2},t_k^n\right)\right.$$

$$\left.-f\left(\tilde{W}^2\tilde{T}(\tilde{W}^2)(\tau_k^n),t_k^n\right)\right)\frac{\tilde{W}^2\tilde{T}(\tilde{W}^2)(t_{k+1}^n)-\tilde{W}^2\tilde{T}(\tilde{W}^2)(t_k^n)}{\hat{T}}$$

$$= \frac{1}{2}\int_0^T \frac{\partial f}{\partial x}(\tilde{W}^2\tilde{T}(\tilde{W}^2)(t),t)dt,$$

i.e.,

$$\int_0^T \hat{f}(\tilde{W}^2\tilde{T}(\tilde{W}^2)(t),t)\circ d\frac{\tilde{W}^2\tilde{T}(\tilde{W}^2)}{\hat{T}} = \int_0^T f(\tilde{W}^2\tilde{T}(\tilde{W}^2)(t),t)d\frac{\tilde{W}^2\tilde{T}(\tilde{W}^2)}{\hat{T}}$$

$$+\frac{1}{2}\int_0^T \frac{\partial f}{\partial x}(\tilde{W}^2\tilde{T}(\tilde{W}^2)(t),t)dt.$$

Definition 6.0.42. *We suppose that the iso-process $\hat{X}(\cdot)$ solves the iso-stochastic iso-integral of the second kind*

$$\hat{X}(t) = \hat{X}(0) + \int_0^t \hat{f}(\hat{X}(s),s)ds + \int_0^t \hat{g}(\hat{X}(s),s)\circ d\frac{\tilde{W}^2\tilde{T}(\tilde{W}^2)}{\hat{T}}, \qquad 0\le t\le T,$$

for $\hat{f}$, $\hat{g}:\mathbb{R}\times[0,T]\longrightarrow\mathbb{R}$. We den write

$$d\hat{X}(t) = \hat{f}(\hat{X}(t),t)dt + \hat{g}(\hat{X}(t),t)\circ d\frac{\tilde{W}^2\tilde{T}(\tilde{W}^2)}{\hat{T}},$$

which is called the iso-Stratonovich iso-stochastic differential of the second kind.

Now we assume that

$$d\hat{X}(t) = \hat{f}(\hat{X}(t),t)dt + \hat{g}(\hat{X}(t),t)\circ d\frac{\tilde{W}^2\tilde{T}(\tilde{W}^2)}{\hat{T}}$$

and $u:\mathbb{R}\times[0,T]\longrightarrow\mathbb{R}$ is smooth. Let also,

$$\hat{Y}(t) = \hat{u}(\hat{X}(t),t).$$

Then

$$d\hat{Y} = \frac{\partial u}{\partial t}dt + \frac{\partial u}{\partial x}\circ d\hat{X}$$

$$= \left(\frac{\partial u}{\partial t} + \hat{f}(\hat{X}(t),t)\right)dt + \frac{\partial u}{\partial x}\hat{g}(\hat{X}(t),t)\circ d\frac{\tilde{W}^2\tilde{T}(\tilde{W}^2)}{\hat{T}}.$$

For 1-dimensional iso-Brownian iso-motion of the third kind

$$\hat{\int}_0^T \frac{\frac{\tilde{W}^3}{\tilde{T}(\tilde{W}^3)}}{\hat{T}} \hat{\times} \hat{d}\frac{\frac{\tilde{W}^3}{\tilde{T}(\tilde{W}^3)}}{\hat{T}}$$

$$:= \lim_{|P^n|\longrightarrow 0} \sum_{k=1}^{m_n-1} \frac{1}{\hat{T}} \frac{\tilde{W}^3}{\tilde{T}(\tilde{W}^3)}(\tau_k^n) \left(\frac{\tilde{W}^3}{\tilde{T}(\tilde{W}^3)}(t_{k+1}^n) - \frac{\tilde{W}^3}{\tilde{T}(\tilde{W}^3)}(t_k^n) \right).$$

We have deducted that

$$\hat{\int}_0^T \frac{\frac{\tilde{W}^3}{\tilde{T}(\tilde{W}^3)}}{\hat{T}} \hat{\times} \hat{d}\frac{\frac{\tilde{W}^3}{\tilde{T}(\tilde{W}^3)}}{\hat{T}} = \frac{\left(\frac{\tilde{W}^3}{\tilde{T}(\tilde{W}^3)}(T)\right)^{\hat{2}}}{2\hat{T}} + \left(\lambda - \frac{1}{2}\right)\frac{T}{\hat{T}}\hat{T}(\Omega).$$

For $\lambda = 0$ we have that

$$\hat{\int}_0^T \frac{\frac{\tilde{W}^3}{\tilde{T}(\tilde{W}^3)}}{\hat{T}} \hat{\times} \hat{d}\frac{\frac{\tilde{W}^3}{\tilde{T}(\tilde{W}^3)}}{\hat{T}} = \frac{\left(\frac{\tilde{W}^3}{\tilde{T}(\tilde{W}^3)}(T)\right)^{\hat{2}}}{2\hat{T}} - \frac{1}{2}\frac{T}{\hat{T}}\hat{T}(\Omega). \tag{6}$$

When $\lambda = 1$ we get

$$\hat{\int}_0^T \frac{\frac{\tilde{W}^3}{\tilde{T}(\tilde{W}^3)}}{\hat{T}} \hat{\times} \hat{d}\frac{\frac{\tilde{W}^3}{\tilde{T}(\tilde{W}^3)}}{\hat{T}} = \frac{\left(\frac{\tilde{W}^3}{\tilde{T}(\tilde{W}^3)}(T)\right)^{\hat{2}}}{2\hat{T}} + \frac{1}{2}\frac{T}{\hat{T}}\hat{T}(\Omega). \tag{7}$$

Definition 6.0.43. *Iso-Stratonovich iso-integral of the third kind for* $\frac{\tilde{W}^3}{\tilde{T}(\tilde{W}^3)}$ *we define as follows*

$$\int_0^T \frac{\tilde{W}^3}{\tilde{T}(\tilde{W}^3)} \circ d\frac{\frac{\tilde{W}^3}{\tilde{T}(\tilde{W}^3)}}{\hat{T}}$$

$$:= \lim_{|P^n|\longrightarrow 0} \sum_{k=0}^{m_n-1} \frac{1}{\hat{T}} \frac{\frac{\tilde{W}^3}{\tilde{T}(\tilde{W}^3)}(t_{k+1}^n) + \frac{\tilde{W}^3}{\tilde{T}(\tilde{W}^3)}(t_k^n)}{2} \left(\frac{\tilde{W}^3}{\tilde{T}(\tilde{W}^3)}(t_{k+1}^n) - \frac{\tilde{W}^3}{\tilde{T}(\tilde{W}^3)}(t_k^n) \right).$$

From the definition, using (6) and (7), we get

$$\int_0^T \frac{\tilde{W}^3}{\tilde{T}(\tilde{W}^3)} \circ d\frac{\frac{\tilde{W}^3}{\tilde{T}(\tilde{W}^3)}}{\hat{T}}$$

$$= \frac{1}{2}\lim_{|P^n|\longrightarrow 0} \sum_{k=0}^{m_n-1} \frac{\frac{\tilde{W}^3}{\tilde{T}(\tilde{W}^3)}(t_{k+1}^n)}{\hat{T}} \left(\frac{\tilde{W}^3}{\tilde{T}(\tilde{W}^3)}(t_{k+1}^n) - \frac{\tilde{W}^3}{\tilde{T}(\tilde{W}^3)}(t_k^n) \right)$$

$$+ \frac{1}{2}\lim_{|P^n|\longrightarrow 0} \sum_{k=0}^{m_n-1} \frac{\frac{\tilde{W}^3}{\tilde{T}(\tilde{W}^3)}(t_k^n)}{\hat{T}} \left(\frac{\tilde{W}^3}{\tilde{T}(\tilde{W}^3)}(t_{k+1}^n) - \frac{\tilde{W}^3}{\tilde{T}(\tilde{W}^3)}(t_k^n) \right)$$

$$= \frac{1}{2}\left(\frac{\left(\frac{\tilde{W}^3}{\tilde{T}(\tilde{W}^3)}(T)\right)^2}{2} - \frac{1}{2}\frac{T}{\hat{T}}\hat{T}(\Omega) \right) + \frac{1}{2}\left(\frac{\left(\frac{\tilde{W}^3}{\tilde{T}(\tilde{W}^3)}(T)\right)^2}{2} + \frac{1}{2}\frac{T}{P\hat{T}}\hat{T}(\Omega) \right)$$

$$= \frac{\left(\frac{\tilde{W}^3}{\tilde{T}(\tilde{W}^3)}(T)\right)^2}{2}.$$

Definition 6.0.44. *Let $\hat{X}$ be an iso-stochastic process with values $\mathbb{R}$. We define*

$$\int_0^T \hat{f}(\hat{X}(t),t)\circ d\frac{\frac{\tilde{W}^3}{\tilde{T}(\tilde{W}^3)}}{\hat{T}}$$

$$:= \lim_{|P^n|\longrightarrow 0}\sum_{k=0}^{m_n-1}\hat{f}\left(\frac{\hat{X}(t_{k+1}^n)+\hat{X}(t_k^n)}{2},t_k^n\right)\frac{\frac{\tilde{W}^3}{\tilde{T}(\tilde{W}^3)}(t_{k+1}^n)-\frac{\tilde{W}^3}{\tilde{T}(\tilde{W}^3)}(t_k^n)}{\hat{T}},$$

where $f:\mathbb{R}\times[0,T]\longrightarrow\mathbb{R}$ is given function.

From the definition it follows

$$\int_0^T \hat{f}(\frac{\tilde{W}^3}{\tilde{T}(\tilde{W}^3)}(t),t)\circ d\frac{\frac{\tilde{W}^3}{\tilde{T}(\tilde{W}^3)}}{\hat{T}} - \int_0^T f(\frac{\tilde{W}^3}{\tilde{T}(\tilde{W}^3)}(t),t)d\frac{\frac{\tilde{W}^3}{\tilde{T}(\tilde{W}^3)}}{\hat{T}}$$

$$= \lim_{|P^n|\longrightarrow 0}\sum_{k=0}^{m_n-1}\hat{f}\left(\frac{\frac{\tilde{W}^3}{\tilde{T}(\tilde{W}^3)}(t_{k+1}^n)+\frac{\tilde{W}^3}{\tilde{T}(\tilde{W}^3)}(t_k^n)}{2},t_k^n\right)\frac{\frac{\tilde{W}^3}{\tilde{T}(\tilde{W}^3)}(t_{k+1}^n)-\frac{\tilde{W}^3}{\tilde{T}(\tilde{W}^3)}(t_k^n)}{\hat{T}}$$

$$-\lim_{|P^n|\longrightarrow 0}\sum_{k=0}^{m_n-1}f\left(\frac{\tilde{W}^3}{\tilde{T}(\tilde{W}^3)}(\tau_k^n),t_k^n\right)\frac{\frac{\tilde{W}^3}{\tilde{T}(\tilde{W}^3)}(t_{k+1}^n)-\frac{\tilde{W}^3}{\tilde{T}(\tilde{W}^3)}(t_k^n)}{\hat{T}}$$

$$= \lim_{|P^n|\longrightarrow 0}\sum_{k=0}^{m_n-1}\left(\hat{f}\left(\frac{\frac{\tilde{W}^3}{\tilde{T}(\tilde{W}^3)}(t_{k+1}^n)+\frac{\tilde{W}^3}{\tilde{T}(\tilde{W}^3)}(t_k^n)}{2},t_k^n\right)\right.$$

$$\left.-f\left(\tilde{W}^2\tilde{T}(\tilde{W}^2)(\tau_k^n),t_k^n\right)\right)\frac{\frac{\tilde{W}^3}{\tilde{T}(\tilde{W}^3)}(t_{k+1}^n)-\frac{\tilde{W}^3}{\tilde{T}(\tilde{W}^3)}(t_k^n)}{\hat{T}}$$

$$= \frac{1}{2}\int_0^T \frac{\partial f}{\partial x}(\frac{\tilde{W}^3}{\tilde{T}(\tilde{W}^3)}(t),t)dt,$$

i.e.,

$$\int_0^T \hat{f}(\frac{\tilde{W}^3}{\tilde{T}(\tilde{W}^3)}(t),t)\circ d\frac{\frac{\tilde{W}^3}{\tilde{T}(\tilde{W}^3)}}{\hat{T}} = \int_0^T f(\frac{\tilde{W}^3}{\tilde{T}(\tilde{W}^3)}(t),t)d\frac{\frac{\tilde{W}^3}{\tilde{T}(\tilde{W}^3)}}{\hat{T}}$$

$$+\frac{1}{2}\int_0^T \frac{\partial f}{\partial x}(\frac{\tilde{W}^3}{\tilde{T}(\tilde{W}^3)}(t),t)dt.$$

Definition 6.0.45. *We suppose that the iso-process $\hat{X}(\cdot)$ solves the iso-stochastic iso-integral of the third kind*

$$\hat{X}(t)=\hat{X}(0)+\int_0^t \hat{f}(\hat{X}(s),s)ds+\int_0^t \hat{g}(\hat{X}(s),s)\circ d\frac{\frac{\tilde{W}^3}{\tilde{T}(\tilde{W}^3)}}{\hat{T}}, \qquad 0\le t\le T,$$

for $\hat{f}$, $\hat{g}:\mathbb{R}\times[0,T]\longrightarrow\mathbb{R}$. We den write

$$d\hat{X}(t)=\hat{f}(\hat{X}(t),t)dt+\hat{g}(\hat{X}(t),t)\circ d\frac{\frac{\tilde{W}^3}{\tilde{T}(\tilde{W}^3)}}{\hat{T}},$$

which is called the iso-Stratonovich iso-stochastic differential of the third kind.

Now we assume that

$$d\hat{X}(t) = \hat{f}(\hat{X}(t),t)dt + \hat{g}(\hat{X}(t),t) \circ d\frac{\frac{\tilde{W}^3}{\hat{T}(\tilde{W}^3)}}{\hat{T}}$$

and $u : \mathbb{R} \times [0,T] \longrightarrow \mathbb{R}$ is smooth. Let also,

$$\hat{Y}(t) = \hat{u}(\hat{X}(t),t).$$

Then

$$d\hat{Y} = \frac{\partial u}{\partial t}dt + \frac{\partial u}{\partial x} \circ d\hat{X}$$

$$= \left(\frac{\partial u}{\partial t} + \hat{f}(\hat{X}(t),t)\right)dt + \frac{\partial u}{\partial x}\hat{g}(\hat{X}(t),t) \circ d\frac{\frac{\tilde{W}^3}{\hat{T}(\tilde{W}^3)}}{\hat{T}}.$$

References

[1] L. Arnold, Stochastic Differential Equations: Theory and Applications, Wiley, 1974.

[2] M. Baxter and A. Rennie, Financial Calculus: An Introduction to Derivative Pricing, Cambridge U. Press, 1996.

[3] A. N. Borodin and P. Salminen, Handbook of Brownian Motion: Facts and Formulae, Birkhauser, 1996.

[4] L. Breiman, Probability, AddisonWesley, 1968.

[5] A. Friedman, Stochastic Differential Equations and Applications, Vol. 1 and 2, Academic Press.

[6] M. Freidlin, Functional Integration and Partial Differential Equations, Princeton U. Press, 1985.

[7] Georgiev, S.(2014). *Foundations of Iso-Differential Calculus*, Vol. 1., Vol. 2, Vol. 3, Nova Science Publishers, Inc.

[8] I. I. Gihman and A. V. Skorohod, Stochastic Differential Equations, Springer, 1972.

[9] D. Gillespie, The mathematics of Brownian motion and Johnson noise, American J. Physics 64 (1996), 225240.

[10] J. Lamperti, Probability, Benjamin.

[11] R. Paley, N. Wiener, and A. Zygmund, Notes on random functions, Math. Z. 37 (1959), 647668.

[12] M. Pinsky, Introduction to Fourier Analysis and Wavelets, Brooks/Cole, 2002.

[13] D. Stroock, Probability Theory: An Analytic View, Cambridge U. Press, 1993.

[14] H. Sussmann, An interpretation of stochastic differential equations as ordinary differential equations which depend on the sample point., Bulletin AMS 83 (1977), 296298.

Index